Taste and Smell

Advances in Oto-Rhino-Laryngology

Vol. 63

Series Editor

W. Arnold *Munich*

KARGER

Taste and Smell

An Update

Volume Editors

Thomas Hummel *Dresden*
Antje Welge-Lüssen *Basel*

33 figures, 1 in color, and 12 tables, 2006

KARGER

Basel · Freiburg · Paris · London · New York ·
Bangalore · Bangkok · Singapore · Tokyo · Sydney

Prof. Thomas Hummel
Smell & Taste Clinic
Department of Otorhinolaryngology
University of Dresden Medical School
Fetscherstrasse 74
DE–01307 Dresden (Germany)

PD Dr. Antje Welge-Lüssen
Department of Otorhinolaryngology
University Hospital Basel
Petersgraben 4
CH–4031 Basel (Switzerland)

Library of Congress Cataloging-in-Publication Data

Taste and smell : an update / editors, Thomas Hummel, Antje Welge-Lüssen.
p. ; cm. – (Advances in oto-rhino-laryngology ; v. 63)
Includes bibliographical references and index.
ISBN 3-8055-8123-8 (hard cover : alk. paper)
1. Taste. 2. Smell. 3. Sense organs. 4. Taste disorders. I. Hummel, Thomas, 1959- II. Welge-Lüssen, Antje. III. Series.
[DNLM: 1. Smell–physiology. 2. Olfaction Disorders. 3. Taste–physiology. 4. Taste Disorders. W1 AD701 v.63 2006 / WV 301 T2148 2006]
QP458.T37 2006
612.8′6–dc22

2006011577

Bibliographic Indices. This publication is listed in bibliographic services, including Current Contents® and Index Medicus.

www.karger.com
Printed in Switzerland on acid-free paper by Reinhardt Druck, Basel
ISSN 0065–3071
ISBN-10: 3–8055–8123–8
ISBN-13: 978-3–8055–8123–3

Contents

Preface

An intact sense of smell and taste allows us to recognize the chemical signals from our environment. By doing this, the chemical senses contribute significantly to the quality of our lives. Despite this fact, and although chemosensory disorders are frequent, they have been 'neglected' in clinical routine for many years. This neglect may be partly explained by traditional difficulties in the diagnosis of chemosensory disorders and the common belief that chemosensory disorders cannot be treated. Having said that, the clinical neglect of chemosensory functions is in sharp contrast to the remarkable interest the chemical senses have received over the last decade, culminating in the 2004 Nobel Prize awarded to two researchers in the sense of smell.

This publication is meant for the clinician confronted with chemosensory disorders. It is supposed to bridge the gap between clinical and basic research. Above all, it is meant to provide an update in this area of research, presented by most distinguished researchers in the field. We do hope that this book will be used by many clinicians in order to improve counselling and treatment of patients with chemosensory dysfunction. Above and beyond this, we would be more than delighted if this book inspires clinical colleagues to perform basic research on the chemical senses where many questions are still open.

Thomas Hummel
Antje Welge-Lüssen

Hummel T, Welge-Lüssen A (eds): Taste and Smell. An Update.
Adv Otorhinolaryngol. Basel, Karger, 2006, vol 63, pp 1–22

Nasal Anatomy and the Sense of Smell

David E. Hornung

St. Lawrence University, Canton, and Neuroscience and Physiology Department,
Upstate Medical University, Syracuse, N.Y., USA

Abstract

As a result of the relative sizes of the various compartments in the nasal cavity, the bulk of the airflow is along the floor of the nasal cavity. The percent of airflow directed to the olfactory region (the superior region of the nasal cavity) is about 10%. Structural changes in the nasal cavity can alter airflow pathways and the characteristics of the airflow (e.g. laminar, mixed or turbulent) within nasal compartments. The relationship between the olfactory response and the stimulus is complex and may vary depending on the physiochemical properties of the odor and the rate at which odorants are delivered to the olfactory receptors. Changes in nasal airflow may impact the various olfactory functions (e.g. identification, differentiation) differently. When there is a nasal obstruction, a decline in olfactory ability may not simply be an access problem, since nasal disease can affect olfactory processing at many levels.

Nasal Anatomy

The internal anatomy of the nose (fig. 1) is divided into two halves along the midline by a bony structure called the nasal septum. The outline of the lateral wall is delineated by the curves of the inferior, middle and superior turbinates [1]. The respiratory and olfactory epithelial cells lining the structures of the internal nose have a very rich blood supply and are covered by a watery mucus that is continually flowing into the back of the throat [2–4]. By altering the blood flow in the dense capillary beds servicing the structures of the internal nose, the size of the airspace can be changed quickly and dramatically [5, 6]. Because these areas can change shape so quickly, they are sometimes referred to as swell spaces. The structural changes in these spaces can alter both the airflow pathways through the nose and the characteristics of the airflow (e.g. laminar, mixed or turbulent) within the various compartments of the nasal cavity [6].

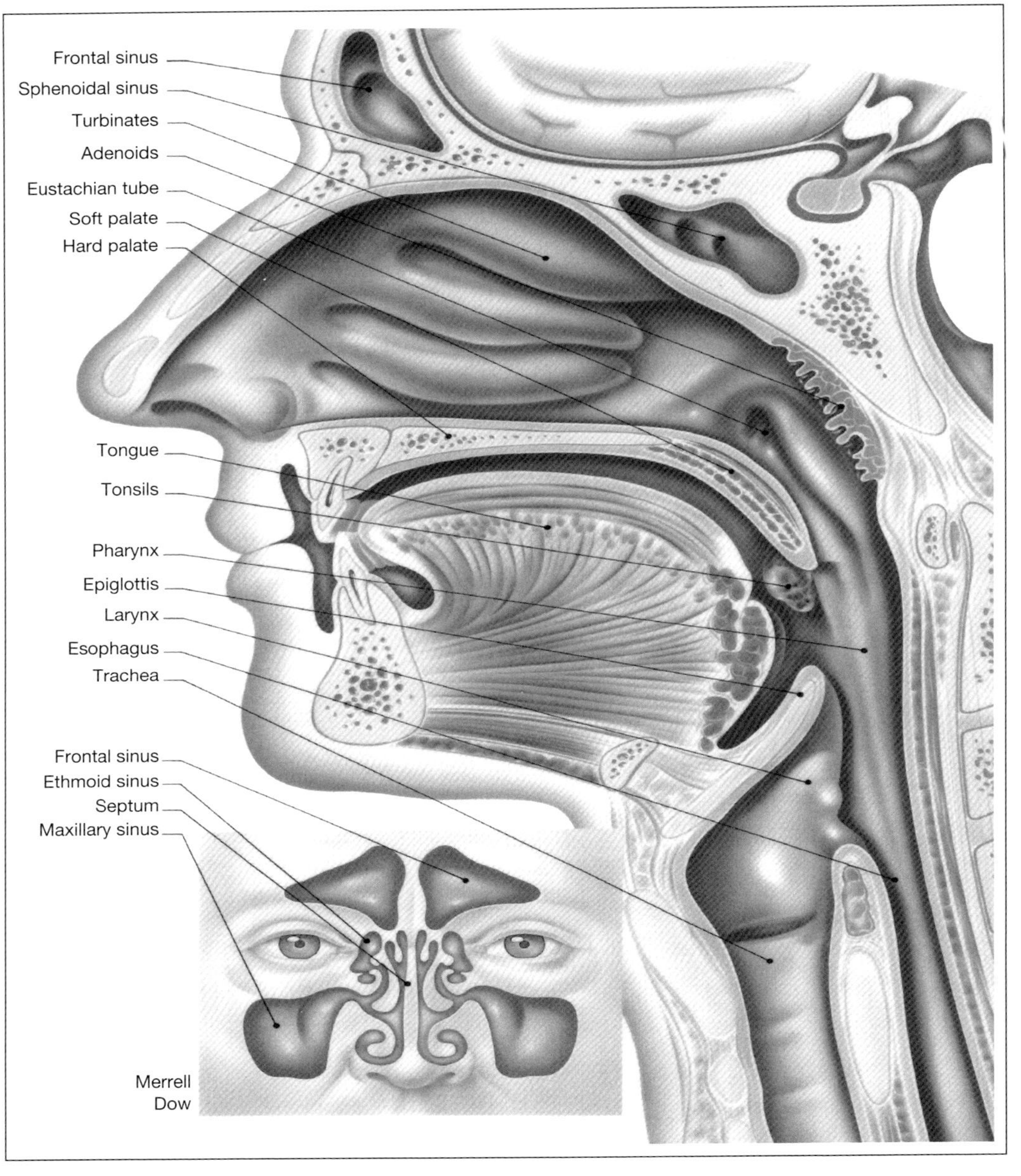

Fig. 1. A cutaway view of the human head. The inferior, middle and superior turbinates are located above the hard palate. The olfactory area is around the superior turbinate. The insert shows a cross-sectional view of the airspaces in the human nose (from Merrell Dow Company).

The cross-sectional area and length of the nasal cavity are the structural constraints, which, in concert with the pressure gradients created by the lungs, determine the amount of airflow in the nose. As will be described in greater detail later in this chapter, the nasal septum and turbinates produce multiple and convoluted flow paths for both the inspired and expired air [7]. Although a subject of some debate, it is generally felt that during breathing, as air moves through some of the more torturous passageways, the flow that is often laminar may become turbulent [8]. Turbulent airflow requires more energy to generate, but because of better mixing also likely makes the various nasal functions more efficient.

By observing the behavior of aerosolized water particles in inspired air, Simmen et al. [8] observed turbulence even at flows in the lower end of those usually seen in the human nose. In addition, as expected, the airflow showed an initial period of acceleration, a period approaching a steady state and a period of deceleration. The higher airflow seen during sniffing may increase the amount of turbulence and so may improve the sensitivity to smells. Turbulent flow may, under some conditions, also improve the efficiency of the nonolfactory functions [9] of the nose including that of providing humidity and temperature control for the inspired and expired air [6]. Additionally, turbulence may facilitate the ability of the mucus to trap foreign material like smoke, dust, bacteria and viruses that are often found in the inspired air [6].

Because of the relative sizes of the various compartments in the nasal cavity, the bulk of the airflow is along the floor of the nasal cavity with the second largest region of airflow being along the middle meatus close to the septum. The percent of airflow directed to the olfactory region (the superior region of the nasal cavity) is about 10% [10–13].

Olfactory Physiology

When air is brought into the nose during breathing or sniffing, odorant molecules pass through the nasal valve area on their way to the headspace above the mucus-coated olfactory receptors. Once in the airspace above the receptors, odorant molecules bind to the receptors on the cilia that are located on the end of the olfactory receptor cells. As will be described in more detail in the next chapter, when odorant molecules bind to the odorant receptors, the structure of the membrane-bound proteins changes in such a way as to allow extracellular calcium ions to enter the cell. This produces a change in the membrane potential at the tip of the olfactory receptor cell that in turn creates an electronic signal that flows along the axons of the olfactory neurons to the olfactory bulb.

Inherent Mucosal Activity Patterns

The olfactory bulb is composed of a number of different types of cells, although the axons of the olfactory receptor cells first make contact with the glomerular cells. Receptors that respond to some common chemical feature (there are approximately 350 different cell types loosely arranged into zones) send their signals to specific glomerular cells. Since most odorants stimulate more than one type of receptor cell, each odorant produces a response pattern across the glomerulus that is unique for that particular odorant. In other words, the olfactory receptor sheet disassembles the odorant molecule into a unique pattern of its functional groups. This disassembled pattern is maintained and perhaps sharpened in the glomerular cell layer of the bulb and is then sent to more central areas (like the primary olfactory cortex) where the patterns are reassembled for processing [14, 15].

Since receptor cells of similar sensitivity are grouped in particular locations along the olfactory receptor sheet, different smells produce different patterns of electrical activity in both the mucosa and in the olfactory bulb. These patterns are perhaps one of the ways the brain can identify a particular smell. The signal created by the specific tuning of olfactory receptor cells is called the 'inherent' mucosal activity pattern [16, 17].

Imposed Mucosal Activity Patterns

As described above, the olfactory receptors are located high in the nose along the septum and in the region of the superior turbinate. Once odorant molecules arrive at the headspace above the olfactory receptors, the molecules distribute themselves along the long axis of the mucosal sheet in patterns reflecting the physical and chemical properties of the odorant molecules themselves. For a chemical that is highly soluble in the mucus, most of the incoming molecules will be trapped early in the flow path, producing a very uneven distribution of odorant molecules along the long axis of the mucosal sheet (fig. 2). On the other hand, for a chemical that is only slightly soluble in the mucus, odorant molecules will be more evenly distributed along the flow path. The specific mucosal odorant distribution pattern created by an odorant's solubility may be another mechanism by which the central nervous system identifies smells. These distribution patterns are called 'imposed' patterns [18–21].

There is electrophysiologic [22–24], radioisotopic [20, 25, 26], and gas chromatographic evidence [27] to support the existence of imposed mucosal activity patterns in a variety of nonhuman animal species. However, for obvious

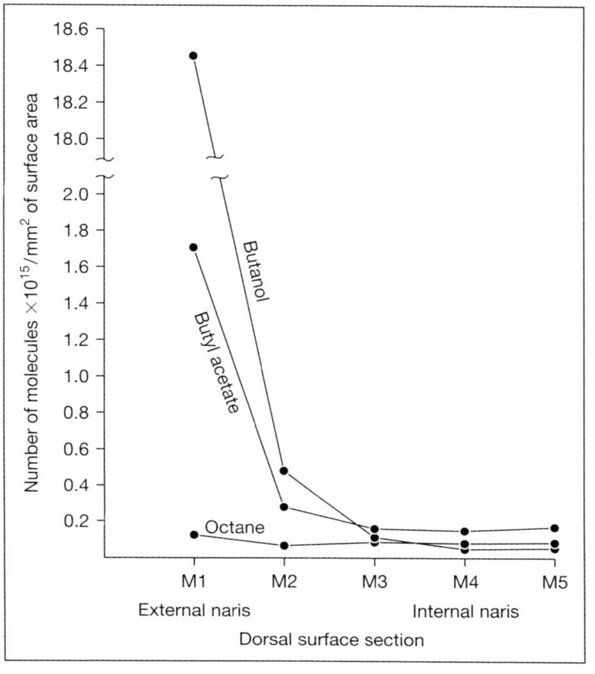

Fig. 2. The distribution of radioactive butanol, butyl acetate and octane across the dorsal surface of the olfactory sac of the bullfrog. Note how for butanol (a very soluble odorant) most of the molecules are sorbed in the mucosal section by the external naris whereas for octane (a poorly sorbed odorant) the sorbed molecules are evenly distributed across the mucosal surface. Butyl acetate, a moderately soluble odorant, has a distribution pattern between the other odorants (from Hornung and Mozell [21]).

reasons, the techniques available to study mucosal distribution patters in non-humans are not appropriate for use in people.

It should be emphasized that imposed and inherent patterns may not be mutually exclusive. They may work in concert to allow humans and other animals to identify a very wide variety of smells. Because they are created by the flow of air across the olfactory receptor sheet, imposed patterns may be more susceptible than inherent patterns to changes in airflow [18, 21].

A Role for Mucosal Patterns in Olfactory Coding

Although a number of investigators have documented the existence of imposed and inherent patterns using a variety of techniques in a number of different species, the role they play in olfactory coding remains a matter of some discussion. To understand the difficulty in determining the role that these patterns play in odor detection (and how the patterns are affected by changes in nasal airflow), one needs to first appreciate the conundrum faced by investigators who are studying the sense of smell. It is relatively easy to ask humans if they smell a particular chemical; at some level, it is even possible to ask them how similar one chemical smells to another. However, it is usually not possible in any systematic and/or permanent way to alter the imposed and inherent patterns produced in the human nose. As a result, the experimental manipulations appropriate in a hypothesis-based exploration of human olfaction are not generally possible. On the other hand, anatomical, biochemical and physiological manipulations are possible in nonhuman animals, but it is very difficult to ask the animals if they smelled anything and it is even more difficult to ask them how similar one chemical is to another. To help solve this conundrum, the five-odorant identification confusion matrix was developed to test the hypothesis that inherent and imposed patterns, at least in nonhuman animals, have some significance in olfactory processing [28].

In the five-odorant confusion matrix, after sampling a target smell, an animal sampled five test ports to determine which port smelled most like the target. The animal indicated his choice by pressing a bar in front of the appropriate port. This technique not only made it possible to measure correct olfactory identification, but by analyzing the off-diagonal response, also made it possible, at least indirectly, for the animal to indicate how similar odors are to each other. The logic is if two odors are often confused they must share some similar perceptual quality and odors not often confused must be more perceptually dissimilar. A multidimensional scaling analysis based on the pattern of errors allowed for a graphical representation of the perceptual relationship among the odors used in the confusion matrix.

By comparing the animal confusion (multidimensional scaling) matrix data with voltage-sensitive dye recordings from the mucosa and the olfactory bulb, Youngentob et al. [29, 32] and Kent et al. [30, 31] were able to demonstrate a relationship between the differential activity patterns and the perceptual characteristics of the odors. That is, odors having similar mucosal activity patterns are more often confused than odors producing very different mucosal activity patterns. In other words, the mucosal activity patterns produced at the mucosal level (a result of imposed and inherent patterns) seem to mirror the psychophysical relationship seen among odors.

If the mucosal patterns play a role in olfactory perception, changing the patterns should change olfactory perception. To test this hypothesis, Youngentob et al. [32] changed the mucosal activity patterns using olfactory marker protein gene depletion. As predicted, as the mucosal activity patterns changed so did olfactory perception.

Table 1. Relationship between primary and derived stimulus variables

Primary variable	Derived variable	Formula
N (number of molecules)	C (concentration)	C = N/V
V (sniff volume)	D (delivery rate)	D = N/T
T (sniff time)	F (flow rate)	F = V/T

So, in summary, as different odorant molecules are delivered to the headspace above the olfactory receptors, different mucosal activity patterns are produced. These patterns are a result of the distribution of the selectively tuned receptors within the olfactory epithelium (inherent patterns) and the physiochemical properties of the odorants as they interact with the various constituents of the mucosa (imposed patterns). Based on nonhuman animal studies, the imposed and inherent patterns seem to mirror olfactory quality perception [32].

Nasal Airflow – Sniff Variables

Obviously, if no odorant molecules get to the receptors, there will be no olfactory response. It therefore seems reasonable to suggest as the number of odorant molecules delivered to the receptor area increases so should the olfactory response. Unfortunately, because of the location of the olfactory receptors in the nasal cavity and because odorants have different physiochemical properties, the relationship between nasal airflow and olfaction is more complex than a simple relationship between the number of molecules and the olfactory response.

Mozell et al. [33] suggested that the olfactory stimulus is defined by three 'primary' variables: the number of molecules (N), the duration of the sniff (T), and the volume of the sniff (V). These primary variables in turn define the three 'derived' variables of concentration (C = N/V), delivery rate (D = N/T) and flow rate (F = V/T). Together these six variables (table 1) characterize the nature of the stimulus delivered to the olfactory receptors.

These variables are not independent of each other. That is, N delivered to the receptors cannot be increased alone, since an increase in N will also increase C and D if T and V remain unchanged. So, if, e.g., increasing N results in an increase in the olfactory response, it will be impossible to attribute the increase to the effect of N alone since C and D also changed. Therefore, Mozell et al. [29] suggest that until there is recognition of the interrelationship of these stimulus variables, the full description of the relationship between nasal airflow and the olfactory response will not be possible.

Still there is a large body of evidence suggesting that the olfactory response (R) is proportional to C such that:

$R \sim C^x$

This relationship suggests R is proportional to the log of C of the stimulus. Since C is derived from the primary variables of N and V, this proportionality can be written as

$R \sim N^x/V^x$

If R is defined only by C, increasing N and V by the same proportion should result in no change in R. If, however, C is not the sole determiner of R, any change in R seen when N and V are increased by the same proportion would reflect the effect of these two primary variables independently of the effect that they have through C. Likewise, the effects of the derived variables of F and D could be studied in relation to the primary variables from which they originate. This is the logic that drove the NVT study described below.

In an effort to parcel out the interrelationship between the six sniff variables, Mozell et al. [34] designed an experiment in which all combinations of two levels of the three primary variables were presented to animals in which the olfactory nerve response was recorded. The levels for each variable were picked so that the levels of volume and time were in the natural range and the lower level of a variable was exactly 50% of the higher level. For example, a lower sniff time was picked from the lower range of what the animal normally produces. The higher sniff time was twice the lower duration but still in the animal's 'physiological' range. The same selection procedure was used in picking the two sniff volumes. The concentrations picked were in the moderate range and the concentration of the larger stimulus was twice the concentration of the smaller stimulus. This strategy resulted in 8 combinations of 'sniffs', which, because of the relationships described above, also resulted in three combinations of each of the derived variables. The odorant used in this study was octane, a chemical that is poorly sorbed by the mucosa and produces an even imposed pattern across the mucosal sheet. The species used was the bullfrog. From an analysis of variance of the log of the olfactory response (the dependent variable), a model using the three primary variables (NVT) was generated that best accounted for the variability in the neural activity. The predicted error variance for this model was 0.0523. This model was:

$R \sim N^{0.35}V^{-0.28}T^{0.22}$

Stated in words, the model proposes that the magnitude of the olfactory nerve response increases as the number of molecules and sniff time increase. Further, the model proposes that the magnitude of the response increases as the sniff volume decreases. This model did very well in explaining the variability in the olfactory nerve response.

However, because of the relationship between the primary and derived variables, it was possible to test other models that included combinations of the primary and derived variables. From this analysis, two additional models emerged that were even better than the NVT model in explaining the variability in the olfactory nerve response. These models were:

$R \sim C^{0.31}T^{0.22}$

$R \sim F^{-0.25}N^{0.35}$

In the NVT model, the three primary variables are independent of each other, but the combined models make a different statement about the olfactory nerve response. For example, in the CT model, the effects of the number of molecules and volume are equal and opposite to each other, while the effect of time is independent of the other two primary variables. Likewise, for the FN model, the effects of volume and time are equal and opposite while the number of molecules is independent of the other two. The other combined models (there were 11) were all very poor in explaining the variability in the olfactory nerve response. Of all the models tested, the FN was the best in explaining the variability in the olfactory nerve response with a predicted error variance of 0.0517.

The negative coefficient for flow rate might be explained in terms of the work of Stuiver [10] who, using a clear plastic nasal model and aluminum particles suspended in water, observed that as flow rate increased the percent of the incoming airstream directed to the olfactory area decreased. However, the modeling studies of Hahn et al. [11] and Keyhani et al. [12] and others have failed to replicate the observation of Stuiver. Most of the modeling studies have suggested that in the absence of a major change in nasal anatomy, the percent of incoming air delivered to the receptors is reasonably constant through a wide range of flow rates.

Rehn [35] used psychophysical techniques to study the effect of flow rate on the human olfactory response by presenting subjects with fixed concentrations of pyridine at different flow rates. Rehn increased flow rate in two ways, increasing sniff volume while holding sniff duration constant and decreasing sniff duration while holding sniff volume constant. The first strategy increased flow rate, delivery rate and the number of molecules. The second strategy increased flow rate and delivery rate while holding the number of molecules constant. The two strategies yielded surprisingly similar results, and so Rehn concluded that as flow rate increased so did the perceived intensity of the olfactory response.

To reconcile the apparent discrepancy between the FN model and the observations of Rehn, Mozell et al. [34] proposed that what appeared to be the flow rate effect was dependent on the physiochemical properties of the odorants. That is, they proposed for odorants only slightly mucosa soluble (like

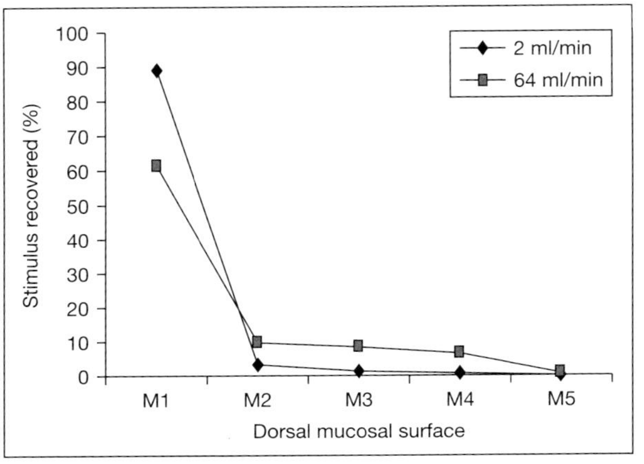

Fig. 3. The distribution of radioactive butanol across the dorsal surface of the olfactory sac of the bullfrog. Note how the distribution patterns wash out as the nasal airflow is increased from 2 to 64 ml/min (from Hornung and Mozell [25]).

octane used in the NVT study described above) that as flow rate increased the response decreased because the likelihood of molecules interacting with the mucosa was decreased. That is, as flow increased, a given molecule spent less time in the headspace above the receptors and so it was less likely to be sorbed by the mucus covering the receptors, decreasing the likelihood of an interaction with the odorant molecule and the olfactory receptor. However, for mucosa-soluble odorants like pyridine, as flow rate increased there were fewer molecules sorbed by the nonolfactory tissue early in the flow path and so more molecules would arrive at the olfactory receptor area. In addition, because of the higher flow rate, more molecules were available further along the flow path of the olfactory receptor sheet. As a result of both of these effects more molecules would be available to stimulate the olfactory receptors and so the olfactory response increased as flow rate increased. Figure 3 shows how the butanol distribution gradient changed across the dorsal surface of the bullfrog mucosa as flow rate increased from 2 to 64 ml/min.

To test the hypothesis that flow rate and the physiochemical properties of the odorants interact, Mozell et al. [34], using a wide rage of odorants and a wide range of flow rates, recorded summated multiunit discharges from two locations along the olfactory flow path in bullfrogs. The results of this study showed, as the FN model predicted, that there was a negative effect of flow rate for the odorants like octane that had very low mucosal solubilities. However, for odorants that were slightly more mucosa soluble than octane, there was less of a negative effect

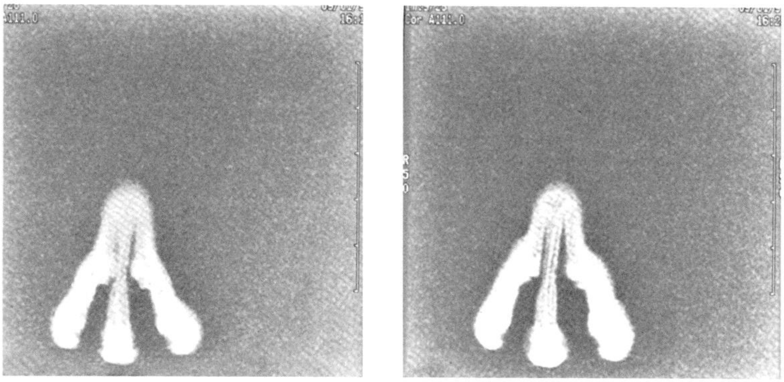

a *b*

Fig. 4. Coronal MRI scan from the soft anterior portion of the nose in a subject without (***a***) and with (***b***) a nasal dilator. Note the expected increase in the size of the nasal airspaces with the dilator present (***b***).

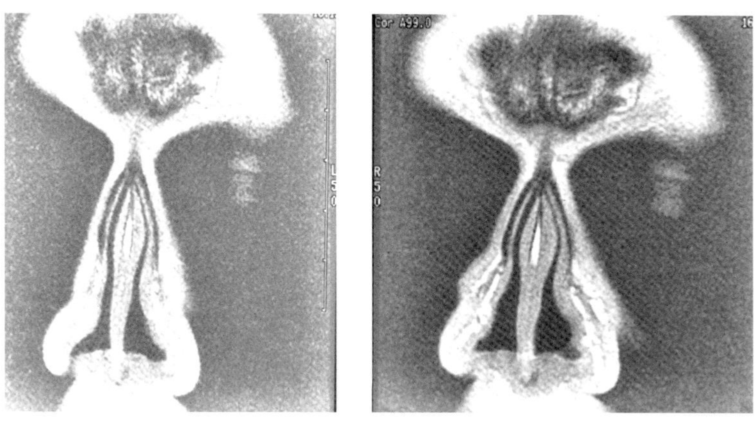

a *b*

Fig. 5. Coronal MRI scan from the bony region of the nose behind which one would expect a direct effect of a nasal dilator. Note the increase in the size of the air passageways with the dilator present (***b***).

of flow rate and for odorants with high mucosal solubilities, the effect of flow rate was positive. In conceptualizing the apparent flow rate effect, it should be remembered that the negative coefficient for flow rate in the FN model is much smaller in absolute magnitude than the positive coefficient for the number of molecules. As a result, an increase in the number of molecules associated with an increase in flow rate could override the negative effect due to the flow rate itself.

The FN model may also explain the observation by Sobel et al. [36] that, when performing an odor threshold test, humans sniff longer when using the nostril with the lower flow rate. (Subjects usually have different flow rates in their two nostrils because of cyclic changes in the size of their nasal cavities.) The FN model suggests increasing the number of molecules rather than flow rate would be a more effective way to improve olfactory performance. In this example, the number of molecules increased by increasing sniff time.

Other Airflow Considerations

The perceptional qualities associated with 'comfortable' breathing are very complicated. Certainly the sensation of being comfortable is related to the degree of nasal openness and the amount of airflow. Also involved is the stimulation of cold receptors in the nasal valve region and trigeminal receptors located throughout the nasal mucosa. In addition, a dry nasal mucosa can influence the sensation of comfortable breathing [37].

Both uninasal and binasal resistance are related to patient complaints about 'uncomfortable' breathing, but there is some debate as to which is more important. Although one might suggest total nasal resistance would be the determiner of complaints, Arbour and Kern [38] suggest the obstruction on the more open side of the nose was the determiner of the presence or absence of nasal complaints.

The activity of many of the central processing centers associated with olfaction has for a long time been known to be phase-linked to respiration. That is, regions of the primary olfactory cortex show much more activity when there is air moving through the nasal cavity than when there is no airflow. Using fMRI, Sobel et al. [39] have been able to show that airflow itself induces a large portion of the neural activity seen in the primary olfactory cortex. In other words, an olfactory sensation is a combination of the central nervous system activity related to the sniff (e.g. the activity of the motor cortex that generated the negative pressure for the sniff) plus the firing of the olfactory receptors. Since airflow through the two nostrils is different (in part because of the nasal cycle), it should follow that complex odors should smell differently to the two nostrils, a hypothesis for which Sobel et al. [39] have been able to generate at least some psychophysical data.

Mathematical Models of Nasal Airflow

Over the past 10 years, mathematical models have been shown to be useful in describing the relationship between nasal airflow and olfactory ability.

Models were first used to describe the mass transport of odorant molecules from the air to the olfactory receptors [11–13, 40]. The output of these models predicted the flow rate, length and thickness of the olfactory mucosa and air/mucosa partitioning of the odorant all were important in determining the intensity of the olfactory response. These models further predicted given 'an adequate mucus surface area', an increase in the flow rate should increase the perceived intensity for odorants with high solubility in the mucus (as reported in the study by Rehn [35] – see above) whereas for poorly sorbed odorants (like those used by Mozell et al. [33] in the NVT study), an increase in flow rate should decrease the perceived intensity. With a reduced surface sorption area, like that seen in some individuals or with certain nasal diseases, increasing flow rate will result in a decreased perceived intensity for all odors.

In general, flow path descriptions by the various modeling studies have yielded similar results. That is, there seems to be a much higher velocity of flow in the nasal valve region and the floor of the nasal cavity and a much lower velocity in the olfactory area [13].

In another modeling approach, Zhao et al. [41], using computational fluid dynamics techniques (CFD), predicted airflow and odorant transport in various locations in the nose as the shape of the cavity itself changed. Numerical finite volume methods were used to evaluate how flow to the olfactory area was altered as the size of critical nasal areas like the olfactory slit and nasal valve region were changed.

For example, Zhao et al. [41] modified the nasal valve area and modeled the total nasal airflow rate and flow to the olfactory area. Although there were only slight changes to the total nasal airflow rate, the changes to airflow through the olfactory region were dramatic, with flow changing by 50% as the size of the airway increased or decreased. Obviously these changes in flow to the olfactory region were accompanied by corresponding changes in odorant uptake. As one example, a small blockage in the nasal valve region (1.45% reduction in local airway volume) resulted in an 18.7% decrease in nasal airflow and a 76.9% decrease in flow to the olfactory area. Since the nasal valve region seems to be the source of most of the nasal resistance [42], the modeling results provided convincing evidence that the nasal valve area is a key in controlling airflow to the olfactory region. In other words, the models support the hypothesis that small changes in critical nasal regions can have profound effects on flow to the olfactory region.

CFD has also been recently combined with an experimental procedure to further describe some of the factors that govern odorant mucosal deposition [43]. The fraction of odorant absorbed by the nasal mucosa was experimentally determined for a number of odorants by measuring the concentration that occurred when an odorant was 'blown' into one nostril and exited the contralateral nostril.

Subjects performed velopharyngeal closure during this procedure. The nasal air/odorant airflows seen during the velopharyngeal closure were modeled using CFD techniques in a model of the human nose [36] and the mucosal odorant uptake was numerically calculated. The comparison between the numerical simulations and the experimental results led to an estimation of the human mucosal odorant solubility. The study suggested that the increase in diffusive resistance of the mucosal layer over that of a thin layer of water seemed to be general and non-odorant-specific. However, the mucus solubility was odorant specific and usually followed the trend that odorants with lower water solubility were more soluble in mucus. The ability of this model to predict odorant movement in the nasal cavity was evaluated by comparison of the model output with known values of odorant mucosal solubility. The results strongly suggest the physical processes involved in olfaction are quite consistent across the various animal species and so it seems reasonable to suggest the results obtained from the many animal studies have some relevance to the understanding of the processes involved in olfactory perception in humans. Combining CFD with the experimental measurements described above provides a technique to measure, in humans, the air/mucosa partition coefficient of any nontoxic odorant.

Nasal Dilators

Nasal dilators (Breathe Right), sometimes worn by athletes, are elastic strips with imbedded plastic springs that fit over the bridge of the nose. Despite their questionable usefulness in improving athletic performance, nasal dilators have been useful as a tool with which to begin to describe the relationship between nasal airflow and olfactory ability [44, 45]. The effect that nasal dilators have on nasal resistance is considerable. For example, Lorion et al. [42] using active posterior rhinometry estimated the presence of an internal nasal dilator reduces nasal resistance by about 40%.

The first step in assessing how dilators affect olfactory function was to determine exactly how dilators produced the dramatic reduction in nasal resistance. Toward that end, MRI and CT were used to determine where and by how much nasal dilators change the anatomy of the nose. In a recent study [46], when subjects wore nasal dilators, the volume of the soft anterior portion of the nose (defined as the nasal cavity from the tip of the nose to the beginning of the turbinates) was increased by 23% compared to the undilated condition. This result agreed well with previously published data [45] reporting dilators produced a 26% increase in the anterior nasal volume. Additionally, the recent studies demonstrate that, at least for 4 h, the anterior nasal volume remained elevated as long as the dilator was worn [46].

Table 2. Effect of nasal dilators on sniff characteristics

Parameter	Without dilator	With dilator
Sniff mean flow rate, l/min	45.3	54.8*
Sniff max. flow rate, l/min	80.0	92.8*
Sniff volume, l	1.09	1.54*
Sniff duration, s	1.4	1.7*

Twelve patients were examined. *$p < 0.05$ (all means were significantly different, paired t test).

In analyzing the volume changes in the bony regions of the nose, the portions behind which one would not expect the dilator to have a direct effect, nasal volume still increased by 9% when the dilator was in place [46]. This agreed well with previously published results [45]. The increase in the volume of the posterior portion of the nose remained as long as the dilator was kept in place. The increase in the volume of the nasal cavity in the bony region must reflect a response of the nose to the dilation of the soft anterior region, a response perhaps mediated by changing blood flow to the swell spaces mentioned above [45].

With the anatomical changes quantified, the next step in determining the effect nasal dilators had on olfactory ability was to determine their effect on the physical characteristics of the sniff. A pneumotachograph measured the sniffing behavior of subjects with and without a dilator in place. As can be seen in table 2, wearing a nasal dilator increased all the characteristics of the sniff compared to the values seen in the undilated condition.

Since wearing the nasal dilator increased the size of the air passageways in the nose and the flow characteristics of a sniff, the parsimonious conclusion is that the pressure generated by the sniff is 'hardwired'. That is, the respiratory muscles seem to be programmed by the nervous system to generate a certain amount of negative pressure. So, when the nasal resistance was decreased (as happens when a nasal dilator is worn), there was an increase in the sniff characteristics [45].

In the mirror image of the nasal dilator experiment, Youngentob et al. [47] studied the effect increasing nasal resistance had on odor intensity. As the nasal resistance increased subjects rated inspired odors as being less intense. At first, this would seem to suggest that with the increase in resistance there was a decrease in the volume and flow rate of the subject's sniff (and so the number of molecules delivered to the receptors was decreased). However, since the sniff flow rate and volume did not change with an increased nasal resistance, Youngentob et al. [47] proposed the concept of a perceptual size constancy model to explain their results. In other words, the olfactory magnitude depends on both the concentration and the perceived effort associated with the sniff. Important to the present discussion is the observation that when presented with an increase in nasal resistance subjects maintained sniff flow rate, volume and time.

On the surface, the nasal dilator data showing an increase in the sniff characteristics seem inconsistent with the observation that when presented with an increase in nasal resistance subjects produce a constant sniff. It seems reasonable that there would be a mechanism to maintain flow as nasal resistance increases since this occurs often in nature. However, a major reduction in nasal resistance rarely happens naturally and so there would be no evolutionary pressure for a neural mechanism to reduce flow in this case. Additionally, as will be described below, olfactory function is apparently not compromised with a major reduction in nasal resistance, and so again there would not be a selective pressure to reduce flow when nasal resistance was suddenly decreased. Therefore, it seems possible that the body maintains minimum sniff characteristics when resistance is increased, but there is less regulation if the nasal resistance is suddenly increased.

The final step was to determine how the nasal dilator induced changes in anatomy and sniff characteristics affected olfactory ability as determined by a number of psychophysical techniques [45]. Since airflow to the olfactory area goes along a channel from the tip of the nose toward the superior turbinate and there was a substantial increase in the sniff characteristics, it was hypothesized that the changes produced by the dilator should have a positive effect on olfactory function, although given some of the constraints of the FN model, other outcomes were possible for at least some types of odorants.

Odorant identification (the ability to name odorants) was evaluated with the odorant confusion matrix, a clinical test developed to assess olfactory function. Since all the subjects tested in the present study had normal senses of smell, the concentrations of the odorants were reduced by 60% to make the test more difficult [48].

Olfactory threshold was evaluated with a 2-interval forced-choice test of phenethyl alcohol (PEA). This compound has a rose-like smell, and is routinely used to measure olfactory threshold. Subjects were presented with two bottles and asked to identify the bottle that contains the odorant. Subjects began with a very low concentration of PEA in the test bottle and each time the subjects picked the incorrect bottle the concentration was increased. This process continued until the subjects correctly picked the bottle containing the odorant five times in a row [49].

Subjects used the Green scale [50] to report their evaluation of the intensity of the 10 odorants in the odorant confusion matrix. Subjects were presented with a bottle containing the smell and asked to indicate how intense the odorant smelled to them. The intensity ratings were compared with and without the dilator [46, 47].

The results of the olfactory testing can be seen in table 3.

Table 3. Effect of nasal dilators on olfactory function

Parameter	Without dilator	With dilator
Correct, %	78	99*
Intensity ratings	7.9	12.0*
Threshold dilution	13.4	15.2*

Twelve patients were examined. *$p < 0.05$ (all means were significantly different, paired t test). For the threshold test, a bigger number means a lower concentration of the PEA. So, a threshold of 15 contained two more binary step dilutions as compared to the concentration seen with a threshold of 13.

The hypothesis described above the perceptual size constancy model in olfaction might account in part for the increase in odor intensity seen when subjects wear nasal dilators. In other words, because subjects had to work less to produce a sniff when presented with a decreased nasal resistance, this could translate into an increase in the perceived intensity of the incoming odorant. Although it is possible the size constancy accounted for some or all of the increase in perceived intensity observed with nasal dilators, it is more likely the contribution that nasal resistance makes to odor intensity is small under the conditions of normal or above-normal nasal airflow.

The nasal dilator data seem to document a relationship between nasal anatomy and olfactory ability. However, how these data fit into the NVT model and the airflow patterns has yet to be fully explained. The relationship of these data to inherent and imposed mucosal distribution patterns is also not clear.

Clinical Considerations

Given the complex nature of the interaction between olfactory function and nasal airflow, it is perhaps not surprising that not all patients who have had surgery to open their nasal passageways show an improvement in olfactory ability. Although this topic will be explored in much more detail in a later chapter, the relationship between nasal surgeries and olfactory ability is explored here as it relates to airflow.

Kimmelman [51] reported that following nasal surgery 66% of the patients had either an improvement or no change in olfactory function whereas 32% showed a decline in their olfactory function (as measured by an identification test). Rowe-Jones and Mackey [52] reported an improvement in olfactory function following surgery in most of their patients and the improvement was correlated with an increase in nasal volume. In a more definitive study, Damm et al. [53] reported an increase in airflow in 87% of their patients following a partial inferior turbinectomy with septoplasty. However, only 80% of these patients showed an improvement in their ability to identify odors and only 54% showed an improvement in their olfactory thresholds.

How the observations of Damm et al. [53] fit into the relationship between nasal airflow and the various olfactory functions has yet to be determined. Perhaps, as the nasal dilator data suggest, airflow has a more pronounced effect on intensity ratings than on olfactory threshold. In the absence of more data, however, it is difficult to even suggest hypotheses concerning the relationship between the specific olfactory functions and nasal airflow.

It is tempting to assume that when there is a nasal obstruction, a decline in olfactory ability is related simply to an access problem. However, Doty and Mishra [54], confirming the observations of Damm et al. [53], report that surgical interventions are often not successful in resolving olfactory losses seen in patients with rhinosinusitis.

When there are other overriding olfactory problems in patients with conductive problems, the task of generating hypotheses concerning the cause of an olfactory loss becomes even more difficult. As one possibility, in some cases nasal disease may be accompanied by changes in the olfactory mucosa, which could, under the right conditions, have a profound impact on the ability to detect and identify odors [55]. Additionally, as nasal airflow decreases, it has been reported there is a decline in mucociliary transport, which, under some circumstances, could influence olfactory ability.

The clinical observations reported above highlight that the relationship between nasal anatomy and olfactory ability is quite complex and clearly has yet to be fully appreciated.

Future Directions

Below are possible directions for future research.

(1) Given the complex nature of the interaction of the sniff parameters and olfactory ability, an attempt needs to be made to describe more fully in humans the interaction of the various sniff variables. What is needed is the human equivalent of the NVT study. Unfortunately, because of the location of the olfactory receptors in the human nose, the design and implementation of this study will not be trivial.

(2) Given the results of Damm et al. [53], the question of the relationship between nasal airflow and specific olfactory functions needs to be much

better described. Questions to be answered include: is the ability to detect the presence of an odorant (threshold) less sensitive to anatomical and flow changes than is the ability to distinguish and recognize odors? What is the relationship between nasal airflow and intensity ratings, and is this relationship the same for all odorants? How do airflow changes affect odor discrimination and recognition tasks?

(3) If airflow affects different olfactory functions differentially, what does this say about the peripheral and central processing mechanisms? Can studies considering the relationship between nasal airflow and specific olfactory functions suggest hypotheses that can be tested in nonhumans?

(4) As models of the airflow patterns become more sophisticated, it may become possible for them to predict how nasal surgery will affect olfactory function in a particular patient. It may even be possible for models to suggest surgical alterations to maximize olfactory ability following a surgical intervention.

Conclusions

(1) Although there seems to be a direct relationship between the olfactory response and stimulus concentration, the olfactory stimulus is actually defined by three 'primary' variables: the number of molecules (N), the duration of the sniff (T), and the volume of the sniff (V). These primary variables in turn define the three 'derived' variables of concentration ($C = N/V$), delivery rate ($D = N/T$) and flow rate ($F = V/T$).

(2) The relationship between the olfactory response and the stimulus may vary depending on the physiochemical properties of the odor.

(3) Nasal airflow may impact the various olfactory functions differently. As one possibility, intensity and discrimination tasks may be more sensitive to flow changes than are measures of odor threshold.

(4) Although it is tempting to assume that when there is a nasal obstruction, a decline in olfactory ability is related simply to an access problem, it should be realized that nasal disease could affect olfactory processing at many levels. As a result, changes in airflow may not always be the cause of an altered olfactory ability.

References

1 Churchill SE, Shackelford LL, Georgi JN, Black MT: Morphological variations in the upper respiratory track and airflow dynamics. Am J Physiol Anthropol Suppl 1966;28:107.

2 Shea BT: Eskimo craniofacial morphology, cold stress and the maxillary sinus. Am J Anthropol 1977;47:289–300.

3 Cole P: Modification in inspired air; in Procter DF, Anderson I (eds): The Nose: Upper Airway Physiology and the Atmospheric Environment. Amsterdam, Elsevier Biomedical Press, 1982, pp 351–375.
4 Barr GS, Tewary AK: Alteration of airflow and mucociliary transport in normal subjects. J Laryngol Otol 1993;107:603–604.
5 Hornung DE: Smell; in Hoagstrom CW (ed): Magill's Encyclopedia of Science: Animal Life. Pasadena, Salem Press, 2002, pp 1514–1516.
6 DeWeese DD, Saunders WH: Textbook of Otolaryngology, ed 3. St Louis, Mosby, 1968.
7 Calhoun KH, House W, Hokanson JA, Quinn FB: Normal nasal airway resistance in noses of different sizes and shapes. Otolaryngol Head Neck Surg 1990;103:605–609.
8 Simmen D, Scherrer JL, Moe K, Heinz B: A dynamic and direct visualization model for the study of nasal airflow. Arch Otolaryngol Head Neck Surg 1999;125:1015–1021.
9 Wolpoff MH: Climatic influence on skeletal nasal aperture. Am J Physiol Anthropol 1968;29: 405–424.
10 Stuiver M: Biophysics of the sense of smell; thesis, Groningen, 1958.
11 Hahn I, Scherer PW, Mozell MM: A mass transport model of olfaction. J Theor Biol 1994;167: 115–128.
12 Keyhani K, Scherer PW, Mozell MM: A numerical model of nasal odorant transport for the analysis of human olfaction. J Theor Biol 1997;186:279–301.
13 Kelly JT, Prasad AK, Wexler AS: Detailed flow patterns in the nasal cavity. J Appl Physiol 2000;89:323–337.
14 Luffingwell JC: Olfaction – a review. Online at http://www.leffingwell.com/olfaction.htm
15 Wilson DA: Synthetic coding of odorant mixtures in rat piriform cortex. Chem Senses 2002; 27:667.
16 Moulton DG: Spatial patterning response to odors in the peripheral olfactory system. Physiol Rev 1976;56:578–593.
17 Mozell MM, Sheehe PR, Hornung DE, Kent PK, Youngentob SL, Murphy SJ: Imposed and inherent mucosal activity patterns: their composite representation of olfactory stimuli. J Gen Physiol 1987;90:625–650.
18 Mozell MM, Sheehe PR, Hornung DE, Kent PK, Youngentob SL, Murphy SJ: The composite representation of olfactory stimuli by imposed and inherent mucosal activities patterns; in Miller I (ed): The Lloyd M Beidler Symposium on Taste and Smell. Winston-Salem, Book Service Associates, 1988, pp 143–158.
19 Hornung DE, Lansing RD, Mozell MM: The distribution of butanol molecules along the olfactory mucosa of the bullfrog. Nature 1975;254:617–618.
20 Mozell MM, Sheehe PR, Hornung DE, Kent PK, Youngentob SL, Murphy SJ: Imposed and inherent mucosal activity patterns: their composite representation of olfactory stimuli. J Gen Physiol 1987;90:625–650.
21 Hornung DE, Mozell MM: Smell – human physiology; in Rivlin RS, Meiselman RH (eds): Human Taste and Smell: Measurements and Uses. New York, Macmillan Publishing, 1986, pp 19–38.
22 Mozell MM: Evidence for sorption as a mechanism of the olfactory analysis of odorants. Nature 1964;203:1181–1182.
23 Kent PF, Mozell MM, Murphy SJ, Hornung DE: The interaction of imposed and inherent olfactory mucosal patterns and their composite representation in a mammalian species using voltage-sensitive dyes. J Neurosci 1996;16:345–353.
24 Kent PF, Youngentob SL, Hornung DE: Mucosal activity patterns and odorant quality perception; in Kurihara K, Suzuki N, Ogawa H (eds): Proceedings of the 11th International Symposium on Olfaction and Taste. Tokyo, Springer, 1994, pp 205–206.
25 Hornung DE, Mozell MM: Factors influencing the differential sorption of odorant molecules across the olfactory mucosa. J Gen Physiol 1977;69:343–361.
26 Hornung DE, Serio JA, Mozell MM: Olfactory mucosa/air partitioning of odorants; in van der Storee H (ed): Olfaction and Taste VII. London, Information Retrieval, 1980, pp 167–170.
27 Mozell MM, Jagodowicz M: Chromatographic separation of odorants by the nose: retention times measured across in vivo olfactory mucosa. Science 1973;181:1247–1249.
28 Youngentob SL, Markert LM, Mozell MM, Hornung DE: A method for establishing a five odorant identification confusion matrix task in rats. Physiol Behav 1990;47:1053–1059.

29 Youngentob SL, Kent PF, Sheehe PR, Schwob JE, Tzournaka E: Mucosal inherent activity patterns in the rat: evidence from voltage-sensitive dyes. J Neurophysiol 1995;73:387–398.
30 Kent PF, Youngentob SL, Sheehe PR: Odorant-specific spatial patterns in mucosal activity predict perceptual differences among odorants. J Neurophysiol 1995;74:1777–1781.
31 Kent PF, Mozell MM, Youngentob SL, Yurco P: Mucosal activity patterns as a basis for olfactory discrimination: comparing behavior and optical recordings. Brain Res 2003;981:1–11.
32 Youngentob SL, Margolis FL, Youngentob LM: OMP gene deletion results in an alteration in odorant quality perception. Behav Neurosci 2001;115:626–631.
33 Mozell MM, Sheehe PR, Swieck SW Jr, Kurtz DB, Hornung DE: A parametric study of the stimulation variables affecting the magnitude of the olfactory nerve response. J Gen Physiol 1984;83: 233–267.
34 Mozell MM, Kent PF, Scherer PW, Hornung DE, Murphy SJ: Nasal airflow; in Getchell TV, Doty RL, Bartoshuk LK, Snow JB Jr (eds): Smell and Taste in Health and Disease. New York, Raven Press, 1991, pp 481–492.
35 Rehn T: Perceived odor intensity as a function of airflow through the nose. Sens Processes 1978;2: 198–205.
36 Sobel N, Khan RM, Hartley CA, Sullivan EV, Gabriele JD: Sniffing longer rather than stronger to maintain olfactory detection threshold. Chem Senses 2000;25:1–8.
37 Pallanch JF, McCaffrey TV, Kern EB: Evaluation of nasal breathing function with objective airway testing; in Cummings CW Jr, Frederickson JM, Harker LA, Richardson M, Schuller DE: Otolaryngology, ed 3. St Louis, Mosby, 1998, pp 799–832.
38 Arbour P, Kern EB: Paradoxical nasal obstruction. Can J Otolaryngol 1975;4:333.
39 Sobel N, Prabhakaran V, Desmond JE, Glover GH, Sullivan EV, Goode RL, Gabrieli JDE: Sniffing and smelling: separate subsystems in the human olfactory cortex. Nature 1998;392: 282–286.
40 Hahn I, Scherer PW, Mozell MM: Velocity profiles measured for airflow through a large-scale model of the human nasal cavity. J Appl Physiol 1994;75:2273–2287.
41 Zhao K, Scherer PW, Hajiloo SA, Dalton P: Effect of anatomy on human nasal airflow and odorant transport patterns: implications for olfaction. Chem Senses 2004;29:365–379.
42 Lorino AM, Lofaso F, Dahan E, Coste A, Harf A, Lorino H: Combined effects of a mechanical nasal dilator and a topical decongestant on nasal airflow resistance. Chest 1999;115:1514–1518.
43 Kurtz DB, Zhao K, Hornung DE, Scherer P: Experimental and numerical determination of odorant solubility in nasal and olfactory mucosa. Chem Senses 2004;29:763–773.
44 Hornung DE, Chin C, Kurtz DB, Kent PF, Mozell MM: Effect of nasal dilators on perceived odor intensity. Chem Senses 1997;22:177–180.
45 Hornung DE, Smith DJ, Kurtz DB, White T, Leopold DA: Effect of nasal dilators on nasal structures, sniffing strategies, and olfactory ability. Rhinology 2001;39:84–87.
46 Lyng GD, Hornung DE, Leopold DA, Irwin SB, Vent J: The long-term effect of nasal dilators on nasal anatomy and olfactory ability (abstracts). 25th Annu Meet Assoc Chemorecept Sci, Sarasota, 2003, p 89.
47 Youngentob SL, Stern NM, Mozell MM, Leopold DA, Hornung DE: Effect of airway resistance on perceived odor intensity. Am J Otolaryngol 1986;7:187–193.
48 Kurtz DB, White TL, Sheehe PR, Hornung DE, Kent PF: Odorant confusion matrix: the influence of patient history on patterns of odorant identification and misidentification in hyposmia. Physiol Behav 2001;72:595–602.
49 Cain WS, Gent J, Catalanato FA, Goodspeed RB: Clinical evaluation of olfaction. Am J Otolaryngol 1983;4:252–256.
50 Green BG, Dalton P, Cowart B, Shaffer G, Rankin K, Higgins J: Evaluating the 'labeled magnitude scale' for measuring sensations of smell and taste. Chem Senses 1996;21:323–334.
51 Kimmelman CP: The risk to olfaction from nasal surgery. Laryngoscope 1994;104:981–988.
52 Rowe-Jones JM, Mackay IS: A prospective study of olfaction following endoscopic sinus surgery with adjuvant medical treatment. Clin Otolaryngol 1997;22:377.
53 Damm M, Eckel HE, Jungehulsing M, Hummel T: Olfactory changes in threshold and suprathreshold levels following septoplasty with partial inferior turbinectomy. Ann Otol Rhinol Laryngol 2003;112:91–97.

54 Doty RL, Mishra A: Olfaction and its alteration by nasal obstruction, rhinitis, and rhinosinusitis. Laryngoscope 2001;111:409–423.
55 Kern RC: Chronic sinusitis and anosmia: pathologic changes in the olfactory mucosa. Laryngoscope 2000;110:1071–1077.

Prof. David E. Hornung
Dana Professor of Biology, St. Lawrence University
Canton, NY 13617 (USA)
E-Mail dhornung@stlawu.edu

Hummel T, Welge-Lüssen A (eds): Taste and Smell. An Update.
Adv Otorhinolaryngol. Basel, Karger, 2006, vol 63, pp 23–43

Transduction and Coding

Nancy E. Rawson, Karen K. Yee

Monell Chemical Senses Center, Philadelphia, Pa., USA

Abstract

Odor transduction and quality coding involves a cascade of events that occur at the level of the olfactory epithelium and olfactory bulbs. Odorants bind to one or a few specific olfactory receptors located in the cilia of olfactory neurons. These olfactory receptor proteins make up the largest gene family discovered and are diverse between and within species. The change of chemical signals to neural signals in the olfactory neurons involves G-coupled proteins and the cascade of second messenger pathways that open ion channels to depolarize the cell and trigger a series of action potentials carried along the receptor cell axon resulting in release of glutamate at synapses with mitral cells within the olfactory bulb. These neural signals in the olfactory bulb produce unique odor maps that play an important role in our ability to detect and discriminate thousands of different odorants. The olfactory neurons are replaced throughout life from a population of slowly dividing basal cells within the epithelium. Disease, infection, injury or aging can interfere with neuronal cell replacement as well as transduction and coding processes, resulting in impairment and distortions of olfactory performance.

This chapter will describe the cellular events occurring in the olfactory epithelium (OE) and at the level of the olfactory bulb that are used to detect, transduce and encode olfactory information. We will present current perspectives on the cell biology of olfaction, emphasizing studies most relevant to understanding normal and diseased olfactory function in humans.

Cellular Anatomy

The OE is a layered structure comprised of neuronal and nonneuronal cell types that resides within the olfactory cleft and extends to varying degrees onto the superior turbinate [1] and superior aspect of the medial turbinate [2] [see chapter 1 by Hornung, this vol, pp 1–22]. The OE is typically well

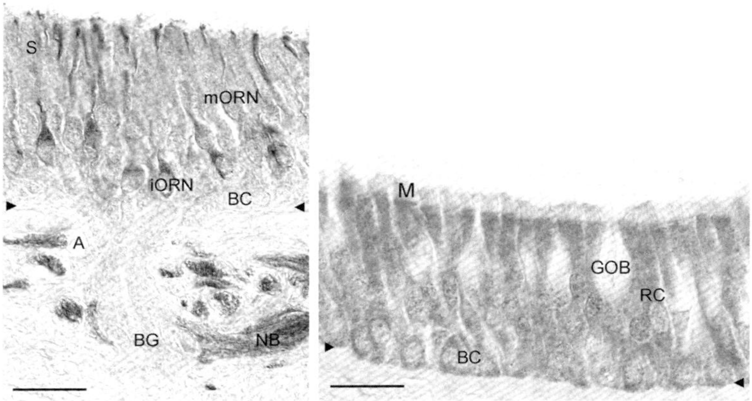

a *b*

Fig. 1. ***a*** Human OE. ***b*** Human RE. Arrow head = Basement membrane; BC = basal cell; BG = Bowman's gland; GOB = goblet cell; iORN = immature ORN; M = microvillar; mORN = mature ORN; NB = olfactory nerve bundle; RC = respiratory cell; S = supporting cell. Scale bar = 20 μm.

organized into a laminar array of cellular elements, but this arrangement is less defined in humans (fig. 1a). A thin basal lamina forms the foundation for a layer of basal cells that give rise to progenitor cells that retain the ability to divide throughout life. These progenitor cells differentiate into neuronal precursor cells that differentiate further into immature and then mature neurons, although the details of this process are not fully delineated [3, 4]. Olfactory receptor neurons (ORNs) are morphologically distinct, comprising a bipolar cell body with a dendrite projecting to the lumenal surface terminating in a swelling called an olfactory knob. Projecting from the knob are thin (approx. 0.1 μm in diameter) cilia that extend into the mucus lining the nasal cavity (fig. 1a). These nonmotile cilia are 5–20 μm long and provide an extensive surface area accessible for interaction with odorant molecules. Odorants that penetrate the mucus interact with receptor proteins present in the ciliary membrane that link odorant binding to a second messenger cascade which leads to excitation (see below). Basally, the neurons extend their unmyelinated axons in bundles through the cribriform plate to synapse with the dendrites of mitral and tufted cells in the olfactory bulb. These synaptic networks form distinct structures called glomeruli in the olfactory bulb (fig. 2). The glomeruli are encircled by periglomerular cells that also synapse with the dendrites of the mitral cells. Tufted cells form intrabulbar communication relays while periglomerular and granule cells contribute to odor quality coding as inhibitory interneurons (see

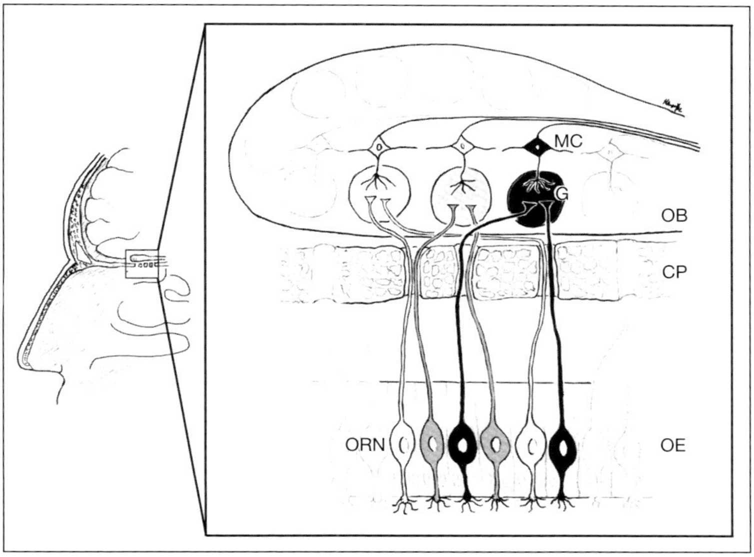

Fig. 2. A schematic illustration of the nasal region showing the projection of ORNs to the olfactory bulb (OB). In the OE, axons of ORNs expressing a specific odor receptor project through openings of the cribriform plate (CP) and innervate a specific glomerulus (G). The olfactory signal is transported to the brain via the mitral cells (MC).

below) [5]. The ORN axons are encircled by nonmyelinating ensheathing glial cells that provide trophic support for axon outgrowth. These glial cells can be purified and cultured and have recently been used to promote axon regeneration in other regions, including the spinal cord [6]. Also present in the epithelium are several types of nonneuronal cells, including sustentacular (supporting) cells and microvillar cells which have nonmotile cilia or microvilli, respectively [7]. The functions of these cells are not clear, but are thought to include detoxification and maintenance of ionic balance. The respiratory epithelium (RE) is marked by cuboid respiratory epithelial cells with motile cilia, goblet cells and basal cells (fig. 1b). RE is demarcated from the sensory areas by the presence of a thicker basement membrane, regular cilia, frequent goblet cells and few nerve bundles [8]. While RE and OE are clearly demarcated in most species studied, in humans, OE may be interrupted by stretches of RE, particularly in the elderly [9–11]. Nerve bundles, vascular components and Bowman's glands reside in the submucosal compartment, and the ducts from Bowman's glands project through the epithelium to secrete a specialized mucus that coats and protects the epithelial surface. The mucus layer thickness ranges from 5 to 30 μm and is comprised of water, electrolytes and a variety of mucopolysaccharides and proteins. A family of odorant-binding proteins have been identified in the mucus that bind distinct classes of hydrophobic odorants and apparently contribute to odor transport [12–14]. These proteins may modulate the mucus concentration of odorants to maintain a level optimal for ORN sensitivity. Mucus secretion is influenced by adrenergic sympathetic fibers from the superior cervical ganglion [15]. Catecholamines, including dopamine and norepinephrine are released from the nerve terminals in the lamina propria, and catecholamines are also released into the mucus in response to activation of the 5th cranial or trigeminal nerve by irritants [16]. These catecholamines have been found to modulate odor sensitivity via D_2 dopamine receptors [17]. Thus, conditions that alter the function of these submucosal tissue components may exert direct and indirect effects on the function of ORNs. A basic understanding of receptor cell function is important to understand the impact of conditions and treatments that may result in olfactory dysfunction.

Odor Detection

Odorants interact with olfactory receptor (OR) proteins present in the ciliary membranes of mature ORNs. These receptor proteins represent the largest gene family yet discovered, and also are among the most diverse both between and within species (for a comprehensive database of ORs, see http://senselab.med.yale.edu/senselab/). The ORs are members of the G-protein-coupled receptor family as they are coupled to GTP-binding regulatory proteins (G-proteins). The OR proteins are typically 300–400 amino acids in length and, like other G-protein-coupled receptors, traverse the membrane seven times. Their sequences contain regions with an unusually high degree of sequence similarity across all of the family members and other regions that are hypervariable (fig. 3). The most highly conserved regions are involved in G-protein binding, while regions that exhibit the most variability are likely to be involved in ligand binding.

A few studies have determined the odor response profiles for specific ORs. These studies insert the gene for that OR into a heterologous cell type including a signaling cascade that allows receptor binding to be detected [18–21]. Data from such studies suggest that each OR is narrowly tuned to bind to a few odorants with particular structural features [22, 23]. Small differences in the amino acid sequence can alter the binding affinity and studies of sequences with defined sequence mutations have identified several regions critical to ligand binding (fig. 3) [24–26]. The vast majority of ORs remain 'orphan' receptors, in that their ligands are not yet known. The task of de-orphanizing

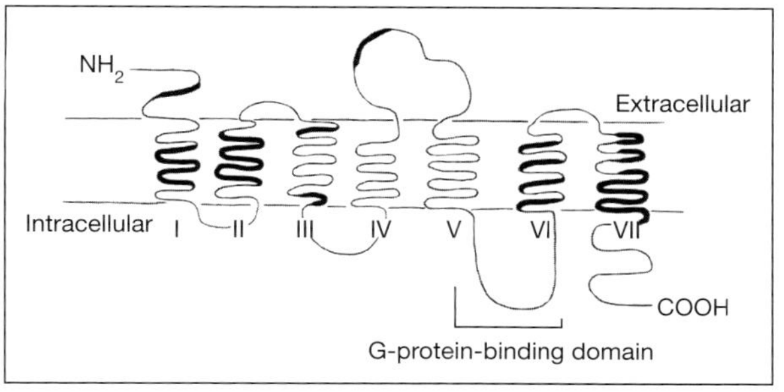

Fig. 3. A schematic diagram of the seven transmembrane mammalian ORs. The thicker lines indicate conserved regions.

these receptors is likely to take some time due to unexpected difficulties in functional expression. Attempts at OR expression often fail due to improper insertion of the protein into the cell membrane [27].

Evidence from studies of many species and sequences indicates that these sequences are subject to a higher than average mutation rate that has selected a large and diverse array of proteins with which to sample our olfactory world [28]. Over 1,000 similar genes have been identified in the human genome, but only a fraction (roughly 1/3) are predicted to be functional, due to the occurrence of a variety of mutations that interrupt the coding regions of many of the sequences preventing them from being translated into a functional protein [29–30]. Different individuals may express different populations of functional receptors due to sequence polymorphisms. While some of these polymorphisms disrupt the gene, others may change the ligand binding profile [31, 32]. Remarkably, even a single amino acid change can alter the preferred ligand of an OR [33]. Thus, each person's genome may contain a slightly different set of functional receptor genes and our perception of the olfactory world is likely to be highly individual.

Studies of the completed genome data from humans and mice suggest that there are about 1/3 as many potentially functional OR sequences in the human genome compared to the mouse genome [34–36]. The precise numbers remain vague due in part to the recent discovery of alternative splicing and large introns separating the start codon from a coding region in sequences previously considered pseudogenes [37–39]. It is not clear whether this evolutionary divergence has had a greater impact on olfactory sensitivity, specificity or breadth. There appears to be a great deal of redundancy within the OR repertoire, and it is possible that the reduction in sequence number has had only a small impact on the number of chemicals we are able to detect. The existence of specific anosmias suggests that mutation of specific ORs could lead to an inability to detect particular odors, but none of these anosmias has yet been linked to a particular OR mutation.

Currently available data suggest that each ORN expresses only a single type of functional OR, and that only one allele of a given OR is expressed [34, 40]. In rodents, ORs of different classes are expressed in several zones across the epithelium [41], but it is not known if similar zones exist in the human. The molecular mechanisms that regulate how each mature ORN selects a particular OR and how their expression pattern is maintained in the face of ongoing ORN replacement remain to be explained. Research in these areas is leading to new insights in our understanding of how the human genome operates, as well as helping us to understand evolutionary processes and the diversity of olfactory perception.

Signal Transduction

Odor signals are generated through a series of intracellular events that are triggered when a volatile chemical binds to the receptor and changes its molecular shape [42, 43]. This conformational change results in dissociation of the corresponding G-protein which then activates enzymes that generate the signaling molecules ('second messengers') needed to open the ion channels in the cell membrane (fig. 4). In mammals, the G-protein G_{olf} activates the enzyme adenylyl cyclase III (AC III), which converts ATP to cyclic AMP [42]. Guanylyl cyclase, which generates cyclic GMP, mediates responses to some odorants in some species [44–46], but its involvement in human ORNs has not been established. Either of these transduction signals can bind to and open the cyclic nucleotide-gated channels (cNcs) [47, 48] (fig. 4). These channels allow positive ions (mainly Na^+ and Ca^{2+}) to enter the cell resulting in depolarization. Ca^{2+} signals in live cells can easily be measured using fluorescence imaging techniques [49] and are commonly used as a metric of an odorant response. Studies with this technique show that mammalian ORNs respond to odorant stimulation with an increase in intracellular calcium and depolarization that is inhibited by pharmacological blockers of the cNcs [2, 49–51]. Transgenic animals lacking functional cNcs are unable to detect most [52], but not all [45] odors. Thus, while the cNc pathway is a major component for odor detection, it is not the sole mechanism by which ORNs respond to odorant stimuli.

Additional pathways may also be activated by odorant stimulation. Biochemical assays have shown that the second messenger inositol-1,4,5-trisphosphate (IP_3) is also produced in response to odorant exposure [53–54],

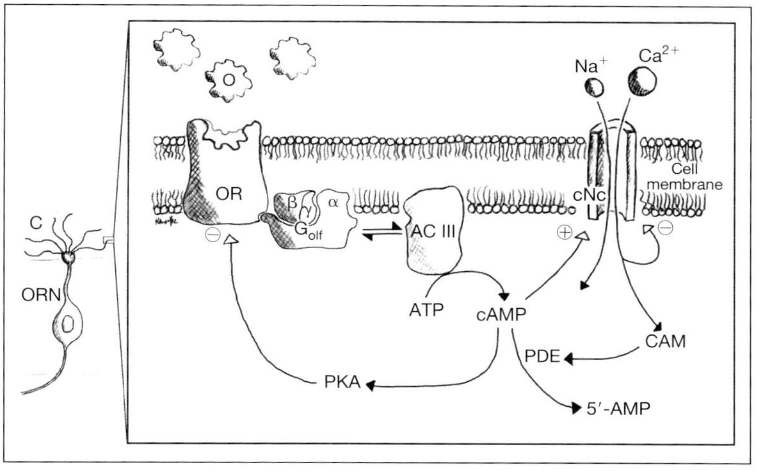

Fig. 4. A schematic illustration of the G-protein signal transduction cascade in the ORN cilia. The binding of an odorant (O) to a G-coupled OR activates G_{olf} and AC III, which leads to an increase in cyclic AMP and the subsequent influx of cations (Na^+ and Ca^{2+}) from cNc. PKA and calcium cations are involved in feedback mechanisms. ATP = Adenosine 5′-triphosphate; C = cilia; CAM = calmodulin; PDE = phosphodiesterase.

and may contribute to the calcium response via opening of membrane channels [55–59]. This pathway has been shown to modulate the sensitivity of the cNc pathway in rodent ORNs [60], and indirect evidence suggests it may play a role in mediating a component of the odorant-stimulated calcium response in human ORNs [2, 61]. While the relative contributions of cyclic AMP and IP_3-linked pathways to odorant signaling remains unclear, several lines of evidence support the existence of multiple pathways for odorant-stimulated ORN activity [45, 56, 60–64]. In most species, the calcium signal may be further transmitted and amplified by voltage-gated calcium channels (VGCCs). Dopamine suppresses VGCC activity and ORN responsiveness [65–67] suggesting that the activity of these channels influences the sensitivity of the ORN. In the axon terminus, VGCCs are activated by the action potential and are necessary for calcium-dependent neurotransmitter release.

The influx of positive ions through any of these mechanisms would be sufficient to trigger an action potential, but under some conditions, additional components may amplify the calcium signal. In many species, the calcium entering via the cNc opens a chloride channel [68–70]. Unlike most cells, the chloride level inside ORNs is higher than in the surrounding compartment, such that opening this channel allows negative chloride ions to exit the cell, further amplifying the depolarizing effect [71–73]. However, the contribution of this channel in human ORNs has not yet been demonstrated.

The second messengers elevated in response to odorant activation lead to other cellular effects [74] (fig. 4). Calcium and cyclic AMP activate various protein kinases (PKs), including PKA and calcium/calmodulin kinase II that are involved in termination of the odorant signal [61, 75–80]. PKs phosphorylate, thereby inactivating, ion channels and other components of the transduction cascade to terminate the odorant signal resulting in short-term adaptation. In catfish, a PKC-sensitive phosphorylation site is present in the ORs [81], and this kinase has also been implicated in adaptation in rat and human ORNs, although its target(s) have yet to be identified [61]. Calcium is removed from the cell via the Na^+/Ca^{2+} exchanger [82] and other mechanisms not yet characterized. These calcium homeostatic pathways have the potential to be influenced by medications and molecules that are elevated in a variety of disease states, either directly or indirectly (see below). As excess or chronic calcium elevation is toxic, such an interaction could shorten ORN life span or impair the function of mature neurons.

Membrane depolarization resulting from cation influx results in generation of an action potential that is transmitted along the axon to trigger release of the neurotransmitter glutamate at the synapses within the olfactory glomeruli (fig. 5). Physiological studies in a variety of species show that, in addition to triggering an increase in action potential firing, odorants may also suppress or inhibit neuronal activity [83–85]. In some studies, odorants have been found to trigger a decrease in intracellular calcium [2, 86] that could account for these inhibitory responses, although data are limited. Studies in toad ORNs suggest these inhibitory receptor potentials may be mediated by opening of Ca^{2+}-activated K^+ channels [62, 87] but similar studies have not been done with human ORNs. The advantage of such a dual mechanism could be to improve the signal to noise ratio. Suppressing the activity of ORNs expressing ORs with lower affinity for that odorant could enhance the impact of the signal from ORNs expressing ORs whose affinity for the stimuli was highest.

Coding and Miscoding

Combinatorial Model

Data acquired through a variety of methods point to a model for odor coding that is based on unique patterns of mitral cell activity. Each odorant can bind to multiple ORs, and each OR can bind multiple odorants. ORNs expressing the same OR project to the same glomeruli in the olfactory bulb, and each mitral cell projects to one glomerulus thereby serving as the central signal

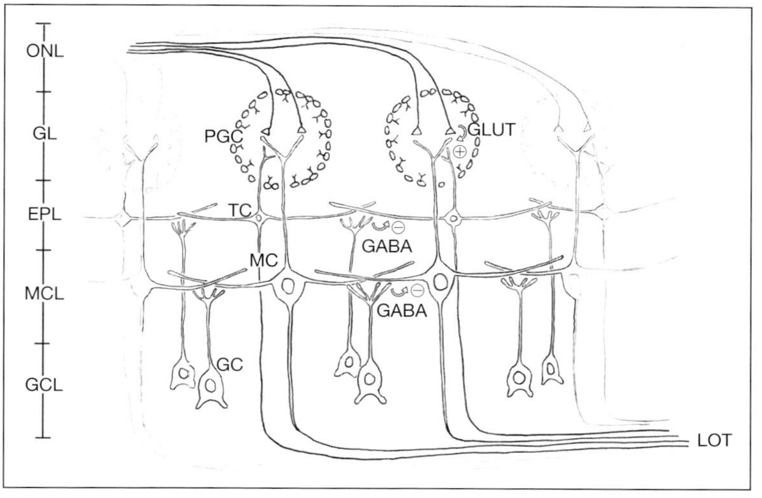

Fig. 5. A schematic illustration of olfactory bulb neurotransmitters. EPL = External plexiform layer; GABA = γ-aminobutyrate; GC = granule cell; GCL = granule cell layer; GL = glomerular layer; GLUT = glutamine; LOT = lateral olfactory tract; MC = mitral cell; MCL = mitral cell layer; ONL = olfactory nerve layer; PGC = periglomerular cell; TC = tufted cell.

detector for odors activating that OR (fig. 2). The activity patterns detected in the glomeruli by the mitral cells are further tuned through the actions of periglomerular, granule and tufted cells. The result is the activation of a unique pattern of mitral cell activity that is triggered by a unique pattern of ORN activity. This activity pattern is decoded in higher brain regions as a particular odor quality, and all components of the pattern are likely to contribute to the code.

Disruption of this pattern due to injury to the epithelium, olfactory bulb or mistargeting of ORN axons to the bulb during recovery from surgery or trauma can result in changes in olfactory perception. As the quality perceived depends on the complete pattern of activity, these changes may include not just reduced sensitivity, but may include qualitative changes in how odors are perceived (for instance, coffee might smell unpleasant or unrecognizable) or discriminated. Experimental and clinical evidence supports this prediction. Behavioral studies of hamsters recovering from complete olfactory nerve transection suggest that perception of odor quality changes following reinnervation [88]. These studies trained animals to discriminate between particular odors using reinforcement and tested the animals 40 days after surgery when olfactory nerves reinnervated the olfactory bulbs. Even when the animals were able to perform simple odor detection tasks, they were not able to discriminate previously learned odors. Thus, following surgery or trauma in which hyposmia or anosmia results from damage to the olfactory nerves, recovery may occur, but the same pattern of glomerular activity may not be reestablished initially due to inaccurate targeting of new axons to glomeruli in the olfactory bulb. With time, it is possible that learning or axon targeting corrections occur and recreate the original patterns or perceptions. In humans, olfactory abnormalities that often precede recovery from anosmia after a head injury might be a result of this type of mistargeting of regenerating axons [89–91; see also chapter 7 by Raviv and Kern, this vol, pp 108–124].

Coding of Mixture Qualities

The vast majority of aromas we encounter in everyday life are actually complex mixtures of tens, hundreds or even thousands of individual volatile chemicals. Our understanding of how these complex stimuli are decoded into a single 'aroma' quality is still very limited. Perfumers have long exploited the ability of certain chemicals to enhance, suppress or alter the quality of other odors to design fragrances. These interactions were traditionally discovered through trial and error, but science has begun to reveal the basis for these phenomena. Referring to the combinatorial model can suggest a basis for some of these interactions.

As mentioned previously, each odor can interact with different affinities with multiple ORs, which recognizes different chemical features (fig. 6A). A high concentration of an odorant may activate the highest affinity or primary receptor as well as lower affinity receptors, thus eliciting a different quality than a lower concentration, which activates only the primary receptor (fig. 6B). Odor suppression or masking may occur when an odorant blocks access to a receptor without activating it, preventing detection of the odor that would be able to activate that receptor [92] (fig. 6C). These predictions are consistent with the results of molecular and behavioral studies indicating that the quality of odor mixtures may be represented either configurally (i.e. the mixture is qualitatively different from the components) or elementally (the components are recognizable) [93, 94]. These data showed that binary mixtures whose components activate overlapping sets of receptors are more likely to be perceived as configural, while components can be discriminated when they activate distinct receptor populations.

While the combinatorial model can provide some insight into the phenomena of odor masking and concentration-dependent changes in odor quality, other predictions of the model fail to account for observed phenomena. For instance, the model predicts that a complex odor should activate more glomeruli than a single compound. Predictions can be made about the number of glomeruli

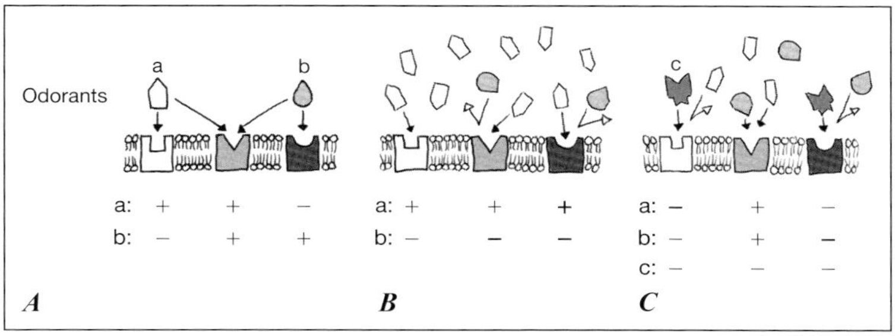

Fig. 6. A schematic diagram illustrating the different models of coding of mixture qualities. ***A*** A single odorant (a or b) can be recognized by multiple ORs. ***B*** High concentration of one odorant can interfere with binding of another odorant (b, bold negative) or change the affinity of the receptor (a, bold positive). ***C*** The receptor site can be blocked by an odorant (c) without activating it (negative).

activated by a mixture containing specified compounds whose glomerular activation patterns have been determined using anatomical methods [95, 96]. However, studies suggest that fewer glomeruli are activated by the mixture than one would predict based on results with individual components of the mixture [96]. The net activity pattern in the olfactory bulb nonetheless remains unique, even when comparing such complex yet similar (to us) odors as the urine from two different mice [96]. This design would enable fine discrimination of mixtures without requiring decomposing the constituents of those mixtures. This finding is consistent with psychophysical studies that document the poor ability of even well-trained human subjects to deconstruct the components of an odor mixture [97, 98]. These data suggest that more complex interneuron and interglomerular signaling is occurring than we currently understand. Studies recording activity at multiple levels within the olfactory bulb are helping to reveal these interactions [99–103]. These studies support the notion that across-fiber and across-glomeruli interactions must eventually be incorporated into any model of olfactory coding before it can be considered complete.

Intensity Coding

Another component of the coding process involves temporal information, which may contribute to both quality and intensity characteristics. Temporal coding may derive from temporally related input at two levels. The first one relates to the initial time of onset of ORN activation that changes across the epithelial sheet due to differences in odorant exposure time as the chemical traverses the nasal cavity [104–106]. This transit time is influenced by sorptive properties of the odor, which are determined primarily by its volatility, water: octanol partition coefficient and solubility. In addition, characteristics of the olfactory organ, such as nasal structure, mucus composition and depth and airflow rate are critical factors influencing the temporal pattern of odor delivery to the ORNs [107]. These latter characteristics can be dramatically influenced by nasal sinus disease, injury or surgery or medications that alter hydration, but these effects are difficult to predict precisely. Studies underway are modeling the impact of airflow and odorant sorptive properties on odor deposition [105, 108] and will help to quantify and predict the impact of these factors on olfaction.

The neural coding of intensity is dependent on temporal characteristics built into the neural firing pattern. Aspects of odorant modulation of firing rates may also contribute to quality coding, but this remains controversial [83, 84]. Nerve recordings indicate that ORNs may exhibit some baseline rate of action potential firing. When stimulated, this baseline rate can increase or decrease and the increase may involve bursts of activity or more tonic increases. The neurophysiological mechanisms responsible for these activity patterns involve the precisely regulated activity of voltage-gated sodium, potassium and calcium channels as well as intermediate transduction elements. Many medications and diseases can alter the functioning of these elements. The implication of this biology is that a variety of conditions may alter olfactory performance more subtly than simply by reducing sensitivity. For instance, changes could also occur in perceived odor quality or rate of adaptation, and effects may be more evident for some odors than others depending on their water solubility or volatility. The clinician should be aware of such potential complex effects that may have a significant impact on the patient's quality of life and olfactory experience, but may be more difficult for the patient to describe and may not be revealed by a simple test of odor sensitivity.

Pathology

A wide variety of conditions can lead to olfactory impairment. These conditions may induce olfactory loss due to conductive changes such as congestion or airflow blockage, and/or through effects on the neural and perineural processes necessary for proper ORN function ('sensorineural' factors).

Aging

Sensory loss is a common age-related complaint, and may be due to changes in the anatomy of the structure (e.g. loss of OR cells) or the environment surrounding the receptor cell (e.g. altered nasal mucus composition). However, aging, as well as age-related diseases and medications may also alter the distribution, density or function of specific receptor proteins, ion channels or signaling molecules that affect the ability of the ORN to function and signal odorant information. For instance, age-related increases in calcium channel density [109] and in the activity of several transduction elements including AC [110], phospholipase C [111, 112], and PKC [113] have been reported, but have not been specifically studied in the olfactory system. Changes such as these, if occurring in the ORNs or olfactory bulb interneurons, could result in chronic activation of calcium-dependent desensitization processes (see above). Studies in our laboratory suggest that with age, individual ORNs become less selective [2, 114], responding to a broader array of odors than do ORNs from younger subjects. The cellular basis for this functional difference is not yet clear, but the impact on perception would be to degrade the signal: noise level at the olfactory bulb and impair odor discrimination. These kinds of neurophysiological changes could also account for faster adaptation and delayed resensitization reported among the elderly [115].

Impact of Inflammation

A variety of pathological conditions such as chronic sinusitis, viral infection, chemical exposure, or allergic rhinitis [116–119] result in inflammation in the nasal cavity that can have acute and long-lasting effects on olfaction. In addition to the acute impact of airway congestion due to injury, secretions and tissue edema ('conductive' factors), a host of chemical signals related to the inflammatory process can have short- and long-term consequences on olfactory epithelial structure and function. At least a few of these inflammatory processes have the potential to directly influence the function of the ORNs or other neural components of the olfactory system.

When faced with injury or infection, a cascade of cellular and molecular events is set into motion to remove damaged cells and stimulate repair. These events may directly or indirectly influence odor detection and signal transduction. White blood cells migrate into the affected area and become activated in response to chemicals released from the injured tissue. The infiltrating cells release a host of pro- and anti-inflammatory chemical mediators including cytokines, chemokines, leukotrienes and lymphokines that act through a complex scenario of signaling pathways to modulate the expression of genes [120] aimed at promoting tissue repair and recovery. Many of these mediators can have effects on a wide array of cell types, including olfactory neurons and related cells in the nasal mucosa, and studies have only begun to explore these effects. For instance, interleukin-1β can stimulate process outgrowth in cultured olfactory neurons [121], while tumor necrosis factor-α triggers apoptosis in the OE [122, 123]. Inflammatory processes also result in elevated production of two gaseous second messengers that may directly affect ORN function: nitric oxide (NO) and carbon monoxide (CO). NO is produced by macrophages in response to activation and can quickly permeate cell membranes [124]. Nasal and expired NO is elevated in chronic rhinosinusitis and has been explored as an index of disease severity in allergic rhinitis and cystic fibrosis [125, 126]. CO is produced by heme oxygenase-1, which confers protection against oxidative stress, and is induced by a variety of lymphokines, including interleukin-1 and tumor necrosis factor-α [127]. Both NO and CO have been shown to stimulate cyclic GMP production in ORNs [46, 128], and to lead to activation of identical types of electrical activity [129]. Thus, both gasses may lead to chronic calcium influx in ORNs via the cNc and thus lead to desensitization or long-term adaptation of the olfactory transduction pathway [130] as well as shortening of ORN life span due to chronic calcium influx. Elevated levels of these inflammatory mediators may linger beyond overt recovery of the nasal epithelium and therapeutic approaches are needed that target the underlying inflammatory process. The success of leukotriene inhibitors in treating asthma has led to exploration of their use in polyposis and allergic and chronic nasal sinus disease [131]. While some data are encouraging, a better understanding of the inflammatory process in the nasal epithelium is needed to identify the most relevant targets for successful drug therapy for the various pathological conditions.

Medications and Olfactory Transduction

A large number of medications have been reported to influence olfactory function [132–134]. Some of these may impact nasal mucus secretion or blood flow, while others may directly influence the neuronal signaling mechanisms. Medications such as calcium channel blockers or dopaminergic drugs can directly alter the functioning of the ORN or the interneurons in the olfactory bulb. Dopamine D_2 receptors are present on ORNs [135] and dopamine can modulate activity of peripheral and central olfactory neurons [136–139]. The excitatory neurotransmitter of the ORN is glutamate, and GABA is an inhibitory neurotransmitter active at both primary and secondary synapses within the olfactory bulb (fig. 5). In addition, receptors for many other neurotransmitters and neuromodulators, including estrogen and insulin, are present in the olfactory bulb [140, 141] and medications influencing their levels in the central nervous system could alter olfactory performance. A variety of medications include olfactory disturbances in their list of side effects (see Doty et al. [134] and Koster et al. [135] for comprehensive reviews). Those listed include ACE inhibitors, calcium channel blockers and a number of antiarrhythmics. These lists are likely to underestimate the true number of medications with olfactory consequences, and do not take into account the compounded effect of multiple medications taken concurrently. Both direct effects on transduction and nerve function, as well as indirect effects due to interference with the epithelial structure or regeneration are possible. There is a clear need for carefully controlled studies using modern in vivo and in vitro methods to study the impact of various classes of medications on olfactory epithelial structure and function, as well as olfactory performance.

Conclusion

Odor detection involves the binding of volatile chemicals to one or more types of receptor proteins present in the cilia of ORNs lining the upper aspects of the nasal cavity. These receptor proteins represent the largest gene family currently known and sequence variations indicate that the olfactory experience is likely to be highly individual. The receptors are coupled to G-proteins that dissociate when activated by conformational changes in the receptor that occur when the odorant binds. The dissociated G-protein subunits activate other cellular processes resulting in activation of the enzymes that produce second messengers which open a membrane channel that allows sodium and calcium into the cell. Other G-protein subunits activate distinct pathways related to adaptation and calcium homeostasis. The entry of positive ions depolarizes the cell triggering an action potential that is carried along the axon to the first synapse in the glomeruli of the olfactory bulb. Glutamate released from the axon terminal activates the associated mitral cells, which relay the activity pattern to higher brain centers. Each odor is thought to activate a particular combination of receptors, and receptor cells expressing the same receptor project to the same glomeruli in the olfactory bulb. Mitral cells project to particular glomeruli and relay the pattern of glomerular activity to the olfactory cortex. Thus, odor quality is determined in part by the particular pattern of ORN and thus mitral cells activated. Each odorant activates multiple OR types and each OR type can bind a number of odorants, generally thought to be related by a particular chemical feature or 'epitope'. In addition to this 'combinatorial' activity pattern, temporal aspects of the neural activation and cross talk among glomeruli mediated by inhibitory interneurons in the olfactory bulb result in a system designed for great sensitivity and specificity that is nonetheless able to encode the astonishing number and diversity of odorous chemicals that we are able to perceive.

A variety of pathological conditions can impair olfactory performance due to direct or indirect effects on these neural transduction and coding mechanisms. Aging, inflammation and medications are common causes of olfactory loss due to both conductive as well as sensorineural effects. Research is needed to better understand how these and other pathological conditions influence the neurophysiology of the olfactory pathways so that improved therapeutic approaches can be developed.

References

1 Lane AP, Gomez G, Dankulich T, Wang H, Bolger WE, Ravoson NE: The superior turbinate as a source of functional human olfactory receptor neurons. Laryngoscope 2002;112: 1183–1189.
2 Rawson NE, Gomez G, Cowart B, Brand JG, Lowry LD, Pribitkin EA, Restrepo D: Selectivity and response characteristics of human olfactory neurons. J Neurophysiol 1997;77:1606–1613.
3 Caggiano M, Kauer JS, Hunter DD: Globose basal cells are neuronal progenitors in the olfactory epithelium: a lineage analysis using a replication-incompetent retrovirus. Neuron 1994;13: 339–352.
4 Calof AL, Hagiwara N, Holcomb JD, Mumm JS, Shou J: Neurogenesis and cell death in olfactory epithelium. J Neurobiol 1996;30:67–81.
5 Mori K, Nagao H, Yoshihara Y: The olfactory bulb: coding and processing of odor molecule information. Science 1999;286:711–715.
6 Ramon-Cueto A, Nieto-Sampedro M: Regeneration into the spinal cord of transected dorsal root axons is promoted by ensheathing glia transplants. Exp Neurol 1994;127:232–244.
7 Menco BP, Morrison EE: Morphology of the mammalian olfactory epithelium: form, fine structure, function, and pathology; in Doty RL (ed): Handbook of Olfactory and Gustation. Basel, Dekker, 2003, pp 17–49.
8 Kern RC: Chronic sinusitis and anosmia: pathologic changes in the olfactory mucosa. Laryngoscope 2000;110:1071–1077.
9 Doty RL, Snow JB Jr: Age-related alterations in olfactory structure and function; in Margolis RL, Getchell TV (eds): Molecular Neurobiology of the Olfactory System. New York, Plenum Press, 1988, pp 355–374.
10 Paik SI, Lehman MN, Seiden AM, Duncan HJ, Smith DV: Human olfactory biopsy: the influence of age and receptor distribution. Arch Otolaryngol Head Neck Surg 1992;118:731–738.
11 Loo AT, Youngentob SL, Kent PF, Schwob JE: The aging olfactory epithelium: neurogenesis, response to damage and odorant-induced activity. Int J Dev Neurosci 1996;14:881–900.
12 Lobel D, Jacob M, Volkner M, Breer H: Odorants of different chemical classes interact with distinct odorant binding protein subtypes. Chem Senses 2002;27:39–44.
13 Briand L, Eloit C, Nespoulous C, Bezirard V, Huet JC, Henry C, Blon F, Trotier D, Pernollet JC: Evidence of an odorant-binding protein in the human olfactory mucus: location, structural characterization, and odorant-binding properties. Biochemistry 2002;41:7241–7252.
14 Pernollet JC, Briand L: Structural recognition between odorants, olfactory-binding proteins and olfactory receptors – First events in odour coding; in Taylor AJ, Roberts DD (eds): Flavor Perception. Oxford, Blackwell, 2004, pp 86–150.
15 Chen Y, Getchell TV, Sparks DL, Getchell ML: Patterns of adrenergic and peptidergic innervation in human olfactory mucosa: age-related trends. J Comp Neurol 1993;334:104–116.
16 Getchell ML, Getchell TV: Fine structural aspects of secretion and extrinsic innervation in the olfactory mucosa. Microsc Res Tech 1992;23:111–127.
17 Vargas G, Lucero MT: Dopamine modulates inwardly rectifying hyperpolarization-activated current (Ih) in cultured rat olfactory receptor neurons. J Neurophysiol 1999;81:149–158.
18 Araneda RC, Kini AD, Firestein S: The molecular receptive range of an odorant receptor. Nat Neurosci 2000;3:1248–1255.
19 Krautwurst D, Yau KW, Reed RR: Identification of ligands for olfactory receptors by functional expression of a receptor library. Cell 1998;95:917–926.
20 Wetzel CH, Oles M, Wellerdieck C, Kuczkowiak M, Gisselmann G, Hatt H: Specificity and sensitivity of a human olfactory receptor functionally expressed in human embryonic kidney 293 cells and *Xenopus laevis* oocytes. J Neurosci 1999;19:7426–7433.

21 Touhara K: Functional cloning and reconstitution of vertebrate odorant receptors. Life Sci 2001;68:2199–2206.
22 Kajiya K, Inaki K, Tanaka M, Haga T, Kataoka H, Touhara K: Molecular bases of odor discrimination: reconstitution of olfactory receptors that recognize overlapping sets of odorants. J Neurosci 2001;21:6018–6025.
23 Touhara K: Functional cloning and reconstitution of vertebrate odorant receptors. Life Sci 2001;68:2199–2206.
24 Singer MS, Shepherd GM: Molecular modeling of ligand-receptor interactions in the OR5 olfactory receptor. Neuroreport 1994;5:1297–1300.
25 Singer MS, Oliveira L, Vriend G, Shepherd GM: Potential ligand-binding residues in rat olfactory receptors identified by correlated mutation analysis. Receptors Channels 1995;3:89–95.
26 Singer MS: Analysis of the molecular basis for octanal interactions in the expressed rat 17 olfactory receptor. Chem Senses 2000;25:155–165.
27 McClintock TS, Landers TM, Gimelbrant AA, Fuller LZ, Jackson BA, Jayawickreme CK, Lerner MR: Functional expression of olfactory-adrenergic receptor chimeras and intracellular retention of heterologously expressed olfactory receptors. Brain Res Mol Brain Res 1997;48:270–278.
28 Dryer L: Evolution of odorant receptors. Bioessays 2000;22:803–810.
29 Sharon D, Glusman G, Pilpel Y, Khen M, Gruetzner F, Haaf T, Lancet D: Primate evolution of an olfactory receptor cluster: diversification by gene conversion and recent emergence of pseudogenes. Genomics 1999;61:24–36.
30 Fuchs T, Glusman G, Horn-Saban S, Lancet D, Pilpel Y: The human olfactory subgenome: from sequence to structure and evolution. Hum Genet 2001;108:1–13.
31 Sharon D, Gilad Y, Glusman G, Khen M, Lancet D, Kalush F: Identification and characterization of coding single-nucleotide polymorphisms within a human olfactory receptor gene cluster. Gene 2000;260:87–94.
32 Gilad Y, Segre D, Skorecki K, Nachman MW, Lancet D, Sharon D: Dichotomy of single-nucleotide polymorphism haplotypes in olfactory receptor genes and pseudogenes. Nat Genet 2000;26:221–224.
33 Gaillard I, Rouquier S, Chavanieu A, Mollard P, Giorgi D: Amino-acid changes acquired during evolution by olfactory receptor 912–93 modify the specificity of odorant recognition. Hum Mol Genet 2004;13:771–780.
34 Mombaerts P: Odorant receptor genes in humans. Curr Opin Genet Dev 1999;9:315–320.
35 Malnic B, Godfrey PA, Buck LB: The human olfactory receptor gene family. Proc Natl Acad Sci USA 2004;101:2584–2589.
36 Godfrey PA, Malnic B, Buck LB: The mouse olfactory receptor gene family. Proc Natl Acad Sci USA 2004;101:2156–2161.
37 Sosinsky A, Glusman G, Lancet D: The genomic structure of human olfactory receptor genes. Genomics 2000;70:49–61.
38 Linardopoulou E, Mefford HC, Nguyen O, Friedman C, van den EG, Farwell DG, Coltrera M, Trask BJ: Transcriptional activity of multiple copies of a subtelomerically located olfactory receptor gene that is polymorphic in number and location. Hum Mol Genet 2001;10:2373–2383.
39 Hoppe R, Weimer M, Beck A, Breer H, Strotmann J: Sequence analyses of the olfactory receptor gene cluster mOR37 on mouse chromosome 4. Genomics 2000;66:284–295.
40 Chess A, Simon I, Cedar H, Axel R: Allelic inactivation regulates olfactory receptor gene expression. Cell 1994;78:823–834.
41 Strotmann J, Wanner I, Helfrich T, Beck A, Breer H: Rostro-caudal patterning of receptor-expressing olfactory neurones in the rat nasal cavity. Cell Tissue Res 1994;278:11–20.
42 Schild D, Restrepo D: Transduction mechanisms in vertebrate olfactory receptor cells. Physiol Rev 1998;78:429–466.
43 Shepherd GM: Discrimination of molecular signals by the olfactory receptor neuron. Neuron 1994;13:771–790.
44 Gibson AD, Garbers DL: Guanylyl cyclases as a family of putative odorant receptors. Annu Rev Neurosci 2000;23:417–439.
45 Lin W, Arellano J, Slotnick B, Restrepo D: Odors detected by mice deficient in cyclic nucleotide-gated channel subunit A2 stimulate the main olfactory system. J Neurosci 2004;24:3703–3710.

46 Breer H, Klemm T, Boekhoff I: Nitric oxide mediated formation of cyclic GMP in the olfactory system. Neuroreport 1992;3:1030–1032.
47 Kurahashi T: The response induced by intracellular cyclic AMP in isolated olfactory receptor cells of the newt. J Physiol 1990;430:355–371.
48 Zufall F, Firestein S, Shepherd GM: Cyclic nucleotide-gated ion channels and sensory transduction in olfactory receptor neurons. Annu Rev Biophys Biomol Struct 1994;23:577–607.
49 Restrepo D, Okada Y, Teeter JH, Lowry LD, Cowart B: Human olfactory neurons respond to odor stimuli with an increase in cytoplasmic Ca^{2+}. Biophys J 1993;64:1961–1966.
50 Restrepo D, Okada Y, Teeter JH: Odorant-regulated Ca^{2+} gradients in rat olfactory neurons. J Gen Physiol 1993;102:907–924.
51 Tareilus E, Noe J, Breer H: Calcium signals in olfactory neurons. Biochim Biophys Acta 1995;1269:129–138.
52 Brunet LJ, Gold GH, Ngai J: General anosmia caused by a targeted disruption of the mouse olfactory cyclic nucleotide-gated cation channel. Neuron 1996;17:681–693.
53 Boekhoff I, Tarelius E, Strotmann J, Breer H: Rapid activation of alternative second messenger pathways in olfactory cilia from rats by different odorants. EMBO J 1990;9:2453–2458.
54 Breer H, Boekhoff I: Odorants of the same odor class activate different second messenger pathways. Chem Senses 1991;16:19–29.
55 Restrepo D, Teeter J, Honda E, Boyle A, Kalinoski DL, Marecek J, Prestwich GD: Evidence for an InsP-gated channel protein in isolated rat olfactory cilia. Am J Physiol 1992;263:C667–C673.
56 Miyamoto T, Restrepo D, Cragoe J, Teeter J: IP3- and cAMP-induced responses in isolated olfactory receptor neurons from the channel catfish. J Membr Biol 1992;127:173–183.
57 Honda E, Teeter JH, Restrepo D: InsP3-gated ion channels in rat olfactory cilia membrane. Brain Res 1995;703:79–85.
58 Kaur R, Zhu XO, Moorhouse AJ, Barry PH: IP3-gated channels and their occurrence relative to CNG channels in the soma and dendritic knob of rat olfactory receptor neurons. J Membr Biol 2001;181:91–105.
59 Okada Y, Fujiyama R, Miyamoto T, Sato T: Comparison of a Ca^{2+}-gated conductance and a second-messenger-gated conductance in rat olfactory neurons. J Exp Biol 2000;203:567–573.
60 Spehr M, Wetzel CH, Hatt H, Ache BW: 3-Phosphoinositides modulate cyclic nucleotide signaling in olfactory receptor neurons: Neuron 2002;33:731–739.
61 Gomez G, Rawson NE, Cowart B, Lowry LD, Pribitkin EA, Restrepo D: Modulation of odor-induced increases in $[Ca^{2+}]_i$ by inhibitors of protein kinases A and C in rat and human olfactory receptor neurons. Neuroscience 2000;98:181–189.
62 Morales B, Bacigalupo J: Chemical reception in vertebrate olfaction: evidence for multiple transduction pathways. Biol Res 1996;29:333–341.
63 Schild D, Lischka FW, Restrepo D: InsP3 causes an increase in apical $[Ca^{2+}]_i$ by activating two distinct current components in vertebrate olfactory receptor cells. J Neurophysiol 1995;73: 862–866.
64 Lischka FW, Teeter JH, Restrepo D: Odorants suppress a voltage-activated K^+ conductance in rat olfactory neurons. J Neurophysiol 1999;82:226–236.
65 Wetzel CH, Spehr M, Hatt H: Phosphorylation of voltage-gated ion channels in rat olfactory receptor neurons. Eur J Neurosci 2001;14:1056–1064.
66 Hegg CC, Lucero MT: Dopamine reduces odor- and elevated-K^+-induced calcium responses in mouse olfactory receptor neurons in situ. J Neurophysiol 2004;91:1492–1499.
67 Okada Y, Miyamoto T, Toda K: Dopamine modulates a voltage-gated calcium channel in rat olfactory receptor neurons. Brain Res 2003;968:248–255.
68 Reisert J, Bauer PJ, Yau KW, Frings S: The Ca-activated Cl channel and its control in rat olfactory receptor neurons. J Gen Physiol 2003;122:349–363.
69 Restrepo D: The ins and outs of intracellular chloride in olfactory receptor neurons. Neuron 2005;45:481–482.
70 Kleene SJ, Gesteland RC: Calcium-activated chloride conductance in frog olfactory cilia. J Neurosci 1991;11:3624–3629.
71 Kurahashi T, Yau KW: Co-existence of cationic and chloride components in odorant-induced current of vertebrate olfactory receptor cells. Nature 1993;363:71–74.

72 Kurahashi T, Yau KW: Olfactory transduction. Tale of an unusual chloride current. Curr Biol 1994;4:256–258.
73 Menini A: Calcium signalling and regulation in olfactory neurons. Curr Opin Neurobiol 1999;9: 419–426.
74 Boekhoff I, Kroner C, Breer H: Calcium controls second-messenger signalling in olfactory cilia. Cell Signal 1996;8:167–171.
75 Conneelly PA, Sisk RB, Schulman H, Garrison JC: Evidence for the activation of the multifunctional Ca^{2+}/calmodulin-dependent protein kinase in response to hormones that increase intracellular Ca^{2+}. J Biol Chem 1987;262:10154–10163.
76 Boekhoff I, Breer H: Termination of second messenger signaling in olfaction. Proc Natl Acad Sci USA 1992;89:471–474.
77 Schleicher S, Boekhoff I, Arriza J, Lefkowitz RJ, Breer H: A beta-adrenergic receptor kinase-like enzyme is involved in olfactory signal termination. Proc Natl Acad Sci USA 1993;90: 1420–1424.
78 Peppel K, Boekhoff I, McDonald P, Breer H, Caron MG, Lefkowitz RJ: G protein-coupled receptor kinase 3 (GRK3) gene disruption leads to loss of odorant receptor desensitization. J Biol Chem 1997;272:25425–25428.
79 Wei J, Zhao AZ, Chan GC, Baker LP, Impey S, Beavo JA, Storm DR: Phosphorylation and inhibition of olfactory adenylyl cyclase by CaM kinase II in neurons: a mechanism for attenuation of olfactory signals. Neuron 1998;21:495–504.
80 Leinders-Zufall T, Ma M, Zufall F: Impaired odor adaptation in olfactory receptor neurons after inhibition of Ca^{2+}/calmodulin kinase II. J Neurosci 1999;19:RC19.
81 Medler KF, Bruch RC: Protein kinase Cbeta and delta selectively phosphorylate odorant and metabotropic glutamate receptors. Chem Senses 1999;24:295–299.
82 Noe J, Tareilus E, Boekhoff I, Breer H: Sodium/calcium exchanger in rat olfactory neurons. Neurochem Int 1997;30:523–531.
83 Rospars JP, Lansky P, Duchamp-Viret P, Duchamp A: Spiking frequency versus odorant concentration in olfactory receptor neurons. Biosystems 2000;58:133–141.
84 Duchamp-Viret P, Duchamp A, Chaput MA: Peripheral odor coding in the rat and frog: quality and intensity specification. J Neurosci 2000;20:2383–2390.
85 Kang J, Caprio J: In vivo responses of single olfactory receptor neurons in the channel catfish, *Ictalurus punctatus*. J Neurophysiol 1995;73:172–177.
86 Delay R, Restrepo D: Odorant responses of dual polarity are mediated by cAMP in mouse olfactory sensory neurons. J Neurophysiol 2004;92:1312–1319.
87 Morales B, Labarca P, Bacigalupo J: A ciliary K^+ conductance sensitive to charibdotoxin underlies inhibitory responses in toad olfactory receptor neurons. FEBS Lett 1995;359:41–44.
88 Yee KK, Costanzo RM: Changes in odor quality discrimination following recovery from olfactory nerve transection. Chem Senses 1998;23:513–519.
89 Christensen MD, Holbrook EH, Costanzo RM, Schwob JE: Rhinotopy is disrupted during the re-innervation of the olfactory bulb that follows transection of the olfactory nerve. Chem Senses 2001;26:359–369.
90 Costanzo RM: Rewiring the olfactory bulb: changes in odor maps following recovery from nerve transection. Chem Senses 2000;25:199–205.
91 Kern RC, Quinn B, Rosseau G, Farbman AI: Post-traumatic olfactory dysfunction. Laryngoscope 2000;110:2106–2109.
92 Kurahashi T, Lowe G, Gold GH: Suppression of odorant responses by odorants in olfactory receptor cells. Science 1994;265:118–120.
93 Araneda RC, Peterlin Z, Zhang X, Chesler A, Firestein S: A pharmacological profile of the aldehyde receptor repertoire in rat olfactory epithelium. J Physiol 2004;555:743–756.
94 Linster C, Cleland TA: Configurational and elemental odor mixture perception can arise from local inhibition. J Comput Neurosci 2004;16:39–47.
95 Johnson BA, Leon M: Modular representations of odorants in the glomerular layer of the rat olfactory bulb and the effects of stimulus concentration. J Comp Neurol 2000;422:496–509.
96 Schaefer ML, Young DA, Restrepo D: Olfactory fingerprints for major histocompatibility complex-determined body odors. J Neurosci 2001;21:2481–2487.

97 Jinks A, Laing DG: A limit in the processing of components in odour mixtures. Perception 1999;28:395–404.
98 Jinks A, Laing DG: The analysis of odor mixtures by humans: evidence for a configurational process. Physiol Behav 2001;72:51–63.
99 Trombley PQ, Shepherd GM: Synaptic transmission and modulation in the olfactory bulb. Cur Opin Neurobiol 1993;3:540–547.
100 Shen GY, Chen WR, Midtgaard J, Shepherd GM, Hines ML: Computational analysis of action potential initiation in mitral cell soma and dendrites based on dual patch recordings. J Neurophysiol 1999;82:3006–3020.
101 Luo M, Katz LC: Response correlation maps of neurons in the mammalian olfactory bulb. Neuron 2001;32:1165–1179.
102 McQuiston AR, Katz LC: Electrophysiology of interneurons in the glomerular layer of the rat olfactory bulb. J Neurophysiol 2001;86:1899–1907.
103 Belluscio L, Katz LC: Symmetry, stereotypy, and topography of odorant representations in mouse olfactory bulbs. J Neurosci 2001;21:2113–2122.
104 Mozell MM, Jagodowicz M: Mechanisms underlying the analysis of odorant quality at the level of the olfactory mucosa. 1. Spatiotemporal sorption patterns. Ann NY Acad Sci 1974;237: 76–90.
105 Hahn I, Scherer PW, Mozell MM: A mass transport model of olfaction. J Theor Biol 1994;167:115–128.
106 Keyhani K, Scherer PW, Mozell MM: A numerical model of nasal odorant transport for the analysis of human olfaction. J Theor Biol 1997;186:279–301.
107 Scott-Johnson PE, Blakley D, Scott JW: Effects of air flow on rat electroolfactogram. Chem Senses 2000;25:761–768.
108 Zhao K, Scherer PW, Hajiloo SA, Dalton P: Effect of anatomy on human nasal air flow and odorant transport patterns: implications for olfaction. Chem Senses 2004;29:365–379.
109 Mattson MP, Rydel RE, Lieberburg I, Smith-Swintosky V: Altered calcium signaling and neuronal injury: stroke and Alzheimer's disease as examples. Ann NY Acad Sci 1993;679:1–21.
110 Yamamoto M, Ozawa H, Saito T, Frolich L, Riederer P, Takahata N: Reduced immunoreactivity of adenylyl cyclase in dementia of the Alzheimer type. Neuroreport 1996;7:2965–2970.
111 Joseph JA, Kowatch MA, Makui T, Roth GS: Selective cross-activation/inhibition of second messenger systems and the reduction of age-related deficits in the muscarinic control of dopamine release from perifused rat striata. Brain Res 1990;537:40–48.
112 Iren N, Pint A, Stef F, Jerz V, Hann M: Increased responsiveness of the cerebral cortical phosphatidylinositol system to noradrenaline and carbachol in senescent rats. Neurosci Lett 1989;107: 195–199.
113 Busquets X, Ventayol P, Sastre M, Garcia-Sevilla JA: Age-dependent increases in protein kinase C-alpha beta immunoreactivity and activity in the human brain: possible in vivo modulatory effects on guanine nucleotide regulatory G(i) proteins. Brain Res 1996;710:28–34.
114 Rawson NE, Gomez G, Cowart B, Restrepo D: The use of olfactory receptor neurons (ORNs) from biopsies to study changes in aging and neurodegenerative diseases. Ann NY Acad Sci USA 1998;855:701–707.
115 Stevens JC, Cain WS, Schiet FT, Oatly MW: Olfactory adaptation and recovery in old age. Perception 1989;18:265–276.
116 Yamagishi M, Fujiwara M, Nakamura H: Olfactory mucosal findings and clinical course in patients with olfactory disorders following upper respiratory viral infection. Rhinology 1994;32:113–118.
117 Yamagishi M, Nakano Y: A re-evaluation of the classification of olfactory epithelia in patients with olfactory disorders. Eur Arch Otorhinolaryngol 1992;249:393–399.
118 Yamagishi M, Nakamura H, Suzuki S, Hasegawa S, Nakano Y: Immunohistochemical examination of olfactory mucosa in patients with olfactory disturbance. Ann Otol Rhinol Laryngol 1990;99:205–210.
119 Church JA, Bauer H, Bellanti JA, Satterly RA, Henkin RI: Hyposmia associated with atopy. Ann Allergy 1978;40:105–109.
120 Baraniuk JN: Mechanisms of rhinitis. Allergy Asthma Proc 1998;19:343–347.

121 Vawter MP, Basaric-Keys J, Li Y, Lester DS, Lebovics RS, Lesch KP, Kulaga H, Freed WJ, Sunderland T, Wolozin B: Human olfactory neuroepithelial cells: tyrosine phosphorylation and process extension are increased by the combination of IL-1beta, IL-6, NGF, and bFGF. Exp Neurol 1996;142:179–194.
122 Farbman AI, Ezeh PI: TGF-alpha and olfactory marker protein enhance mitosis in rat olfactory epithelium in vivo. Neuroreport 2000;11:3655–3658.
123 Suzuki Y, Farbman AI: Tumor necrosis factor-alpha-induced apoptosis in olfactory epithelium in vitro: possible roles of caspase 1 (ICE), caspase 2 (ICH-1), and caspase 3 (CPP32). Exp Neurol 2000;165:35–45.
124 Moncada S: Nitric oxide in the vasculature: physiology and pathophysiology. Ann NY Acad Sci 1997;811:60–69.
125 Lefevere L, Willems T, Lindberg S, Jorissen M: Nasal nitric oxide. Acta Otorhinolaryngol Belg 2000;54:271–280.
126 Gilain L, Bedu M, Jouaville L, Guichard C, Advenier D, Mom T, Laurent S, Caillaud D: Analysis of nasal and exhaled nitric oxide concentration in nasal polyposis. Ann Otolaryngol Chir Cervicofac 2002;119 234–242.
127 Terry CM, Clikeman JA, Hoidal JR, Callahan KS: Effect of tumor necrosis factor-alpha and interleukin-1 alpha on heme oxygenase-1 expression in human endothelial cells. Am J Physiol 1998;274:H883–H891.
128 Ingi T, Chiang G, Ronnett GV: The regulation of heme turnover and carbon monoxide biosynthesis in cultured primary rat olfactory receptor neurons. J Neurosci 1996;16:5621–5628.
129 Lischka FW, Schild D: Effects of nitric oxide upon olfactory receptor neurones in *Xenopus laevis*. Neuroreport 1993;4:582–584.
130 Zufall F, Leinders-Zufall T: Identification of a long-lasting form of odor adaptation that depends on the carbon monoxide/cGMP second-messenger system. J Neurosci 1997;17:2703–2712.
131 Parnes SM: The role of leukotriene inhibitors in allergic rhinitis and paranasal sinusitis. Curr Allergy Asthma Rep 2002;2:239–244.
132 Elsner RJ: Environment and medication use influence olfactory abilities of older adults. J Nutr Health Aging 2001;5:5–10.
133 Ackerman BH, Kasbekar N: Disturbances of taste and smell induced by drugs. Pharmacotherapy 1997;17:482–496.
134 Doty RL, Philip S, Reddy K, Kerr KL: Influences of antihypertensive and antihyperlipidemic drugs on the senses of taste and smell: a review. J Hypertens 2003;21:1805–1813.
135 Koster NL, Norman AB, Richtand NM, Nickell WT, Puche AC, Pixley SK, Shipley MT: Olfactory receptor neurons express D_2 dopamine receptors. J Comp Neurol 1999;411:666–673.
136 Hsia AY, Vincent JD, Lledo PM: Dopamine depresses synaptic inputs into the olfactory bulb. J Neurophysiol 1999;82:1082–1085.
137 Brunig I, Sommer M, Hatt H, Bormann J: Dopamine receptor subtypes modulate olfactory bulb gamma-aminobutyric acid type A receptors. Proc Natl Acad Sci USA 1999;96:2456–2460.
138 Zhang JJ, Okutani F, Yagi F, Inoue S, Kaba H: Facilitatory effect of ritanserin is mediated by dopamine D_1 receptors on olfactory learning in young rats. Dev Psychobiol 2000;37:246–252.
139 Berkowicz DA, Trombley PQ: Dopaminergic modulation at the olfactory nerve synapse. Brain Res 2000;855:90–99.
140 Macrides F, Schoenfeld TA, Marchand JE, Clancy AN: Evidence for morphologically, neurochemically and functionally heterogeneous classes of mitral and tufted cells in the olfactory bulb. Chem Senses 1985;10:175–202.
141 Halasz N, Shepherd GM: Neurochemistry of the vertebrate olfactory bulb. Neuroscience 1983;10: 579–619.

Nancy E. Rawson, PhD
Monell Chemical Senses Center
3500 Market Street
Philadelphia, PA 19104–3308 (USA)
Tel. +1 215 898 0943, Fax +1 215 898 2084, E-Mail rawson@monell.org

Hummel T, Welge-Lüssen A (eds): Taste and Smell. An Update.
Adv Otorhinolaryngol. Basel, Karger, 2006, vol 63, pp 44–69

Smell: Central Nervous Processing

Jay A. Gottfried

Northwestern University Feinberg School of Medicine, Cognitive Neurology and Alzheimer's Disease Center, Chicago, Ill., USA

Abstract

This chapter focuses on central olfactory processing in the human brain. As the psychophysiology of human olfactory function is important for appreciating its underlying neurophysiology, the chapter will begin with a brief overview of what the human nose can do, contesting notions that human olfaction is a second-rate system. It will be followed by an anatomical survey of the principal recipients of olfactory bulb input, with some comments on the unique organizing properties that distinguish olfaction from other sensory modalities. The final section will cover the neural correlates of human olfactory function, including aspects of basic chemosensory processing (odor detection, sniffing, intensity, valence) and higher-order olfactory operations (learning, memory, crossmodal integration), with particular emphasis on functional imaging data, though human lesion studies and intracranial recordings will also be discussed.

Historically the clinical science of smell has received scant attention compared to other sensory modalities. In neurology, the olfactory system goes virtually without mention, and textbooks are quick to point out that the first cranial nerve is of little diagnostic utility. For many practitioners in otolaryngology, their acquaintance with olfactory structures is limited to the visible surface of the olfactory epithelial sheet, having no reason to tread beyond the cribriform plate. *Grinker's Neurology* [1], a classic textbook of its day (1943), introduced the topic of human olfaction by emphasizing 'the extremely rudimentary sense of smell and the insignificant role it plays in man's existence' (p 337). Several decades later, the conventional wisdom had barely changed: 'From a clinical point of view the importance of the olfactory system is slight, just as the sense of smell is of relatively minor importance in the normal life of civilized man' [2] (p 509).

This clinical state of affairs is echoed by an evolutionary snub. So the theory goes, human olfaction became a subpar 'microsmatic' sense precisely at the moment long ago when we first learned how to walk upright. By bringing our noses up off the odor-rich ground, our bipedal station deprived us of nature's best scents, and a phylogenetic regression ensued (discussed in Shepherd [3]). The relatively smaller sizes of human olfactory structures, along with the decline in functioning human olfactory receptor genes [4, 5], are cited as biological consequences of this evolutionary upheaval. In contrast, for 'macrosmatic' quadrupeds like dogs, cats, and rodents, the sense of smell remained a biological imperative, governing a wide variety of behaviors important for survival, namely those related to food, sex, and threat. As such, evolutionary pressure has ensured olfactory preeminence in these ground-dwelling species.

Part of the historical indifference to human olfaction has been due to the technical challenges of working with odorous stimuli and of measuring brain activity with sufficient spatiotemporal detail. However, with recent experimental advances, many of these problems have become soluble, and a whole new set of research questions, more ambitious and more sophisticated, can now be applied to the olfactory domain. As a result, it has become evident that the olfactory system is well-suited to addressing key neurobiological questions. Importantly, the fact that many neurodegenerative disorders, including Alzheimer's disease and Parkinson's disease, are typified in their early stages by olfactory deficits [6–8] means that a scientific knowledge of the functional organization of central olfaction in healthy subjects may enlighten our understanding of these conditions.

This chapter focuses on central olfactory processing in the human brain. As the psychophysiology of human olfactory function is important for appreciating its underlying neurophysiology, the chapter will begin with a brief and biased overview of what the human nose can do, contesting notions that human olfaction is a second-rate system. It will be followed by an anatomical survey of the principal recipients of olfactory bulb input, with some comments on the unique organizing properties that distinguish olfaction from other sensory modalities. The final section will cover the neural correlates of human olfactory function, including aspects of basic chemosensory processing (odor detection, sniffing, intensity, valence) and higher-order olfactory operations (learning, memory, crossmodal integration), with particular emphasis on functional imaging data, though human lesion studies and intracranial recordings will also be discussed.

What Can Your Nose Do?

Perception and Discrimination

Human olfactory perception is surprisingly good, and in certain instances better than in species with their noses closer to the ground. A comparison of olfactory detection thresholds across several mammalian species indicates that for certain simple monomolecular compounds, human thresholds are lower (more sensitive) than the corresponding thresholds in rats, as well as in nonhuman primates [9]. Notably, human detection thresholds for many odors are commonly in the parts-per-billion range [10]. Moreover, a series of elegant psychophysical studies by Laska and Seibt [9], Laska and Teubner [11], Laska et al. [12] and Laska and Hubener [13] has shown that humans can readily discriminate between two different odors that differ by a single molecular component. For example, humans have no problem distinguishing aliphatic aldehydes of a 4- vs. 5-carbon chain length, otherwise matched for intensity, and in the absence of other physicochemical differences [11]. Finally, it is estimated that humans can distinguish thousands of different smells, although such powers of discrimination are generally not matched by verbal skills, and when a human subject is asked to name an odor, performance tends to break down [14]. Nevertheless, this verbal limitation does not invalidate the idea that human olfactory perception is highly sensitive and specific.

Behavioral Modulation

In the animal kingdom, behavior is inevitably shaped by the biological salience of stimuli encountered in the environment. Odors can have a powerful impact on behavioral and motivational states, by virtue of their associations with threat (predators), food, and sexual gratification. Bombykol, a volatile pheromone secreted by female silkworm moths, will drive potential mates miles upwind in hot pursuit [15]. Despite an apparent phylogenetic decline, there is evidence to suggest that odors are also capable of modulating human behavior. Human infants discriminate their own mothers' odors by postnatal day 6 [16] and are more likely to initiate sucking specifically in response to these stimuli by 6 weeks of age [17]. The use of perfume can affect social interactions by influencing subjective impressions of its wearer by others [18], and odor signals may be favored over visual cues in dictating food preferences [19]. Odor components in a woman's axillary sweat can synchronize the ovulatory cycles of other female subjects [20, 21], suggesting that neuroendocrine states are sensitive to olfactory cues. A related work indicates that the major histocompatibility complex genotype determines an individual's body odor and body odor preference [22, 23], and moreover that a woman's preference of male body odors is defined by major histocompatibility complex haplotype inheritance patterns [24], all of which may influence mating choice in humans [25].

Integration and Plasticity

Most everyday smells are composed of a mixture of many different airborne components, but tend to be experienced as unitary percepts. The smell of chocolate contains hundreds of volatile organic compounds [26], yet the olfactory system synthesizes this complex mixture seamlessly into one odor. As a result of the integrative nature of odor perception, we are notoriously poor at identifying discrete components within an odor blend, and even those with professional training (e.g. wine tasters, perfumists) have no major discriminatory advantage [27, 28]. In general, odor perception is highly plastic and depends on sensory context and past experience. For example, identification of single odors is poor, but improves when relevant semantic (e.g. verbal) information is available [14]. In the absence of visual information, a group of blindfolded enology students was unable to determine the color of a test wine from its odor alone [29]. Even elementary aspects of olfactory processing, including detection thresholds, adaptation rates, and intensity judgments, are strongly modulated by visual, perceptual and cognitive factors [30–33].

The power of suggestion (as another contextual cue) plays an equally important role. A classroom of students was convinced that a bottle filled with distilled water actually contained a 'strong and peculiar' odor that slowly spread from the front to the back of the room [34]. When a radio station informed its listeners that a certain auditory tone would recreate the physiological experience of a 'pleasant country smell', many people reported perceiving such an odor [35]. Learning and experience are also critical in human olfactory identification and discrimination (reviewed in Wilson and Stevenson [36]). Two unfamiliar odors that have been paired together in a mixture come to acquire each other's perceptual qualities [37], and odors experienced in the presence of sweet or sour tastes take on those attributes [38]. Taken together, these findings indicate that an individual's olfactory viewpoint is profoundly shaped by higher-order operations, likely to be mediated via central olfactory processes.

Anatomy

This section describes the anatomical organization of central olfactory structures in the human brain [detailed in 39–44]. Where appropriate, reference will be made to animal data, though the interested reader is referred to several excellent reviews for data regarding other species [45–49].

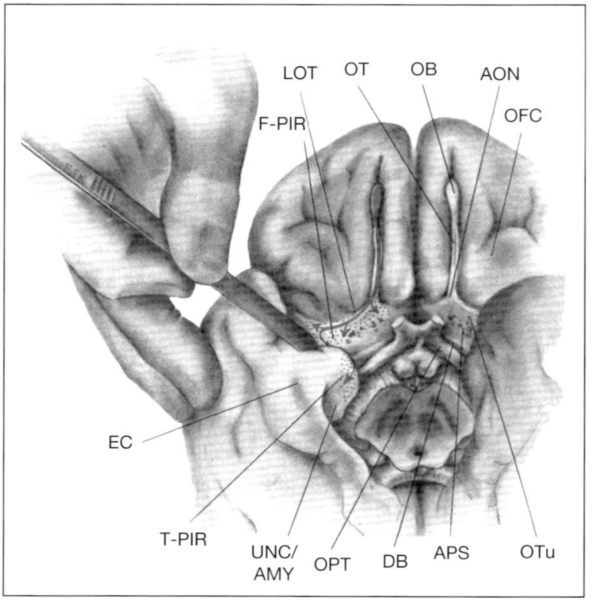

Fig. 1. Anatomical illustration of the human basal forebrain and medial temporal lobes, depicting the olfactory tract, its principal projections, and surrounding nonolfactory structures. DB = Diagonal band; EC = entorhinal cortex; F-PIR = frontal piriform cortex; OB = olfactory bulb; OPT = optic tract; OT = olfactory tract; OTu = olfactory tubercle; T-PIR = temporal piriform cortex; UNC/AMY = uncus with amygdala situated beneath. (Reproduced and modified from figure 159 of Heimer [41]; copyright 1983 by Springer, and with permission of the author.)

General Principles

As outlined in chapter 1 by Hornung [this vol, pp 1–32] and in chapter 2 by Rawson and Yee [this vol, pp 23–43], odor-evoked responses are initially conducted from first-order neurons at the nasal mucosa toward the olfactory bulb, where olfactory sensory axons make contact with second-order (mitral and tufted cell) dendrites within discrete glomeruli. Axons of the mitral and tufted cells of each bulb coalesce to form the olfactory tract, one on each side. This structure lies in the olfactory sulcus of the basal forebrain, immediately lateral to the gyrus rectus, and conveys olfactory information ipsilaterally to a wide number of brain areas within the orbital surface of the frontal lobe and the dorsomedial surface of the temporal lobe (fig. 1). Collectively these projection sites comprise the 'primary olfactory cortex', signifying all of the brain regions receiving direct bulbar input [50, 51]. Notably, the highly ordered chemotopic organization in the olfactory bulb [see chapter 2 by Rawson and Yee, this vol, pp 23–43] does not appear to be systematically maintained in central brain regions. In rats, a given area of the olfactory bulb projects extensively throughout the olfactory cortex, and a given area of the cortex receives projections from widespread areas of the bulb [52]. On the other hand, a recent genetic tracer study in rodents indicates that a given olfactory receptor subtype projects to discrete neuronal clusters within the olfactory cortex, suggesting a certain topographical preservation [53].

As the olfactory tract courses posteriorly, collateral branches peel off and synapse upon the anterior olfactory nucleus (AON), a collection of cell clusters scattered along the caudal extent of the tract and the caudomedial orbital cortex. While histologically variable in humans, this region appears to be a target of pathology in both Parkinson's disease [54] and multiple system atrophy [55]. In animals, projections between one AON and its contralateral partner, via the anterior commissure, provide the major route of olfactory information transfer between hemispheres, though such contralateral pathways have not been documented in humans. Upon nearing the entry point to the brain (at the so-called olfactory trigone), the olfactory tract is generally thought to divide into lateral, intermediate, and medial olfactory striae. Although well-documented in animals, the intermediate and medial branches are extremely rudimentary in humans and seldom evident histologically [39, 56]. Thus, the lateral olfactory tract (LOT) apparently provides the only source of bulbar afferents to the human brain [40].

The LOT deviates laterally around the rostral edge of the anterior perforated substance (APS), and then makes a sharp bend caudally onto the medial temporal surface (uncus). Principal recipients include the piriform cortex, amygdala, and rostral entorhinal cortex (fig. 1), all of which are substantially interconnected via associational intracortical fiber systems (fig. 2). Another target of the LOT input is the olfactory tubercle, which in animals forms a prominent bump, but in humans is difficult to visualize. It is thought to be situated along the posterior-most segment of the medial orbital cortex within the APS and may be a derivative of the pallial striatum [57]. Other regions of the basal forebrain, such as the taenia tecta, indusium griseum, anterior hippocampal continuation, and the nucleus of the horizontal diagonal band, have been shown to receive direct bulbar input in animal models [45, 58], but whether similar connections are preserved in humans is unknown.

Higher-order projections arising from each of these olfactory structures converge on the orbital prefrontal cortex, agranular insula, other amygdala subnuclei, thalamus, hypothalamus, basal ganglia, and hippocampus [47] (fig. 2). Together this complex network of connections provides the basis for odor-guided regulation

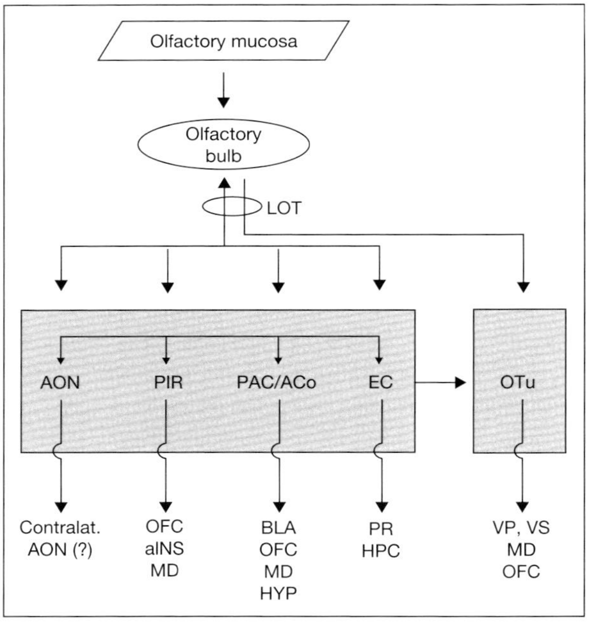

Fig. 2. Schematic diagram of the major olfactory pathways [based on refs. 39, 40, 45–48]. Regions in gray together represent the primary olfactory cortex. Projections between the olfactory bulb and most areas of the primary olfactory cortex are bidirectional, with the exception of the olfactory tubercle (OTu). Similarly, associational connections between the primary olfactory cortex subregions are reciprocal, apart from OTu. The downstream targets of the primary olfactory cortex represent some of the major projection sites (bottom of figure), many of which provide feedback to the primary olfactory cortex (not shown), but these connections are not meant to be comprehensive or all-inclusive. While broadly illustrative of the human olfactory system, this diagram is largely based on information obtained from animal models, due to the scarcity of human data. ACo = Anterior cortical nucleus of the amygdala; aINS = agranular insula; BLA = basolateral nucleus of the amygdala; EC = entorhinal cortex; HPC = hippocampus; HYP = hypothalamus; MD = mediodorsal thalamus; PAC = periamygdaloid cortex; PIR = piriform cortex; PR = perirhinal cortex; VP = ventral pallidum; VS = ventral striatum.

of behavior, feeding, emotion, autonomic states, and memory. In addition, each region of the primary olfactory cortex (apart from the olfactory tubercle) sends dense feedback projections to the olfactory bulb [45], supplying numerous physiological routes for central or 'top-down' modulation of olfactory information processing as early as the second-order neuron in the olfactory hierarchy. What follows below is a more detailed anatomical description regarding several key regions involved in human olfaction.

Piriform Cortex

Because the piriform cortex is the major recipient of inputs from the olfactory bulb, the term 'primary olfactory cortex' is sometimes used interchangeably. Piriform means 'pear-shaped', on account of its gross appearance in certain animal species. The piriform cortex is the largest of the central olfactory areas and spans the anatomical junction between the frontal and temporal lobe, effectively defining two subdivisions. The frontal piriform (or 'prepiriform') cortex extends from the frontotemporal junction anteriorly to the caudal-most extent of the orbitofrontal cortex (OFC) [39, 43]. This region lies lateral to the olfactory tubercle and subcallosal gyrus and medial to the insular cortex. The temporal piriform cortex extends along the dorsomedial surface of the uncus, capping the amygdala which lies just interior to this structure, and merging posteriorly into the anterior cortical nucleus of the amygdala. These two piriform subregions appear to be histologically identical, both consisting of a 3-layer allocortex (paleocortex), reflecting its ancient evolutionary origins. There is recent evidence to suggest that the human frontal and temporal piriform cortex are functionally distinct (see next section), which accords with animal models demonstrating anatomical and physiological heterogeneity along the rostral-caudal axis of the piriform cortex [47, 59, 60].

Amygdala

Olfactory bulb projections terminate in several discrete amygdala subnuclei of the corticomedial group, including the periamygdaloid region, anterior and posterior cortical nuclei, the nucleus of the LOT, and the medial nucleus [45]. These structures are situated along the dorsomedial margin of the amygdala, and rostrally the cytoarchitectonic transition from the olfactory amygdala to the temporal piriform cortex is poorly demarcated. Neurophysiological recordings in animals [61, 62] and humans [63, 64] suggest that the amygdala is highly responsive to odor stimulation. Interestingly, from an evolutionary perspective, the physical expansion of the primate amygdala paralleled increases in the paleocortex, mostly comprising the piriform cortex, and consequently much of the amygdala was committed to olfactory processing [65]. In addition to sending projections back to the bulb, the olfactory portions of the amygdala provide direct input to the other major subdivisions, including lateral, basolateral, and central amygdaloid nuclei [66, 67], as well as basal ganglia, thalamus, hypothalamus, and prefrontal cortex.

Orbitofrontal Cortex

The OFC represents the main neocortical projection of the olfactory cortex. This structure consists of a 5-layer agranular or dysgranular neocortex, with an absent or poorly developed layer 4 [68]. It is located along the basal surface of the caudal frontal lobes, including the gyrus rectus medially and the agranular insula laterally, which wraps onto the caudal orbital surface. Apart from the olfactory tubercle, direct afferent inputs arrive from all primary olfactory areas, including the piriform cortex, amygdala, and entorhinal cortex, in the absence of an obligatory thalamic relay. In turn, the OFC provides direct feedback projections to each of these regions. While the majority of available data is derived from nonhuman primate tracer studies [45], there is general agreement that the anatomical organization and cytoarchitecture of human and primate OFC closely correspond [69]. Within the ventral prefrontal cortex, primate areas in the agranular insula (including Iam and Iapm) and the OFC (including area 13a) receive the most substantial olfactory inputs, and each of these regions appears to have an anatomical counterpart in the human brain [70]. In addition, electrical stimulation of the olfactory bulb in anesthetized macaques elicits short- to medium-latency action potentials in the same neocortical areas [45], further supporting the idea that these orbital targets are among the initial stages of olfactory information processing within the brain. Finally, it is important to note that adjacent, nonoverlapping regions of the OFC receive sensory input from gustatory and visual centers, as well as information about visceral states, providing a neural substrate for associative learning and cross-modal integration [71–74], all in the service of promoting feeding-related and odor-guided behaviors.

Unique Properties

From the above description it is apparent that the central organization of the olfactory system has several unique anatomical features that distinguish it from other sensory modalities. These include the ipsilateral nature of central projections, the absence of a thalamic intermediary, and the intimate overlap with 'limbic' regions of the brain.

Ipsilateral Olfactory Projections

Apart from the possibility of minor interhemispheric exchanges via the anterior commissure, odor processing remains ipsilateral all the way from the nasal periphery to the primary olfactory cortex. This stands in sharp contrast, for example, to the visual or auditory systems, which already quite early in the processing stream assemble information from both sides of the head (visually at the optic chiasm, auditorily at the superior olive) into integrated sensory codes. Complementary lines of evidence demonstrating the influence of nasal airflow on odor adsorption patterns and olfactory discrimination may provide a possible explanation for this unique arrangement. First, unilateral swelling of the nasal turbinates routinely alternates between the two nostrils every few hours through the course of the day and determines whether airflow is high or low through each side of the nose [75]. Second, psychophysical data show that in general, high flow rates favor sorption of hydrophilic compounds, whereas low flow rates favor sorption of hydrophobic compounds [76]. Third, recent work indicates that when subjects sniff through one nostril or the other, their ability to identify the components of a binary 2-odor mixture (containing equal proportions of hydrophilic and hydrophobic compounds) significantly depends on local airflow through the tested side [77]. In other words, the hydrophobic odor within the mixture was more likely to be perceived through the low-flow nostril, whereas the hydrophilic odor was better perceived through the high-flow nostril. Thus, restricting the stream of olfactory information to the same side of nasal stimulation would allow for the possibility that different odor 'percepts' are presented through each nostril to each hemisphere. In this manner, the olfactory cortex would be able to make bilateral odor comparisons, potentially enhancing its discriminative capacities [78]. Such an arrangement could also be useful for providing differential access to odor memories [79].

No Thalamic Intermediary

The conduction of odor-evoked signals to central brain regions, including the primary olfactory cortex and neocortical (prefrontal) areas, is achieved without an obligatory thalamic relay. This stands in contrast to the transmission of sensory information across all other modalities, whereby an incoming signal undergoes thalamic modulation prior to being delivered to the sensory-specific cortex. The most parsimonious explanation for this anatomical variation is an evolutionary one: as primitive paleocortex, the olfactory circuitry simply developed long before the emergence of a thalamic module. The implication is that the olfactory bulb and cortex are capable of carrying out many of the functions otherwise supported by the thalamus, such as sharpening sensory receptive fields and repackaging sensory representations into information streams containing different physical qualities (e.g. the parvo- and magnocellular layers in the lateral geniculate nucleus). The absence of a thalamic node in the processing hierarchy has the added advantage of preserving the fidelity of the original olfactory percept. This may be of particular importance to the olfactory system, which has to cope with the challenge of distinguishing an odor in the context of unpredictable shifts in stimulus concentration, background smells, and respiratory patterns. Of course, an alternative possibility is that the olfactory system, in the absence of any meaningful spatial topographical codes (unlike the spatial primacy of visual and somatosensory receptive fields), has no major need for the sensory refinements conferred by a thalamic nucleus.

'Limbic' Overlap

It is clear that an intimate structural overlap exists between the olfactory-related regions described above and those devoted more generally to human emotional processing. Anecdotally this comes as no surprise, as odors appear to have an unusual capacity to evoke highly vivid and emotional autobiographical memories, much more so than other mnemonic sensory triggers, and timelessly exemplified in Proust's tea-soaked madeleines [80]. A consideration of the role of olfaction in animals may shed light on this issue. For many animal species, odorous stimuli are the primary means of motivating almost every aspect of their behavior. Maternal bonding, kinship recognition, food search, mate selection, predator avoidance, and territorial marking are all guided by smells. The ultimate manifestation of these behaviors relies on the coordination of hormonal, visceral, autonomic, emotional, and mnemonic states, all of which are precisely under the control of primary and secondary olfactory structures, namely, the amygdala, entorhinal cortex, orbital cortex, striatum, hypothalamus, and hippocampus. As outlined by Carpenter [81], the human brain has usurped this same set of structures to promote emotional reactions and to motivate behavior, driven not only by olfactory cues (recall the 'microsmatic' human), but also by a host of other secondary reinforcers, such as money, good grades, or a Chicago Cubs baseball ticket, which have acquired biological salience via learning and experience. As it happens, the same neural systems appear to be involved in both the motivational expression of behavior and the formation of novel stimulus-reinforcer associations.

Function

It has been recognized for over 100 years that the temporal lobe contributes to the human experience of smell. Hughlings-Jackson and Beevor [82] and Hughlings-Jackson and Stewart [83] described the occurrence of olfactory auras in patients with certain types of epilepsy and attributed these phenomena to ictal discharges in the medial temporal lobe ('uncinate fits'). Half a century later, Penfield and Jasper [84] discovered that focal electrical stimulation of the olfactory bulb, uncus, or amygdala in awake patients could evoke olfactory impressions, frequently described as smelling unpleasant or 'burnt' in quality. More recently, focal lesion studies have highlighted the importance of the mediotemporal and orbitofrontal lobes to human olfactory perception (reviewed in West and Doty [85]). In patients who have undergone anterior temporal lobectomy for intractable epilepsy, deficits of odor detection [86], identification [87–91], naming [92, 93], quality discrimination [87, 93, 94], matching [87, 88, 95], and memory [88, 89, 92, 93, 96–100] have all been described. Partial excisions of the prefrontal lobe, either as a result of tumor, hematoma, or intractable epilepsy, are also associated with impairments of odor identification [90], quality discrimination [94, 101], and memory [98]. However, the large size and spatial extent of such lesions precludes careful anatomical delineation, particularly given the close proximity of so many critical olfactory structures within the temporal lobe (fig. 1). Thus, despite the abundance of clinical data implicating the ventral frontal and temporal lobes in human olfaction, a more precise functional organization has not been elucidated.

The advent of modern neuroimaging techniques, such as positron emission tomography (PET) and functional magnetic resonance imaging (fMRI), has led to important advances in our understanding of central olfactory function in the normal human brain (for recent reviews, see Savic [102], Sobel et al. [103] and Royet and Plailly [104]). While both PET and fMRI permit simultaneous data acquisition from the whole brain at high spatial resolution, fMRI has several important advantages: it is safe and noninvasive (not requiring the intravenous injection of radioactive tracers), is easily repeated within subjects across different sessions, and is ideal for complex experiments involving the delivery of multiple different odors. It is important to note that the fMRI signal reflects local activity-dependent changes in hemodynamic state (technically, blood oxygenation level-dependent contrast) [105–107], and therefore provides a surrogate marker of neural activity that is temporally constrained by the intrinsic time lag of neurovascular coupling. Thus, the spatial benefits of fMRI (and PET) come at a certain expense of temporal resolution, on the order of seconds.

Human neuroimaging research has begun to identify a network of brain structures important for olfactory processing [108–119]. While many of these studies identified odor-evoked neural responses within primary and secondary areas of the olfactory cortex, one notable feature was the inconsistent activation of the piriform cortex. This is now thought to reflect at least two factors. First, conventional fMRI sequences are associated with signal loss (susceptibility artifact) at air-tissue interfaces, reducing image quality in olfactory-specific areas of the ventral temporal and basal frontal lobes [120]. Second, olfactory habituation occurs with prolonged odor exposure in the rodent piriform cortex [121], and analogous phenomena have been confirmed with fMRI in humans [119, 122]. Because many previous olfactory neuroimaging studies used blocked designs, with constant odor presentation over 30–60 s, habituation has been an unavoidable confound. The recent introduction of event-related designs to limit the duration of odor exposure [123–125], as well as the development of specialized imaging protocols [126, 127] have helped improve fMRI signal detection in the olfactory cortex.

In keeping with the earlier anatomical focus on the piriform cortex, amygdala, and orbitofrontal cortex, the remaining section will provide a selective survey of functional imaging data from these three regions. This decision is motivated by the fact that: (1) there is a relative abundance of PET and fMRI data for these structures, in comparison to other areas, and (2) regions like the AON and olfactory tubercle are too small to be identified with confidence, given the spatial limitations of fMRI. For a discussion of other olfactory structures, including areas not traditionally thought to be olfactory in function (e.g. cerebellum), the reader is referred to Sobel et al. [103]. In addition, while cortical lateralization of olfactory function is not considered here, the topic is covered in detail in a recent review [104]. Finally, other sections of the book are specifically devoted to nasal trigeminal function [see chapter 10 by Breslin and Huang, this vol, pp 152–190] and vomeronasal function [see chapter 11 by Small, this vol, pp 191–220], which are therefore not treated here.

Piriform Cortex

Despite some initial difficulties in imaging the piriform cortex (as highlighted above), several early studies demonstrated that human piriform activity could be evoked by smelling of an odor [108–110, 115], consistent with the idea that this region participates in basic olfactory processing. One important finding to emerge was that the piriform cortex responded not only to smells, but to the act of sniffing itself, even in the absence of odor [115]. In this study, sniffing of odorless air, as well as artificial sniffing induced by air puffs into the nostrils, activated the piriform cortex, whereas partial physical occlusion of the nostrils, or topical anesthesia to the nasal passages, reduced this activity (fig. 3). These results indicate that sniff-induced piriform activity is not simply due to the motor act of sniffing, but rather to the physical sensation of airflow across the nasal mucosa, and are compatible with animal data suggesting that the sniff may prime the piriform cortex for optimal reception of a smell [128, 129].

Until recently, it had been unclear whether the human piriform cortex basically served as a passive relay of olfactory information, or whether it was capable of more complex operations. The first hint that the human piriform cortex was functionally complex, and moreover that fMRI techniques were capable of resolving subregional piriform differences, came from a study on olfactory processing and odor valence [123], in which human subjects smelled three odors that differed in valence (pleasant, neutral, unpleasant). The posterior (temporal) piriform cortex was significantly activated bilaterally by each odor, independent of valence (fig. 4a, b), suggesting that this region mediates basic odor perception. Such a role complements theories derived from animal models suggesting that the piriform cortex is broadly tuned to odors [62, 72] and conforms to recent neuroimaging studies demonstrating similar activation patterns in response to low-level olfactory processes [108, 115, 118, 122]. In contrast, activations

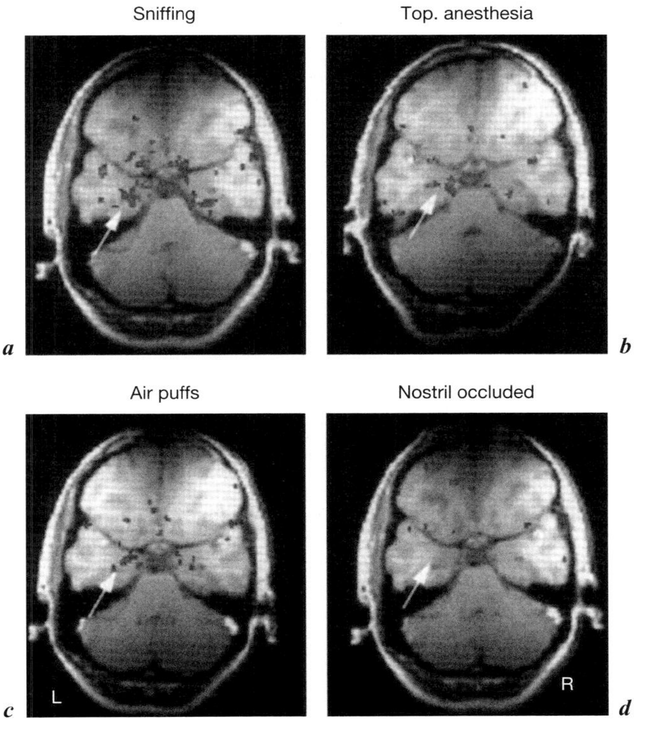

Fig. 3. Sniffing activates the piriform cortex (***a***). By comparison, the sniff-induced fMRI activity is reduced when topical anesthesia is applied to the nasal membranes (***b***) or when airflow is blocked via nostril occlusion (***d***). Piriform activity is also present during artificial puffs of odorless air into the nose (***c***). L = Left; R = right. (Reproduced and modified from figure 3 of Sobel et al. [115]; copyright 1998 by the Nature Publishing Group, and with permission of the authors.)

were also observed within the anterior (frontal) piriform cortex in response to unpleasant and pleasant, but not neutral, odors (fig. 4c, d). One plausible hypothesis is that the anterior piriform cortex is receptive to hedonic quality, especially at extremes of odor valence. The identification of functional heterogeneity in the human piriform cortex accords with animal models demonstrating anatomical and physiological distinctions along its rostral-caudal axis [47, 59, 60] and suggests that this region likely transcends the role of mere sensory intermediary.

The piriform cortex also appears to be involved in olfactory learning and memory, again in keeping with the animal literature. In two PET studies, short- and

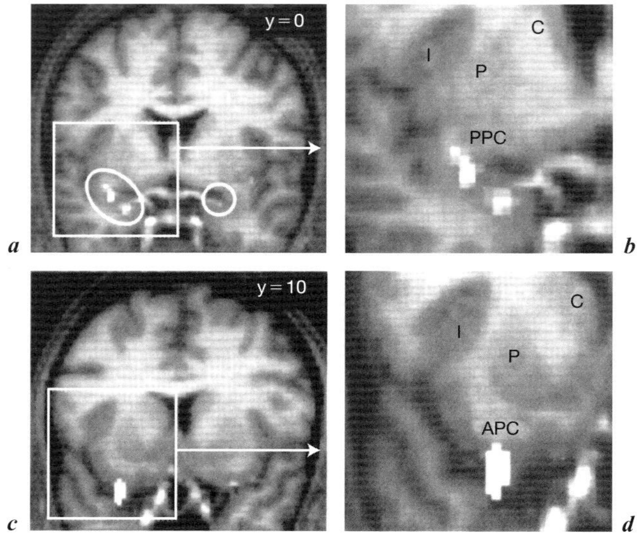

Fig. 4. Functional heterogeneity in the piriform cortex. ***a*** The posterior piriform cortex is mutually activated by pleasant, neutral, and unpleasant odors, characteristic of its role in low-level olfactory processing. The area outlined by a rectangle is magnified in (***b***) to illustrate the anatomy more clearly. ***c*** In turn, the anterior piriform cortex is sensitive to unpleasant and pleasant (but not neutral) odors, suggesting this region encodes information about extremes of odor valence. The fMRI activation shown here was specifically evoked by unpleasant odor, and is magnified in (***d***). APC = Anterior piriform cortex; C = caudate; I = insula; P = putamen; PPC = posterior piriform cortex. The right side of the figure matches the right side of the brain. (Reproduced and modified from figures 3 and 4 of Gottfried et al. [123]; copyright 2002 by the Society for Neuroscience.)

long-term odor recognition memory was associated with enhanced piriform cortex activity when compared to odorless baseline scans [99, 118]. More recently, an fMRI study of visual-olfactory associative learning (Pavlovian conditioning) between a neutral visual stimulus and a pleasant food odor showed that the conditioned visual cue, in the absence of odor, elicited neural activity in the piriform cortex [130]. In this same experiment, after subjects consumed a food (corresponding to the odor) to satiety, the learning-evoked piriform responses decreased in accordance with their current appetite state, lending credence to the idea that the piriform cortex is a site of learning-induced plasticity. Other work has focused on episodic memory retrieval of olfactory context [131]. In an initial study phase, subjects were given combinations of smells and pictures and asked to imagine a link or association between the two stimuli. The aim of this session was to encourage episodic memory encoding of odor-picture

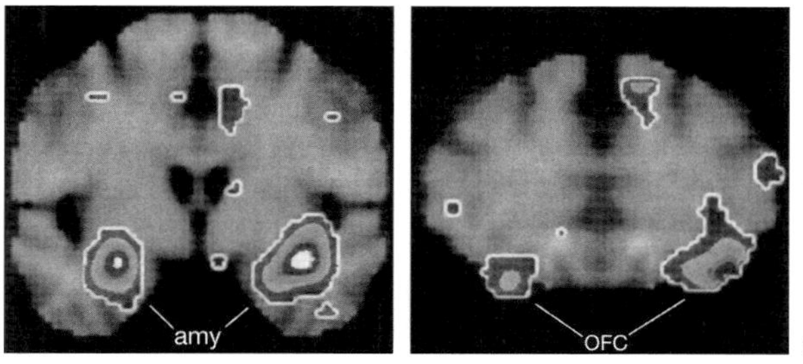

a b

Fig. 5. Highly aversive odors activate the amygdala. This PET image demonstrates bilateral increases in regional cerebral blood flow within the amygdala (amy) (***a***), as well as in the OFC (***b***), in response to unpleasant odors. The right side of the figure matches the left side of the brain. (Reproduced and modified from figure 1 of Zald and Pardo [112]; copyright 1997 by the National Academy of Sciences, USA.)

pairs. In a subsequent recognition memory test, subjects had to decide whether they were viewing a study (old) or novel (new) picture, in the absence of any odor cues. Comparison of correctly remembered items to correct rejections ('old/new' effect) revealed significant memory-related activity in the piriform cortex. Critically, odor absence during the memory session ruled out the possibility that piriform activity was merely being driven by olfactory stimulation. In addition, this effect was specific to the retrieval of olfactory context, as a nonolfactory control experiment (otherwise identical to the primary study) failed to activate this region. These findings suggest that the retrieval-related responses in the piriform cortex provide evidence for the incidental retrieval of olfactory context (experienced at encoding). The data further imply that the piriform cortex, in sustaining sensory-specific traces of a crossmodal memory, must be encoding high-order representations of odor quality or identity.

Amygdala

The amygdala is commonly implicated in emotional processing [132, 133], and one plausible hypothesis is that neural representations of odor valence (pleasantness) are maintained in this region. The first study to test this idea was by Zald and Pardo [112], who showed bilateral amygdala activation in response to highly aversive (compared to minimally aversive) smells (fig. 5). However, as the highly aversive stimuli were also more intense, it was difficult to conclude whether the amygdala responded to odor valence per se or to perceived odor intensity. Follow-up experiments have been conflicted in various ways by these intensity-valence confounds. One study showed greater amygdala activity for unpleasant than pleasant odors, otherwise matched for intensity, suggesting a valence-specific effect [134]. Another study suggested the amygdala was basically insensitive to valence, being similarly activated by pleasant, neutral, and unpleasant odors [123], while a third group demonstrated correlations between perceived intensity and neural activity in temporal structures adjacent to the amygdala [135]. In an effort to resolve these issues, Anderson et al. [125] dissociated intensity and valence within a single experiment, by presenting one pleasant odor (citral: lemon smell) and one unpleasant odor (valeric acid: sweaty sock smell), each at low and high intensity. In this manner, the amygdala was significantly activated by intensity (high vs. low), but not valence (unpleasant vs. pleasant, or vice versa), suggesting odor intensity coding in this region.

Even then, recent findings from our laboratory suggest this story is still unfinished [135a]. Here, intensity and valence were again dissociated within one experimental design, but with the inclusion of a neutrally valenced odor condition, in order to estimate amygdala activity across a more comprehensive valence spectrum. That is, high and low intensity versions of pleasant, neutral, and unpleasant odors were delivered to subjects. We predicted that if the amygdala responds to intensity irrespective of valence, then all three odor types should elicit similar levels of activation. However, if the amygdala only responds to intensity at the extremes of valence, then pleasant and unpleasant odors, but not neutral odor, should elicit activity. Our findings are in keeping with this latter prediction, suggesting that the interaction between intensity and valence, reflecting the overall behavioral salience of an odor, is what matters most to the amygdala.

As described above for the piriform cortex, the amygdala is also involved in associative learning between visual stimuli and olfactory reinforcers [130, 136]. Recent data suggest this region is preferentially involved in the formation of new associations, but does not maintain these representations over time [136]. This latter mnemonic function may be reserved for regions such as the OFC (see below). In addition, there is evidence to suggest a role for the amygdala in the evocation of emotional odor memories [137]. Five subjects were presented with one of four different stimulus conditions in the fMRI scanner: odors (perfumes) that elicited a pleasant, personal memory; visual pictures of those odors (perfume bottles); control odors; and control pictures. Comparison of emotional odor to the other three conditions (or to the control odor alone) was associated with parahippocampal activation extending into the amygdala. While these findings do not permit a distinction between memory retrieval, valence processing, and stimulus salience, they suggest that behaviorally relevant odor cues may be a more potent activator of emotional circuitry than nonolfactory stimuli. These functional imaging data are supported by lesion studies of patients with selective bilateral amygdala damage [100, 138], as well as by stereotactic intracranial recordings [134], all of which further implicate the amygdala in various tasks related specifically to olfactory memory.

Orbitofrontal Cortex

Human olfactory imaging has tentatively revealed dissociations of function along two separate anatomical dimensions in the OFC. The first of these is a caudal-rostral distinction. Odor-evoked neural activity in the caudal OFC is typically associated with low-level aspects of olfactory processing, such as passive smelling and odor detection [108, 112, 123, 139], and probably represents the initial neocortical projection site from the primary olfactory cortex. The location of these activations roughly corresponds to the so-called 'central-posterior orbitofrontal cortex' identified by Yarita et al. [140], who considered it a broadly tuned area of the primate olfactory association cortex. According to Carmichael et al. [45], the central-posterior orbitofrontal cortex is roughly homologous to the orbital areas 13m, 13a, and Iam, the primary prefrontal locus of olfactory input (see the Anatomy section above). By comparison, more rostral areas of the OFC are engaged in higher-order olfactory computations, including associative learning [124, 130, 136], working memory [141], and short- and long-term odor recognition memory [99, 118]. This caudal-rostral division is bolstered by animal findings suggesting an anatomical hierarchy of orbitofrontal specialization, whereby caudal regions (such as the OFC) converge on medial and anterior territories to permit more complex information processing [69, 142].

The OFC also exhibits regional differences along a medial-lateral axis of functional specialization. Numerous studies increasingly show that pleasant odors evoke activity in the medial OFC and ventromedial prefrontal cortex, whereas unpleasant odors evoke activity in the lateral OFC and adjacent inferior prefrontal cortex [123, 125, 135]. Similar dissociations have been identified in visual-olfactory crossmodal paradigms of episodic memory [131] and associative learning [136]. These valence-specific patterns have emerged in other experiments spanning a variety of modalities. Thus, processing of tastes [143], faces [144], and abstract monetary reinforcers [145] have all revealed medial-lateral response differences in the OFC that vary according to the degree of pleasantness. On neuroanatomical grounds, these orbital regions can be regarded as distinct functional units with unique sets of cortical and subcortical connections [142]. Notably, projections between the OFC and amygdala are reciprocal [142]. It is thus plausible that differences in input patterns from the amygdala, e.g., might contribute to the expression of positive and negative value in medial and lateral orbital subdivisions, respectively.

As a major recipient of projections not only from the primary olfactory cortex, but also from gustatory, visual, visceral, and thalamic centers, the OFC is certain to participate in a wide variety of complex olfactory functions related to multimodal integration, reward processing, and goal-directed learning and behavior [see chapter 13 by Landis and Lacroix, this vol, pp 242–254, for an extensive discussion on neural correlates of flavor and feeding]. The role of the OFC as a site of sensory convergence has been documented across several different sensory combinations, including odor/taste [146–148] and odor/vision [149]. Moreover, through manipulations of semantic correspondence between odors and tastes [147], or between odors and pictures [149], OFC activity increased with increasing subjective congruency ratings. These observations underscore the general idea that prior learning and experience can profoundly modulate the central processing of sensory information. Such mechanisms may also help to resolve the inherent ambiguity in olfactory perception [149]. Interestingly, as discussed above, unimodal odor stimuli appear to be processed in more caudal regions of the OFC than the corresponding bimodal stimulus pairs [147, 149].

The human OFC also provides a substrate for the encoding of primary and secondary (learned) value of olfactory stimuli. In a study of sensory-specific satiety, subjects were delivered two different food odors, both before and after a feeding session designed to decrease the pleasantness of one of the odors [117]. Following satiety, the neural activity in the OFC declined in parallel with the decrease in food pleasantness ratings, suggesting that reward value of a food odor is encoded and updated in this structure. Our own work consistently demonstrates robust participation of the OFC in classical conditioning paradigms of olfactory learning [124, 130, 136]. Specifically, after an arbitrary visual picture is repetitively paired with either a pleasant or unpleasant odor, presentation of the visual cue by itself elicits activation in the OFC, suggesting that this region is involved in the establishment of picture-odor contingencies. Moreover, when the current affective value of the olfactory reinforcer is either decreased (via selective satiety) or increased (via odor inflation), cue-evoked OFC responses are modulated in parallel with the behavioral manipulation. The implication is that a predictive cue has direct access to central representations of value in the OFC, and that these representations are flexibly updated according to an individual's motivational state.

Finally, it should be mentioned that a variety of explicit cognitive tasks influences responses in the OFC [104]. Intensity judgments [150], familiarity judgments [116, 139], hedonicity judgments [139], and quality discrimination tasks [118] are all associated with orbitofrontal activity, irrespective of specific perceptual features of the odors themselves (e.g. their intensity, familiarity, or pleasantness). These activations are usually accompanied by regional responses in large portions of the frontal, temporal, parietal, and occipital cortex (frequently in the absence of the primary olfactory cortex), implying the involvement of nonolfactory networks in mediating higher-level olfactory decision-making. It remains to be established how each of these brain areas specifically contributes to the cognitive components of the high-level tasks outlined here.

Conclusions

This chapter has provided an overview of central olfactory processing in the human brain, with particular emphasis on the anatomical and functional features of those primary and secondary olfactory structures receiving the bulk of afferent inputs from the nasal periphery. Critical to our present understanding of human olfaction has been the development of modern imaging techniques, in parallel with the emergence of increasingly sophisticated experimental paradigms and research questions. These findings indicate that: (1) the piriform cortex is no mere sensory intermediary, but is functionally heterogeneous and participates in numerous aspects of olfactory learning and memory; (2) the amygdala is also functionally complex, encoding the emotionality of an odor stimulus and helping establish links between environmental cues and biologically salient smells, and (3) the OFC, as the principal neocortical target of the primary olfactory cortex, performs a wide assortment of higher-level operations related to multisensory integration, reward processing, and associative learning. The biological complexity of central olfactory processing underscores the perceptual acuity and aptitude of the human sense of smell, as briefly touched on here, and makes a compelling case for the advanced functional capabilities of human olfaction, even when constrained by ecological and evolutionary factors.

We are still only on the cusp of deciphering olfactory processes in the human brain, and the field of olfactory functional imaging, which had less than 10 publications in the 5 years between 1992 and 1997, has seen that number quadruple since that time. The improvement of existing techniques, or the development of new ones, will lead to improvements in spatial and temporal resolution, which will be critical if we hope to achieve a more fine-grained understanding of structure-function relationships in the basal forebrain. One crucial issue will be to determine how peripheral and central processes interact to create perceptual representations of smell, a goal that will best succeed via multidisciplinary collaboration across the fields of neuroscience, neurology, psychology, and otorhinolaryngology. From a purely clinical viewpoint, the clarification of central olfactory processing in the healthy human brain may ultimately lead to diagnostic and treatment interventions in neurodegenerative disorders such as Alzheimer's disease and Parkinson's disease, both of which have smell impairments early in the course of illness, sometimes preceding the onset of other neurological symptoms [6–8]. To this end, progress in basic olfactory neuroscience should encourage reciprocal translations between laboratory and clinic. Knowledge gained about the functional organization of olfaction in the healthy brain will inform our understanding of neurological disease, whereas insights gained in the clinical population will serve as a useful constraint on hypotheses tested back in the laboratory.

References

1 Grinker RR: Grinker's Neurology. Springfield, Thomas, 1943.

2 Brodal A: Neurological Anatomy in Relation to Clinical Medicine. London, Oxford University Press, 1969.

3 Shepherd GM: The human sense of smell: are we better than we think? PLoS Biol 2004;2: E146.

4 Rouquier S, Friedman C, Delettre C, van den Engh G, Blancher A, Crouau-Roy B, Trask BJ, Giorgi D: A gene recently inactivated in human defines a new olfactory receptor family in mammals. Hum Mol Genet 1998;7:1337–1345.

5 Gilad Y, Man O, Paabo S, Lancet D: Human specific loss of olfactory receptor genes. Proc Natl Acad Sci USA 2003;100:3324–3327.

6 Mesholam RI, Moberg PJ, Mahr RN, Doty RL: Olfaction in neurodegenerative disease: a meta-analysis of olfactory functioning in Alzheimer's and Parkinson's diseases. Arch Neurol 1998;55:84–90.

7 Murphy C: Loss of olfactory function in dementing disease. Physiol Behav 1999;66:177–182.

8 Hawkes C: Olfaction in neurodegenerative disorder. Mov Disord 2003;18:364–372.

9 Laska M, Seibt A: Olfactory sensitivity for aliphatic esters in squirrel monkeys and pigtail macaques. Behav Brain Res 2002;134:165–174.

10 Devos M, Patte F, Rouault J, Laffort P, Van Gemert LJ: Standardized Human Olfactory Thresholds. Oxford, IRL Press at Oxford University Press, 1990.

11 Laska M, Teubner P: Olfactory discrimination ability for homologous series of aliphatic alcohols and aldehydes. Chem Senses 1999;24:263–270.

12 Laska M, Ayabe-Kanamura S, Hubener F, Saito S: Olfactory discrimination ability for aliphatic odorants as a function of oxygen moiety. Chem Senses 2000;25:189–197.

13 Laska M, Hubener F: Olfactory discrimination ability for homologous series of aliphatic ketones and acetic esters. Behav Brain Res 2001;119:193–201.

14 Cain WS: To know with the nose: keys to odor identification. Science 1979;203:467–470.

15 Carde RT, Mafra-Neto A: Mechanisms of flight of male moths to pheromone; in Carde RT, Minks AK (eds): Insect Pheromone Research: New Directions. New York, Chapman and Hall, 1997, pp 275–290.

16 MacFarlane A: Olfaction in the development of social preferences in the human neonate. Ciba Found Symp 1975;103–117.

17 Russell MJ: Human olfactory communication. Nature 1976;260:520–522.

18 Baron RA: Perfume as a tactic of impression management in social and organizational settings; in Van Toller S, Dodd GG (eds): Perfumery: The Psychology and Biology of Fragrance. London, Chapman and Hall, 1988, pp 91–104.

19 Beauchamp GK, Maller O: The development of flavor preferences in humans: a review; in Kare MR, Maller O (eds): The Chemical Senses and Nutrition. New York, Academic Press, 1977, pp 291–311.

20 McClintock MK: Menstrual synchrony and suppression. Nature 1971;229:244–245.

21 Stern K, McClintock MK: Regulation of ovulation by human pheromones. Nature 1998;392: 177–179.

22 Wedekind C, Seebeck T, Bettens F, Paepke AJ: MHC-dependent mate preferences in humans. Proc Biol Sci 1995;260:245–249.
23 Wedekind C, Furi S: Body odour preferences in men and women: do they aim for specific MHC combinations or simply heterozygosity? Proc Biol Sci 1997;264:1471–1479.
24 Jacob S, McClintock MK, Zelano B, Ober C: Paternally inherited HLA alleles are associated with women's choice of male odor. Nat Genet 2002;30:175–179.
25 Ober C, Weitkamp LR, Cox N, Dytch H, Kostyu D, Elias S: HLA and mate choice in humans. Am J Hum Genet 1997;61:497–504.
26 Counet C, Callemien D, Ouwerx C, Collin S: Use of gas chromatography-olfactometry to identify key odorant compounds in dark chocolate. Comparison of samples before and after conching. J Agric Food Chem 2002;50:2385–2391.
27 Laing DG, Francis GW: The capacity of humans to identify odors in mixtures. Physiol Behav 1989;46:809–814.
28 Livermore A, Laing DG: Influence of training and experience on the perception of multicomponent odor mixtures. J Exp Psychol Hum Percept Perform 1996;22:267–277.
29 Morrot G, Brochet F, Dubourdieu D: The color of odors. Brain Lang 2001;79:309–320.
30 Zellner DA, Kautz MA: Color affects perceived odor intensity. J Exp Psychol Hum Percept Perform 1990;16:391–397.
31 Dalton P: Odor perception and beliefs about risk. Chem Senses 1996;21:447–458.
32 Dalton P, Doolittle N, Nagata H, Breslin PA: The merging of the senses: integration of subthreshold taste and smell. Nat Neurosci 2000;3:431–432.
33 Distel H, Hudson R: Judgement of odor intensity is influenced by subjects' knowledge of the odor source. Chem Senses 2001;26:247–251.
34 Slosson EE: A lecture experiment in hallucinations. Psychol Rev 1899;6:407–408.
35 O'Mahony M: Smell illusions and suggestion: reports of smell contingent on tones played on television and radio. Chem Senses Flav 1978;3:183–189.
36 Wilson DA, Stevenson RJ: The fundamental role of memory in olfactory perception. Trends Neurosci 2003;26:243–247.
37 Stevenson RJ: Associative learning and odor quality perception: how sniffing an odor mixture can alter the smell of its parts. Learn Motiv 2001;32:154–177.
38 Stevenson RJ: The acquisition of odour qualities. Q J Exp Psychol 2001;54:561–577.
39 Eslinger PJ, Damasio AR, Van Hoesen GW: Olfactory dysfunction in man: anatomical and behavioral aspects. Brain Cogn 1982;1:259–285.
40 Price JL: Olfactory system; in Paxinos G (ed): The Human Nervous System. San Diego, Academic Press, 1990, pp 979–998.
41 Heimer L: The Human Brain and Spinal Cord: Functional Neuroanatomy and Dissection Guide. New York, Springer, 1983.
42 Nolte J: The Human Brain: An Introduction to Its Functional Anatomy. St Louis, Mosby, 2002, pp 319–336.
43 Mai JK, Assheuer J, Paxinos G: Atlas of the Human Brain. San Diego, Academic Press, 1997.
44 Martin JH: Neuroanatomy: Text and Atlas. New York, McGraw-Hill, 2003, pp 207–226.
45 Carmichael ST, Clugnet MC, Price JL: Central olfactory connections in the macaque monkey. J Comp Neurol 1994;346:403–434.
46 Shipley MT, Ennis M: Functional organization of olfactory system. J Neurobiol 1996;30: 123–176.
47 Haberly LB: Olfactory cortex; in Shepherd GM (ed): The Synaptic Organization of the Brain. New York, Oxford University Press, 1998, pp 377–416.
48 Cleland TA, Linster C: Central olfactory structures; in Doty RL (ed): Handbook of Olfaction and Gustation, ed 2. New York, Dekker, 2003, pp 165–180.
49 Wilson DA, Sullivan RM: Sensory physiology of central olfactory pathways; in Doty RL (ed): Handbook of Olfaction and Gustation, ed 2. New York, Dekker, 2003, pp 181–201.
50 Price JL: An autoradiographic study of complementary laminar patterns of termination of afferent fibers to the olfactory cortex. J Comp Neurol 1973;150:87–108.
51 de Olmos J, Hardy H, Heimer L: The afferent connections of the main and the accessory olfactory bulb formations in the rat: an experimental HRP-study. J Comp Neurol 1978;181:213–244.

52 Haberly LB, Price JL: Association and commissural fiber systems of the olfactory cortex of the rat. J Comp Neurol 1978;178:711–740.
53 Zou Z, Horowitz LF, Montmayeur JP, Snapper S, Buck LB: Genetic tracing reveals a stereotyped sensory map in the olfactory cortex. Nature 2001;414:173–179.
54 Pearce RK, Hawkes CH, Daniel SE: The anterior olfactory nucleus in Parkinson's disease. Mov Disord 1995;10:283–287.
55 Kovacs T, Papp MI, Cairns NJ, Khan MN, Lantos PL: Olfactory bulb in multiple system atrophy. Mov Disord 2003;18:938–942.
56 Montemurro DG, Bruni JE: The Human Brain in Dissection. New York, Oxford University Press, 1988.
57 Millhouse OE, Heimer L: Cell configurations in the olfactory tubercle of the rat. J Comp Neurol 1984;228:571–597.
58 Shipley MT, Adamek GD: The connections of the mouse olfactory bulb: a study using orthograde and retrograde transport of wheat germ agglutinin conjugated to horseradish peroxidase. Brain Res Bull 1984;12:669–688.
59 Litaudon P, Mouly AM, Sullivan R, Gervais R, Cattarelli M: Learning-induced changes in rat piriform cortex activity mapped using multisite recording with voltage sensitive dye. Eur J Neurosci 1997;9:1593–1602.
60 Chabaud P, Ravel N, Wilson DA, Mouly AM, Vigouroux M, Farget V, Gervais R: Exposure to behaviourally relevant odour reveals differential characteristics in rat central olfactory pathways as studied through oscillatory activities. Chem Senses 2000;25:561–573.
61 Cain DP, Bindra D: Responses of amygdala single units to odors in the rat. Exp Neurol 1972;35:98–110.
62 Tanabe T, Iino M, Takagi SF: Discrimination of odors in olfactory bulb, pyriform-amygdaloid areas, and orbitofrontal cortex of the monkey. J Neurophysiol 1975;38:1284–1296.
63 Hughes JR, Andy OJ: The human amygdala. 1. Electrophysiological responses to odorants. Electroencephalogr Clin Neurophysiol 1979;46:428–443.
64 Hudry J, Ryvlin P, Royet JP, Mauguiere F: Odorants elicit evoked potentials in the human amygdala. Cereb Cortex 2001;11:619–627.
65 Barton RA, Aggleton JP: Primate evolution and the amygdala; in Aggleton JP (ed): The Amygdala: A Functional Analysis, ed 2. New York, Oxford University Press, 2000, pp 479–508.
66 Amaral DG, Price JL, Pitkänen A, Carmichael ST: Anatomical organization of the primate amygdaloid complex; in Aggleton JP (ed): The Amygdala, ed 1. New York, Wiley, 1992, pp 1–67.
67 Pitkänen A: Connectivity of the rat amygdaloid complex; in Aggleton JP (ed): The Amygdala: A Functional Analysis, ed 2. Oxford, Oxford University Press, 2000, pp 31–116.
68 Carmichael ST, Price JL: Architectonic subdivision of the orbital and medial prefrontal cortex in the macaque monkey. J Comp Neurol 1994;346:366–402.
69 Ongur D, Price JL: The organization of networks within the orbital and medial prefrontal cortex of rats, monkeys and humans. Cereb Cortex 2000;10:206–219.
70 Ongur D, Ferry AT, Price JL: Architectonic subdivision of the human orbital and medial prefrontal cortex. J Comp Neurol 2003;460:425–449.
71 Rolls ET, Baylis LL: Gustatory, olfactory, and visual convergence within the primate orbitofrontal cortex. J Neurosci 1994;14:5437–5452.
72 Schoenbaum G, Eichenbaum H: Information coding in the rodent prefrontal cortex. 1. Single-neuron activity in orbitofrontal cortex compared with that in pyriform cortex. J Neurophysiol 1995;74: 733–750.
73 Rolls ET, Critchley HD, Mason R, Wakeman EA: Orbitofrontal cortex neurons: role in olfactory and visual association learning. J Neurophysiol 1996;75:1970–1981.
74 Schoenbaum G, Chiba AA, Gallagher M: Orbitofrontal cortex and basolateral amygdala encode expected outcomes during learning. Nat Neurosci 1998;1:155–159.
75 Hasegawa M, Kern EB: The human nasal cycle. Mayo Clin Proc 1977;52:28–34.
76 Mozell MM, Kent PF, Murphy SJ: The effect of flow rate upon the magnitude of the olfactory response differs for different odors. Chem Senses 1991;16:631–649.
77 Sobel N, Khan RM, Saltman A, Sullivan EV, Gabrieli JD: The world smells different to each nostril. Nature 1999;402:35.

78 Wilson DA, Sullivan RM: Respiratory airflow pattern at the rat's snout and an hypothesis regarding its role in olfaction. Physiol Behav 1999;66:41–44.
79 Kucharski D, Hall WG: New routes to early memories. Science 1987;238:786–788.
80 Proust M: Swann's Way, 1913 (translated by CKS Moncrieff and T Kilmartin). New York, Modern Library, 1956.
81 Carpenter RHS: Neurophysiology. London, Arnold Publishers, 2003.
82 Hughlings-Jackson J, Beevor CE: Case of tumour of the right temporo-sphenoidal lobe bearing on the localization of the sense of smell and on the interpretation of a particular variety of epilepsy. Brain 1890;12:346–357.
83 Hughlings-Jackson J, Stewart P: Epileptic attacks with a warning of a crude sensation of smell and with the intellectual aura (dreamy state) in a patient who had symptoms pointing to gross organic disease of the right temporo-sphenoidal lobe. Brain 1899;22:534–549.
84 Penfield W, Jasper H: Epilepsy and the Functional Anatomy of the Human Brain. Boston, Little Brown & Co, 1954.
85 West SE, Doty RL: Influence of epilepsy and temporal lobe resection on olfactory function. Epilepsia 1995;36:531–542.
86 Rausch R, Serafetinides EA: Specific alterations of olfactory function in humans with temporal lobe lesions. Nature 1975;255:557–558.
87 Eichenbaum H, Morton TH, Potter H, Corkin S: Selective olfactory deficits in case HM. Brain 1983;106:459–472.
88 Eskenazi B, Cain WS, Novelly RA, Friend KB: Olfactory functioning in temporal lobectomy patients. Neuropsychologia 1983;21:365–374.
89 Eskenazi B, Cain WS, Novelly RA, Mattson R: Odor perception in temporal lobe epilepsy patients with and without temporal lobectomy. Neuropsychologia 1986;24:553–562.
90 Jones-Gotman M, Zatorre RJ: Olfactory identification deficits in patients with focal cerebral excision. Neuropsychologia 1988;26:387–400.
91 Jones-Gotman M, Zatorre RJ, Cendes F, Olivier A, Andermann F, McMackin D, Staunton H, Siegel AM, Wieser HG: Contribution of medial versus lateral temporal-lobe structures to human odour identification. Brain 1997;120:1845–1856.
92 Carroll B, Richardson JT, Thompson P: Olfactory information processing and temporal lobe epilepsy. Brain Cogn 1993;22:230–243.
93 Martinez BA, Cain WS, de Wijk PA, Spencer DD, Novelly RA, Sass KJ: Olfactory functioning before and after temporal lobe resection for intractable seizures. Neuropsychology 1993;7:351–363.
94 Zatorre RJ, Jones-Gotman M: Human olfactory discrimination after unilateral frontal or temporal lobectomy. Brain 1991;114:71–84.
95 Abraham A, Mathai KV: The effect of right temporal lobe lesions on matching of smells. Neuropsychologia 1983;21:277–281.
96 Henkin RI, Comiter H, Fedio P, O'Doherty D: Defects in taste and smell recognition following temporal lobectomy. Trans Am Neurol Assoc 1977;102:146–150.
97 Rausch R: Cognitive strategies in patients with unilateral temporal lobe excisions. Neuropsychologia 1977;15:385–395.
98 Jones-Gotman M, Zatorre RJ: Odor recognition memory in humans: role of right temporal and orbitofrontal regions. Brain Cogn 1993;22:182–198.
99 Dade LA, Zatorre RJ, Jones-Gotman M: Olfactory learning: convergent findings from lesion and brain imaging studies in humans. Brain 2002;125:86–101.
100 Buchanan TW, Tranel D, Adolphs R: A specific role for the human amygdala in olfactory memory. Learn Mem 2003;10:319–325.
101 Potter H, Butters N: An assessment of olfactory deficits in patients with damage to prefrontal cortex. Neuropsychologia 1980;18:621–628.
102 Savic I: Imaging of brain activation by odorants in humans. Curr Opin Neurobiol 2002;12: 455–461.
103 Sobel N, Johnson BN, Mainland J, Yousem DM: Functional neuroimaging of human olfaction; in Doty RL (ed): Handbook of Olfaction and Gustation, ed 2. New York, Dekker, 2003, pp 251–273.
104 Royet JP, Plailly J: Lateralization of olfactory processes. Chem Senses 2004;29:731–745.
105 Ogawa S, Lee TM, Kay AR, Tank DW: Brain magnetic resonance imaging with contrast dependent on blood oxygenation. Proc Natl Acad Sci USA 1990;87:9868–9872.

106 Kwong KK, Belliveau JW, Chesler DA, Goldberg IE, Weisskoff RM, Poncelet BP, Kennedy DN, Hoppel BE, Cohen MS, Turner R, et al: Dynamic magnetic resonance imaging of human brain activity during primary sensory stimulation. Proc Natl Acad Sci USA 1992;89:5675–5679.

107 Ogawa S, Tank DW, Menon R, Ellermann JM, Kim SG, Merkle H, Ugurbil K: Intrinsic signal changes accompanying sensory stimulation: functional brain mapping with magnetic resonance imaging. Proc Natl Acad Sci USA 1992;89:5951–5955.

108 Zatorre RJ, Jones-Gotman M, Evans AC, Meyer E: Functional localization and lateralization of human olfactory cortex. Nature 1992;360:339–340.

109 Koizuka I, Yano H, Nagahara M, Mochizuki R, Seo R, Shimada K, Kubo T, Nogawa T: Functional imaging of the human olfactory cortex by magnetic resonance imaging. ORL J Otorhinolaryngol Relat Spec 1994;56:273–275.

110 Levy LM, Henkin RI, Hutter A, Lin CS, Martins D, Schellinger D: Functional MRI of human olfaction. J Comput Assist Tomogr 1997;21:849–856.

111 Yousem DM, Williams SC, Howard RO, Andrew C, Simmons A, Allin M, Geckle RJ, Suskind D, Bullmore ET, Brammer MJ, Doty RL: Functional MR imaging during odor stimulation: preliminary data. Radiology 1997;204:833–838.

112 Zald DH, Pardo JV: Emotion, olfaction, and the human amygdala: amygdala activation during aversive olfactory stimulation. Proc Natl Acad Sci USA 1997;94:4119–4124.

113 Dade LA, Jones-Gotman M, Zatorre RJ, Evans AC: Human brain function during odor encoding and recognition. A PET activation study. Ann NY Acad Sci 1998;855:572–574.

114 Fulbright RK, Skudlarski P, Lacadie CM, Warrenburg S, Bowers AA, Gore JC, Wexler BE: Functional MR imaging of regional brain responses to pleasant and unpleasant odors. AJNR Am J Neuroradiol 1998;19:1721–1726.

115 Sobel N, Prabhakaran V, Desmond JE, Glover GH, Goode RL, Sullivan EV, Gabrieli JD: Sniffing and smelling: separate subsystems in the human olfactory cortex. Nature 1998;392:282–286.

116 Royet JP, Koenig O, Gregoire MC, Cinotti L, Lavenne F, Le Bars D, Costes N, Vigouroux M, Farget V, Sicard G, Holley A, Mauguiere F, Comar D, Froment JC: Functional anatomy of perceptual and semantic processing for odors. J Cogn Neurosci 1999;11:94–109.

117 O'Doherty J, Rolls ET, Francis S, Bowtell R, McGlone F, Kobal G, Renner B, Ahne G: Sensory-specific satiety-related olfactory activation of the human orbitofrontal cortex. Neuroreport 2000;11:893–897.

118 Savic I, Gulyas B, Larsson M, Roland P: Olfactory functions are mediated by parallel and hierarchical processing. Neuron 2000;26:735–745.

119 Sobel N, Prabhakaran V, Zhao Z, Desmond JE, Glover GH, Sullivan EV, Gabrieli JD: Time course of odorant-induced activation in the human primary olfactory cortex. J Neurophysiol 2000;83:537–551.

120 Ojemann JG, Akbudak E, Snyder AZ, McKinstry RC, Raichle ME, Conturo TE: Anatomic localization and quantitative analysis of gradient refocused echo-planar fMRI susceptibility artifacts. Neuroimage 1997;6:156–167.

121 Wilson DA: Habituation of odor responses in the rat anterior piriform cortex. J Neurophysiol 1998;79:1425–1440.

122 Poellinger A, Thomas R, Lio P, Lee A, Makris N, Rosen BR, Kwong KK: Activation and habituation in olfaction – An fMRI study. Neuroimage 2001;13:547–560.

123 Gottfried JA, Deichmann R, Winston JS, Dolan RJ: Functional heterogeneity in human olfactory cortex: an event-related functional magnetic resonance imaging study. J Neurosci 2002;22:10819–10828.

124 Gottfried JA, O'Doherty J, Dolan RJ: Appetitive and aversive olfactory learning in humans studied using event-related functional magnetic resonance imaging. J Neurosci 2002;22:10829–10837.

125 Anderson AK, Christoff K, Stappen I, Panitz D, Ghahremani DG, Glover G, Gabrieli JD, Sobel N: Dissociated neural representations of intensity and valence in human olfaction. Nat Neurosci 2003;6:196–202.

126 Deichmann R, Gottfried JA, Hutton C, Turner R: Optimized EPI for fMRI studies of the orbitofrontal cortex. Neuroimage 2003;19:430–441.

127 Wilson JL, Jezzard P: Utilization of an intra-oral diamagnetic passive shim in functional MRI of the inferior frontal cortex. Magn Reson Med 2003;50:1089–1094.

128 Adrian ED: Olfactory reactions in the brain of the hedgehog. J Physiol 1942;100:459–473.
129 Ueki S, Domino EF: Some evidence for a mechanical receptor in olfactory function. J Neurophysiol 1961;24:12–25.
130 Gottfried JA, O'Doherty J, Dolan RJ: Encoding predictive reward value in human amydata and orbitofrontal cortex. Science 2003;301:1104–1107.
131 Gottfried JA, Smith APR, Rugg MD, Dolan RJ: Remembrance of odors past: human olfactory cortex in crossmodal recognition memory. Neuron 2004;42:687–695.
132 LeDoux JE: Emotion circuits in the brain. Annu Rev Neurosci 2000;23:155–184.
133 Dolan RJ: Emotion, cognition, and behavior. Science 2002;298:1191–1194.
134 Hudry J, Perrin F, Ryvlin P, Mauguiere F, Royet JP: Olfactory short-term memory and related amygdala recordings in patients with temporal lobe epilepsy. Brain 2003;126:1851–1863.
135 Rolls ET, Kringelbach ML, De Araujo IE: Different representations of pleasant and unpleasant odours in the human brain. Eur J Neurosci 2003;18:695–703.
135a Winston JS, Gottfried JA, Kilner JM, Dolan RJ: Integrated neural representations of odor intensity and affective valence in human amydata. J Neurosci 2005;25:8903–8907.
136 Gottfried JA, Dolan RJ: Human orbitofrontal cortex mediates extinction learning while accessing conditioned representations of value. Nat Neurosci 2004;7:1144–1152.
137 Herz RS, Eliassen J, Beland S, Souza T: Neuroimaging evidence for the emotional potency of odor-evoked memory. Neuropsychologia 2004;42:371–378.
138 Markowitsch HJ, Calabrese P, Wurker M, Durwen HF, Kessler J, Babinsky R, Brechtelsbauer D, Heuser L, Gehlen W: The amygdala's contribution to memory – A study on two patients with Urbach-Wiethe disease. Neuroreport 1994;5:1349–1352.
139 Royet JP, Hudry J, Zald DH, Godinot D, Gregoire MC, Lavenne F, Costes N, Holley A: Functional neuroanatomy of different olfactory judgments. Neuroimage 2001;13:506–519.
140 Yarita H, Iino M, Tanabe T, Kogure S, Takagi SF: A transthalamic olfactory pathway to orbitofrontal cortex in the monkey. J Neurophysiol 1980;43:69–85.
141 Dade LA, Zatorre RJ, Evans AC, Jones-Gotman M: Working memory in another dimension: functional imaging of human olfactory working memory. Neuroimage 2001;14:650–660.
142 Carmichael ST, Price JL: Connectional networks within the orbital and medial prefrontal cortex of macaque monkeys. J Comp Neurol 1996;371:179–207.
143 Small DM, Zatorre RJ, Dagher A, Evans AC, Jones-Gotman M: Changes in brain activity related to eating chocolate: from pleasure to aversion. Brain 2001;124:1720–1733.
144 O'Doherty J, Winston JS, Critchley H, Perrett D, Burt DM, Dolan RJ: Beauty in a smile: the role of medial orbitofrontal cortex in facial attractiveness. Neuropsychologia 2003;41:147–155.
145 O'Doherty J, Kringelbach ML, Rolls ET, Hornak J, Andrews C: Abstract reward and punishment representations in the human orbitofrontal cortex. Nat Neurosci 2001;4:95–102.
146 Cerf-Ducastel B, Murphy C: fMRI activation in response to odorants orally delivered in aqueous solutions. Chem Senses 2001;26:625–637.
147 De Araujo IE, Rolls ET, Kringelbach ML, McGlone F, Phillips N: Taste-olfactory convergence, and the representation of the pleasantness of flavour, in the human brain. Eur J Neurosci 2003;18:2059–2068.
148 Small DM, Voss J, Mak YE, Simmons KB, Parrish T, Gitelman D: Experience-dependent neural integration of taste and smell in the human brain. J Neurophysiol 2004;92:1892–1903.
149 Gottfried JA, Dolan RJ: The nose smells what the eye sees: crossmodal visual facilitation of human olfactory perception. Neuron 2003;39:375–386.
150 Zatorre RJ, Jones-Gotman M, Rouby C: Neural mechanisms involved in odor pleasantness and intensity judgments. Neuroreport 2000;11:2711–2716.

Jay A. Gottfried, MD, PhD
Northwestern University Feinberg School of Medicine
Cognitive Neurology and Alzheimer's Disease Center
320 East Superior Street, Searle 11–453
Chicago, IL 60611 (USA)
Tel. +1 312 503 1834, Fax +1 312 908 8789, E-Mail j-gottfried@northwestern.edu

Hummel T, Welge-Lüssen A (eds): Taste and Smell. An Update.
Adv Otorhinolaryngol. Basel, Karger, 2006, vol 63, pp 70–83

Structure and Function of the Vomeronasal Organ

Martin Witt[a], *Witold Woźniak*[b]

[a]Smell and Taste Clinic, Department of Otorhinolaryngology, and Department of Anatomy, University of Technology, Dresden, Germany; [b]Department of Anatomy, Medical University, Poznań, Poland

Abstract

The vomeronasal organ (VNO) is a complex of different structures that forward specific chemical signals commonly called pheromones to the central nervous system. In some macrosmatic animals, e.g. rodents, the VNO consists of vomeronasal receptor neurons located in a sensory epithelium of the vomeronasal duct, their afferent axons connecting the duct with the accessory olfactory bulb, associated glands and ganglionic cells in the nasal septal mucosa. The organ's main task is to influence mating and social behavior. In humans, the VNO does not exist, at least not in its complexity. Although developed in early fetal life, all structures except the vomeronasal duct undergo regression. The orifice of this duct can be easily observed by nasal endoscopy. Histochemically, it is lined with a remarkable pseudostratified epithelium, the nature and significance of which are still unclear. Recent studies indicate that pheromone-like compounds are most likely registered at the level of olfactory receptor cells, rendering the chemical information system more independent of specific organ structures.

In vertebrates, there are three different olfactory subsystems, which perceive chemical stimuli from the external environment: the main olfactory system [see chapter by Rawson et al., this vol, pp 23–43], the vomeronasal system and the septal organ (organ of Masera). As a common feature, all of them develop from the olfactory placode. However, while the embryonic origin is almost constant in most species, there is considerable variation in the differentiation and maintenance of these organs during adulthood.

This chapter will give a survey of development, differentiation, and preservation of the vomeronasal organ (VNO) in humans. It will be preceded by some short historical notes on its discovery and interpretations. Lastly, since only few

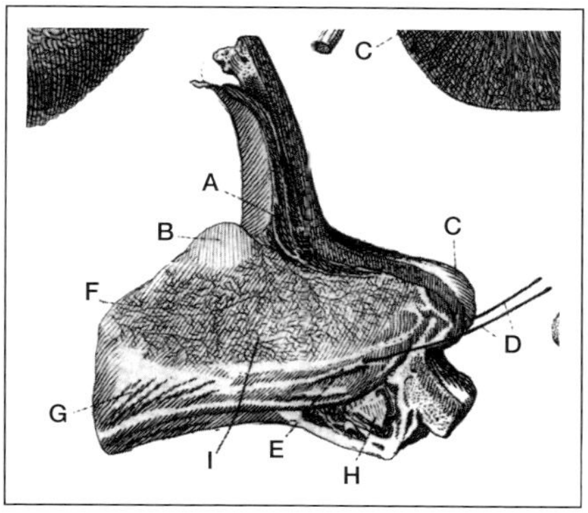

Fig. 1. First description of the human VND by F. Ruysch [1]. Lateral view of the nasal septum, nasal tip to the right (C). The duct is marked by a syringe (D) (Ref. SLUB Dresden Anat.A.216).

organs share the privilege of sometimes flourishing interpretations, which developed at least in part from false analogies with nonhuman vertebrates in view of their behavioral and sexual attitudes, we will have to deal with the question if typical 'VNO-related tasks' may be accomplished by other chemical receptor organs in humans.

History

In 1703, the anatomist and botanist Frederic Ruysch [1] discovered a small bilateral canal ('canalis nasalis') in the anterior lower part of the nasal septum of a human infant (fig. 1). However, skeptical about the function of the canal, he allocated a presumably secretory function ('…de cuius usu et existentia nil apud auctores legi: inservire muco excernendo existimo'). More than 100 years later, Ludwig Jacobson [2, 3] provided more systematic observations of the VNO in several animals, but stated erroneously that humans seem to be the only vertebrates in which such an organ is missing. Until then, most investigators were concerned about the existence or nonexistence of a communication between the nasal and oral cavities, named after Stenson (nasopalatine canal). Jacobson, however, described an independent, 'novel' duct opening into the nasal cavity. This canal impresses as a blind-ending sac, accompanied by a dense network of nerves, which ascend obliquely and pass through holes of the cribriform plate, differing from the course of olfactory nerves proper. Based on his observations in various mammals, Jacobson considered a secretory function and, maybe, also an engagement in some sensory processes (for excellent historical reviews and translations, see also Bhatnagar and Reid [4] and Bhatnagar and Smith [5]). In 1877, Kölliker [6] first described the histological human

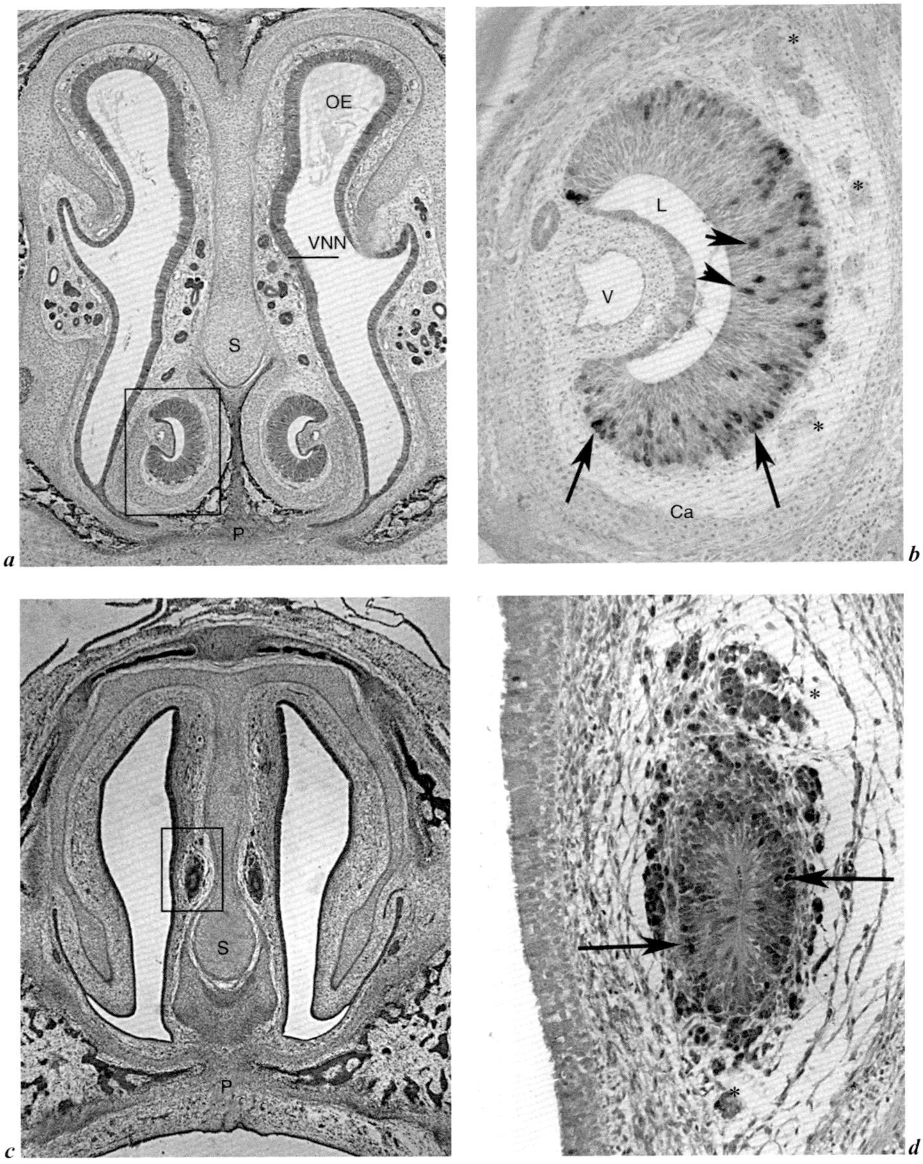

Fig. 2. VNO in the newborn rat (***a***, ***b***) and in a 8-week-old human fetus (***c***, ***d***). Frontal section through the cartilage of the nasal septum (S). ***a*** The u-shaped VND outlined with a rectangle is shown in detail in (***b***). OE = Olfactory epithelium; P = palate; VNN = fibers of the vomeronasal nerve. Iron-hematoxylin stain. ***b*** The lumen (L) of the VND is lined with a lateral nonsensory epithelium (left) belonging to the mushroom body that also contains the

VNO (as organ of Jacobson) both during development and in adults. He developed the idea that the adult human VNO should not be considered as a rudimental, possibly nonfunctional organ, rather as an underdeveloped embryonic structure, similar to the male mammary gland. Contemporary authors until today have described the VNO in most adult individuals as inconstantly occurring remnants, but its function has remained obscure [7–9].

The Vomeronasal Organ in Vertebrates

The VNO as a complete and independent chemosensory organ occurs in most reptiles, amphibians, and mammals, but is lacking in some phylogenetically successful representatives such as crocodiles, birds, or marine mammals. The anatomy of the VNO varies considerably among the different classes [10, 11]. Bhatnagar and Meisami [12] proposed the term 'vomeronasal organ complex' addressing the fact that the organ is composed of different constituents, such as the vomeronasal duct (VND), seromucous glands, the vomeronasal nerve, a vomeronasal cartilage, and a venous pumping system (fig. 2a, b). In rodents and many macrosmatic animals, the VND is a tunnel-shaped blind-ending channel lined with a sensory epithelium on its medial side (fig. 2b). This epithelium consists of elongated bipolar receptor neurons, sustentacular (supporting) cells and basal cells. The axons of receptor neurons project via the vomeronasal nerve to the accessory olfactory bulb, which in turn projects to the medial amygdala and then to hypothalamic areas, where controlling of neuroendocrine functions and social behavior is regulated [13]. The cellular organization of the VNO is related to that of the main olfactory system, but in some important issues distinct. For example, the epithelium is much thicker and receptor cells possess microvilli instead of cilia [14–16]. Meanwhile, there are many studies that have examined the relationship between vomeronasal stimuli in rodents, so-called 'pheromones' (see below), and the specific physiological responses [17–22].

vomeronasal pumping system. The vomeronasal epithelium proper is on the medial of the VND. Two subpopulations of vomeronasal receptor neurons, superficially and basally located, are marked with arrowheads and arrows, respectively. Their axons (asterisks) project to different regions of the accessory olfactory bulb and belong to different receptor families. Ca = Vomeronasal cartilage; V = vein. Immunohistochemical localization of vomeronasal receptor cells with PGP 9.5, counterstain with hematoxilin. ***c*, *d*** The anlage of the VND in fetal humans (rectangle in ***c***) is lined with a uniform epithelium consisting of many PGP-positive neurons (arrows). Putative axons harbor sets of ganglionic cells (asterisks), which are probably intermingled LHRH neurons. The lumen is usually very narrow and barely visible here. Immunohistochemistry for PGP 9.5, counterstain with hematoxylin.

Derivatives of the Olfactory Placode: Olfactory Nerve, Vomeronasal Nerve, Terminal Nerve, and Luteinizing Hormone-Releasing Hormone Neurons

According to recent reevaluations of staged human embryos by Müller and O'Rahilly [23], first vomeronasal formations are observed between weeks 4 and 5 (stages 13–15; 'terminal-vomeronasal crest'), which appear slightly before olfactory thickenings in the nasal disk. The dorsolateral epithelium of the so-called olfactory placode gives rise to the olfactory nerves, the medial part of the VNO, the terminal nerve, and luteinizing hormone-releasing hormone (LHRH) neurons [23, 24]. Vomeronasal nerve fibers together with ensheathing cells and those of the terminal nerve (see below) begin to sprout from the epithelium at stage 16 and reach the central nervous system (CNS; anlagen of the olfactory bulb and the median eminence, respectively) around stages 17 and 18. Shortly thereafter, the nasal septum divides both pits from each other to form bilateral pouches or recesses. Vomeronasal and terminal ganglia arise during stages 18 and 19. Around stage 22 (week 7.5) [23, 25, 26], the VND is clearly detectable (fig. 2c, d). In contrast to rodents, humans do not form a clearly separate accessory olfactory bulb [23, 27]; i.e., vomeronasal fibers seem to project parallel to olfactory receptor neurons into the 'main' olfactory bulb. Precise projection patterns are well described in mice, but remain obscure in humans.

Some fibers of the vomeronasal nerve serve as guiding structures for migratory gonadotropin (LHRH-like) neurons from the olfactory placode to the forebrain [28]. Another derivative of the olfactory placode constitutes the terminal nerve, whose fibers project directly to the median eminence of the hypothalamus bypassing the laterally located olfactory-associated structures. Although its components are not very clearly defined [29], a small subset (about 10%) of its fibers also carries LHRH neurons, which migrate transiently to the forebrain [29–31]. LHRH neurons are believed to originate from the medial olfactory placode [24], though there is recent evidence of an earlier migratory route of precursor cells from the neural crest [29]. It appears that LHRH neurons use vomeronasal and, more intensely, terminal nerve fibers as scaffolds for their migratory route to the lamina terminalis and the median eminence of the diencephalon [24].

Much less is known about the regression of the vomeronasal system in humans. Controversial data are caused by the scarce material investigated after the embryonic development. Parts of the VNO, such as vomeronasal nerve or VND, have been described until the 6th fetal month [32]. According to Schaeffer [33] and Kallius [34], Jacobson's organ attains its highest development in the 20th week, if it does not degenerate earlier. In a very careful investigation, Humphrey [27] observed 'vestigial remains' of the VNO until fetal week 18.5, which also included lamination of the accessory olfactory bulb, but

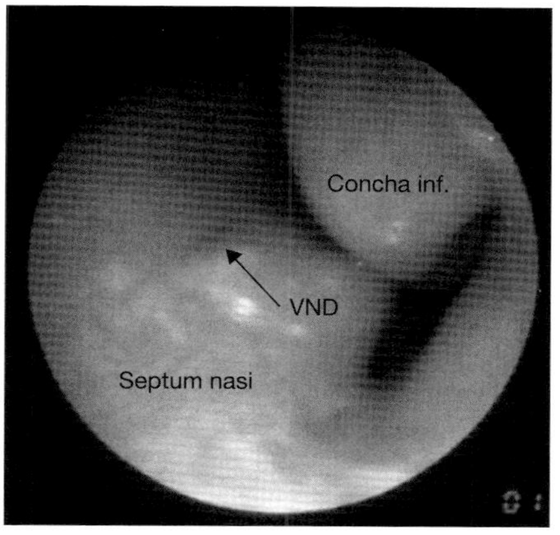

Fig. 3. Nasal endoscopy showing the opening of the VNO in the anterior nasal septum. (From Knecht et al. [35].)

these observations were irregular and not always bilaterally expressed. As a sign of early degeneration, Kallius [34] noted the closure of the VND and the transient formation of a cyst-like structure in the nasal septum. In most adults, however, the duct remains open to several degrees, but the timing of the 'disconnection' process of the nerve from epithelial duct cells has not yet been described. Also, there may be some uncertainty about the differentiation between vomeronasal and terminal nerves, which run largely parallel to each other.

Vomeronasal Structures in Adult Humans

As mentioned above, most constituents of the VNO complex are no more detectable at birth. The only structure, generally referred to as 'remnant' or 'vestigial relict', is the VND located in the anterior nasal septum. Therefore, when addressing vomeronasal functions, we will use the term vomeronasal duct (VND) or vomeronasal epithelium instead of vomeronasal organ (VNO).

The orifice of the duct appears as a pit less than 2 mm in diameter, which sometimes exhibits a yellow-brownish pigment (fig. 3) [35]. It is generally described as a blind-ending duct, or a mucosal pouch located in the anterior nasal septum [36, 37] (fig. 5); however, there is one report of a VNO with an approx. 6-cm-long(!) tubular mucosal structure with anterior and posterior openings [38]. Typically, its length varies between 3 and 22 mm [39]. Considerable confusion has arisen by its occasional vicinity to the incisive canal/nasopalatine duct [40]. However, the incisive canal is usually closed serving only as gateway for the nasopalatine nerve. Rarely, it may be misinterpreted as the VND, when the nasopalatine duct does not close during development [40, 41]. Thus, the human VND seems to exhibit considerable variability in

Table 1. Frequency of the VND in adult humans

Reference	n	Frequency
Potiquet [8]	200	25%
Johnson et al. [43]	100	patients: 39% (endoscopy) postmortem specimens: 70% (histology)
Stensaas et al. [71]	410	93%
Moran et al. [72]	200	100% (bilateral)
Garcia-Velasco and Mondragon [73]	1,000	90%
Trotier et al. [44]	1,842	patients: 26% (endoscopy)
Gaafar et al. [74]	200	patients: 76% (endoscopy)
Won et al. [75]	78	36%
Knecht et al. [76]	173	65% (41% bilateral, 24% unilateral; endoscopy)
Witt et al. [26]	25	65%

size, shape, and detectability/presence, which is dependent on the technique of investigation and the criteria set by the investigators. In general, histological studies reveal a higher percentage of VNDs and a more reliable basis than endoscopical investigations [26, 35, 42], which provide differences even in the same individual when observed at different times. A comprised survey about the frequency of the VND in humans is given in table 1.

Nevertheless, the fact that 'functional' vomeronasal structures are not found in man does not automatically imply that humans lack the capacity to present reaction patterns similar to those of animals that possess a complete VNO. The only working hypothesis may be that tasks fulfilled in animals by vomeronasal structures may be accomplished by different structures in humans.

Function of the Human Vomeronasal Organ

Histochemistry

The VND is lined with irregularly composed epithelia, sometimes resembling respiratory epithelium with ciliated cells and goblet cells, but there are also areas of stratified epithelium and sensory-like formations [26, 43–46]. Contrasting the aspect in rodents, both sides of the VND are more or less similarly lined with various epithelia; there is no mediolateral polarity of 'sensory' and 'nonsensory' epithelium [26, 43–45]. On the other hand, there are also morphologically unique epithelial areas: long, bipolar cells similar to a certain degree to olfactory epithelium, though considerably less high, restricted to two-layered pseudostratified epithelium (fig. 4b). These bipolar cells, however, have

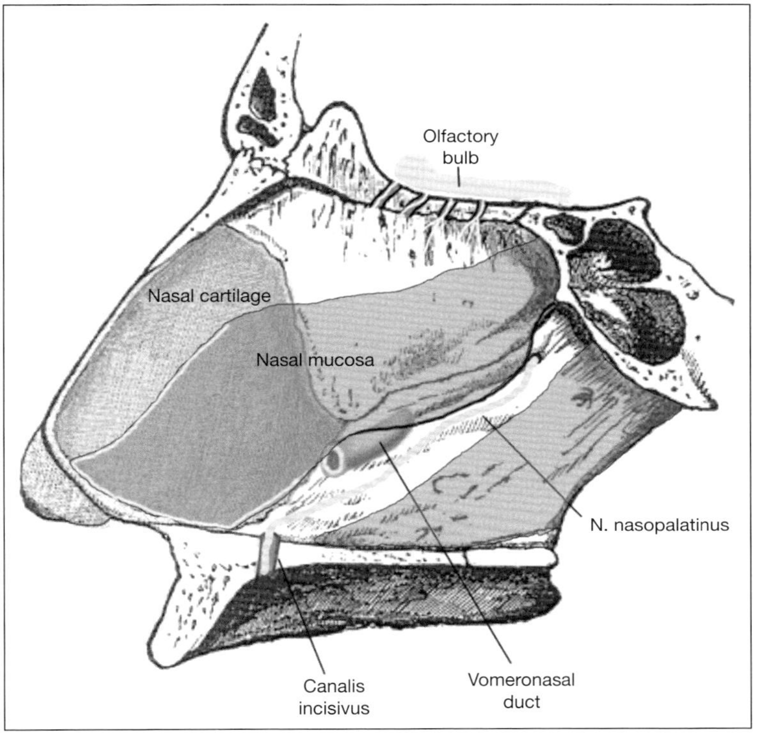

Fig. 4. Schematic drawing of the location of the VND. Paramedian-sagittal section through the nasal cavity. The septal mucosa is partially split off to demonstrate the obliquely ascending VND. Underneath runs the nasopalatine nerve which enters the oral cavity by traversing through the incisive canal. There are no connections with the olfactory bulb.

no connection to nerve fibers, thus they do not seem to be able to transfer potential information, if any, via the classical neuronal way to the CNS. Moreover, bipolar, sensory-like cells of the vomeronasal epithelium are keratin-positive, i.e., they provide intermediate filaments typical of epithelial cells, and not of differentiated neuroepithelial cells, which do not possess keratin filaments [26, 44]. Generally, the characteristic vomeronasal cells do not express neuronal markers such as protein gene product 9.5 (PGP 9.5), olfactory marker protein, or neuronal tubulin, though a very small subset of cells may contain one of these proteins [26, 47]. Instead, there are some remarkable proteins, which point to a specialized function in terms of cell signalling such as caveolins, which were described in olfactory cells [48]. β1-Integrin also seem to play a role, though a specific function of these proteins has not yet been shown. Nevertheless, surrounding epithelia, e.g. of the nasal septum as well as

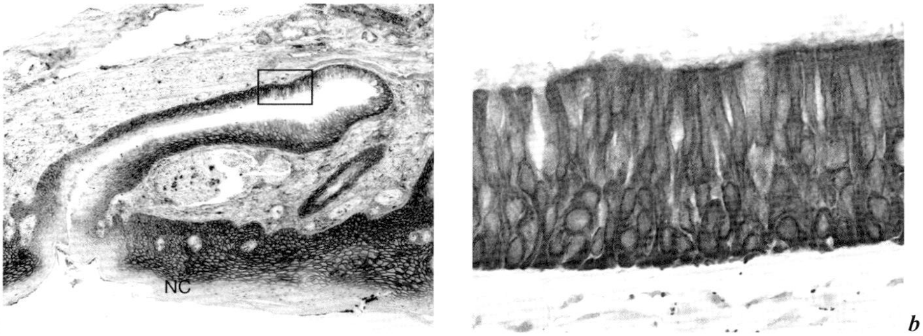

a b

Fig. 5. VND in the adult human. ***a*** The VND appears as a blind-ending duct originating from the anterior nasal cavity (NC), which is lined with squamous cell epithelium. ***b*** A high-power enlargement of the region similar to the rectangle in (***a***). Immunohistochemical detection of cytokeratin (MNF16 clone) demonstrates the epithelial rather than neuronal nature of slender bipolar cells organized in a pseudostratified epithelium that is different from respiratory epithelium.

squamous cell epithelium or respiratory-like epithelium of the VND did not contain caveolins or β1-integrin. Although a series of interesting proteins are expressed in sensory-like portions of the VND, a specific vomeronasal function in the adult human does not seem likely, especially because there are no afferent projection fibers to the CNS. However, the epithelium displays a mitotic activity, also in regions in which bipolar epithelial cells arise [26].

What Are Human Pheromones?

The chemical compounds which mediate social communication in animals through the vomeronasal system are usually called pheromones [49, 50]. The pheromone concept was originally developed by Karlson and Lüscher [51], who investigated the mating behavior in the silk moth *Bombyx mori*, notably, an insect that does not possess the classical VNO found in vertebrates. According to their concept, pheromones are 'substances which are secreted to the outside by an individual and received by a second individual of the same species, in which they release a specific reaction, e.g., a definite behavior or a developmental process'. Undoubtedly, there are numerous mainly behavioral experiments that support the original concept in nonprimate vertebrates [11]. For humans, four types of pheromones have been defined, i.e. primers, signalers, modulators, and releasers. One well-known source of pheromone candidates is the axilla, from which many chemically different volatile compounds have been described [52–54]. As a well-known example, the pivotal experiments by Stern and McClintock [55] established the influence of axillary secretions to the synchronization of menstrual cycles among female roommates, whereas exposure to male axillary extracts gives more regular menstrual cycles [56]. Hypothalamic activity can be measured after exposure of subjects to sex-hormone-like substances such as androstadien-3-one (similar to testosterone) or oestra-1,3,5(10),16-tetraen-3-ol, similar to estrogen [57]. Quantitatively, unsaturated acids such as (E)-3-methyl-2-hexenoic acid play a greater role than previously thought, e.g., most often cited 'human-related', androstenone, androstenol, and 4,16-androstadien-3-one [53, 58]. More recently, Meredith [50] proposed to restrict the definition for pheromones by inclusion of mutual benefit to sender and receiver. We will not discuss the particular chemistry in detail (for excellent reviews, see Wysocki and Preti [59] and Halpern and Martinez-Marcos [11]). Of some importance is the chemical similarity between pheromone candidates in humans and those of other animals [59] and the relationship of complex 'odor prints' with the immune system, e.g. the major histocompatibility complex. Axillary volatiles collected from T-shirts allowed individuals to identify their own odor and that of closely related persons, e.g. spouses [60]. These 'signaler pheromones', represented by paternally inherited human lymphocyte antigen alleles, may be responsible for mate choice [61]. In humans, the receptor region for pheromone candidates such as androstenone does not seem to be the VND, as own studies with experimentally covered VND entrance did not affect olfactory function or androstenone sensitivity [62].

What and Where Are the Receptors for Human Pheromone Candidates?
The observation that a series of particular compounds can exert specific behavioral and physiological effects requires the presence and genetic expression of specific receptors. The superfamily of thyroid hormones, approx. 300 vomeronasal receptors in mice, are structurally different from odorant receptors [63], but they share the common 7-transmembrane domain structure. Most of the genes encoding the vomeronasal receptors belong to at least two different families (VR1 and VR2), but 95% of all V1R genes are pseudogenes, i.e. are nonfunctional in man [64]; receptor proteins present in many animals [13, 65, 66] have not yet been identified in the human vomeronasal epithelium. There is no intact human V2R gene [67]. At least one of them is expressed at the mRNA level in the olfactory mucosa [68]. On the other hand, several studies have established that not all stimuli, which were believed to be VNO related, require a complete VNO [69, 70]. Also, major histocompatibility complex-related odor type recognition does not seem to be a monopoly of the VNO [70]. Thus, the most likely binding sites for human pheromone candidates are receptor cells within the olfactory epithelium.

Acknowledgements

The authors are grateful for the expert immunohistochemical work of Mrs. Anja Neisser and the artwork of Mrs. I. Beck.

References

1 Ruysch F: Thesaurus anatomicus tertius. Amstelaedami, Wolters, 1703, p 49.

2 Jacobson L: Description anatomique d'un organe observé dans les mammifères (reported by M Cuvier; translated into English by Bhatnagar and Reid, 1996). Ann Mus Hist Nat (Paris) 1811;18: 412–424.

3 Jacobson L: Anatomisk beskrivelse over et nyt organ i huusdyrenes naese. Veter Salesk Skrift 1813;2:209–246.

4 Bhatnagar KP, Reid KH: The human vomeronasal organ. 1. Historical perspectives. A study of Ruysch's (1703) and Jacobson's (1811) reports on the vomeronasal organ with comparative comments and English translations. Biomed Res 1996;7:219–229.

5 Bhatnagar KP, Smith TD: The human vomeronasal organ. 5. An interpretation of its discovery by Ruysch, Jacobson, or Kolliker, with an English translation of Kolliker (1877). Anat Rec 2003;270B: 4–15.

6 Kölliker A: Ueber die Jacobsonschen Organe des Menschen. Leipzig, Engelmann, 1877.

7 Merkel F: Jacobson'sches Organ und Papilla palatina beim Menschen. Festschrift zum fünfzigjährigen Doktorjubiläum des Herrn Geheimrat A v Kölliker. Anat Hefle, Wiesbaden I., 1892, pp 213–232.

8 Potiquet A: Du canal de Jacobson. De la possibilité de le reconnaître sur le vivant et de son rôle probable dans la pathogénie de certaines lésions de la cloison nasale. Rev Laryngol Otol Rhinol 1891;24:737–753.

9 Zuckerkandl E: Normale und pathologische Anatomie der Nasenhöhle und ihrer pneumatischen Anhänge. Wien, Braumüller, 1893.

10 Garrosa M, Gayoso MJ, Esteban FJ: Prenatal development of the mammalian vomeronasal organ. Microsc Res Tech 1998;41:456–470.

11 Halpern M, Martinez-Marcos A: Structure and function of the vomeronasal system: an update. Prog Neurobiol 2003;70:245–318.

12 Bhatnagar KP, Meisami E: Vomeronasal organ in bats and primates: extremes of structural variability and its phylogenetic implications. Microsc Res Tech 1998;43:465–475.

13 Luo M, Katz LC: Encoding pheromonal signals in the mammalian vomeronasal system. Curr Opin Neurobiol 2004;14:428–434.

14 Yoshida-Matsuoka J, Osada T, Mori Y, Ichikawa M: A developmental study using three antibodies (VOBM1, VOBM2, and VOM2): immunocytochemical and electron microscopical analysis of the luminal surface of the rat vomeronasal sensory epithelium. Anat Embryol (Berl) 1999;199:215–224.

15 Höfer D, Shin DW, Drenckhahn D: Identification of cytoskeletal markers for the different microvilli and cell types of the rat vomeronasal sensory epithelium. J Neurocytol 2000;29: 147–156.

16 Menco BP, Carr VM, Ezeh PI, Liman ER, Yankova MP: Ultrastructural localization of G-proteins and the channel protein TRP2 to microvilli of rat vomeronasal receptor cells. J Comp Neurol 2001;438:468–489.

17 Mombaerts P: Seven-transmembrane proteins as odorant and chemosensory receptors. Science 1999;286:707–711.

18 Takami S: Recent progress in the neurobiology of the vomeronasal organ. Microsc Res Tech 2002;58:228–250.

19 Zufall F, Kelliher KR, Leinders-Zufall T: Pheromone detection by mammalian vomeronasal neurons. Microsc Res Tech 2002;58:251–260.

20 Zhang J, Webb DM: Evolutionary deterioration of the vomeronasal pheromone transduction pathway in catarrhine primates. Proc Natl Acad Sci USA 2003;100:8337–8341.

21 Zhang J-X, Ni J, Ren X-J, Sun L, Zhang Z-B, Wang Z-W: Possible coding for recognition of sexes, individuals and species in anal gland volatiles of *Mustela eversmanni* and *M. sibirica*. Chem Senses 2003;28:381–388.

22 Restrepo D, Arellano J, Oliva A, Schaefer M, Lin W: Emerging views on the distinct but related roles of the main and accessory olfactory systems in responsiveness to chemosensory signals in mice. Horm Behav 2004;46:247–256.

23 Müller F, O'Rahilly R: Olfactory structures in staged human embryos. Cells Tissues Organs 2004;178:93–116.

24 Schwanzel-Fukuda M: Origin and migration of luteinizing hormone-releasing hormone neurons in mammals. Microsc Res Tech 1999;44:2–10.

25 Kreutzer EW, Jafek BW: The vomeronasal organ of Jacobson in the human embryo and fetus. Otolaryngol Head Neck Surg 1980;88:119–123.

26 Witt M, Georgiewa B, Knecht M, Hummel T: On the chemosensory nature of the vomeronasal epithelium in adult humans. Histochem Cell Biol 2002;117:493–509.

27 Humphrey T: The development of the olfactory and the accessory olfactory formations in human embryos and fetuses. J Comp Neurol 1940;73:431–468.

28 Cummings DM, Brunjes PC: Migrating luteinizing hormone-releasing hormone (LHRH) neurons and processes are associated with a substrate that expresses S100. Brain Res Dev Brain Res 1995;88:148–157.

29 von Bartheld CS: The terminal nerve and its relation with extrabulbar 'olfactory' projections: lessons from lampreys and lungfishes. Microsc Res Tech 2004;65:13–24.

30 Kjaer I, Fischer Hansen B: The human vomeronasal organ: prenatal developmental stages and distribution of luteinizing hormone-releasing hormone. Eur J Oral Sci 1996;104:34–40.

31 Schwanzel-Fukuda M, Crossin KL, Pfaff DW, Bouloux PM, Hardelin JP, Petit C: Migration of luteinizing hormone-releasing hormone (LHRH) neurons in early human embryos. J Comp Neurol 1996;366:547–557.

32 McCotter RE: The connection of the vomero-nasal nerves with the accessory olfactory bulb in the opossum and other mammals. Anat Rec 1912;6:299–318.

33 Schaeffer JP: The Nose, Parasinal Sinuses, Nasolacrimal Passageways, and Olfactory Organ in Man. Philadelphia, Blakiston, 1920.

34 Kallius E: Geruchsorgan (organon olfactus). Bardeleben's Handbuch der Anatomie des Menschen. Jena, Fischer, 1905.

35 Knecht M, Kuhnau D, Hüttenbrink KB, Witt M, Hummel T: Frequency and localization of the putative vomeronasal organ in humans in relation to age and gender. Laryngoscope 2001;111: 448–452.

36 Smith TD, Buttery TA, Bhatnagar KP, Burrows AM, Mooney MP, Siegel MI: Anatomical position of the vomeronasal organ in postnatal humans. Ann Anat 2001;183:475–479.

37 Smith TD, Bhatnagar KP, Shimp KL, Kinzinger JH, Bonar CJ, Burrows AM, Mooney MP, Siegel MI: Histological definition of the vomeronasal organ in humans and chimpanzees, with a comparison to other primates. Anat Rec 2002;267:166–176.

38 Mangakis M: Ein Fall von Jacobson'schen Organen beim Erwachsenen. Anat Anz 1902;21: 106–109.

39 Anton W: Beiträge zur Kenntnis des Jacobson'schen Organes bei Erwachsenen. Z Heilk 1895;16:355–372.

40 Jacob S, Zelano B, Gungor A, Abbott D, Naclerio R, McClintock MK: Location and gross morphology of the nasopalatine duct in human adults. Arch Otolaryngol Head Neck Surg 2000;126: 741–748.

41 Knecht M, Kittner T, Beleites T, Hüttenbrink KB, Hummel T, Witt M: Morphological and radiologic evaluation of the human nasopalatine duct. Ann Otol Rhinol Laryngol 2005;114:229–232.

42 Zbar RI, Zbar LI, Dudley C, Trott SA, Rohrich RJ, Moss RL: A classification schema for the vomeronasal organ in humans. Plast Reconstr Surg 2000;105:1284–1288.

43 Johnson A, Josephson R, Hawke M: Clinical and histological evidence for the presence of the vomeronasal (Jacobson's) organ in adult humans. J Otolaryngol 1985;14:71–79.

44 Trotier D, Eloit C, Wassef M, Talmain G, Bensimon JL, Doving KB, Ferrand J: The vomeronasal cavity in adult humans. Chem Senses 2000;25:369–380.

45 Bhatnagar KP, Smith TD: The human vomeronasal organ. 3. Postnatal development from infancy to the ninth decade. J Anat 2001;199:289–302.

46 Bhatnagar KP, Smith TD, Winstead W: The human vomeronasal organ. 4. Incidence, topography, endoscopy, and ultrastructure of the nasopalatine recess, nasopalatine fossa, and vomeronasal organ. Am J Rhinol 2002;16:343–350.

47 Takami S, Getchell ML, Chen Y, Monti-Bloch L, Berliner DL, Stensaas LJ, Getchell TV: Vomeronasal epithelial cells of the adult human express neuron-specific molecules. Neuroreport 1993;4:375–378.

48 Schreiber S, Fleischer J, Breer H, Boekhoff I: A possible role for caveolin as a signaling organizer in olfactory sensory membranes. J Biol Chem 2000;275:24115–24123.
49 Beauchamp GK, Doty RL, Moulton DG, Mugford RA: The pheromone concept in mammalian chemical communication: a critique; in Doty RL (ed): Mammalian Olfaction, Reproductive Processes, and Behavior. New York, Academic Press, 1976, pp 143–160.
50 Meredith M: Human vomeronasal organ function: a critical review of best and worst cases. Chem Senses 2001;26:433–445.
51 Karlson P, Lüscher M: Pheromones: a new term for a class of biologically active substances. Nature 1959;183:55–56.
52 Spielman AI, Zeng XN, Leyden JJ, Preti G: Proteinaceous precursors of human axillary odor: isolation of two novel odor-binding proteins. Experientia 1995;51:40–47.
53 Zeng C, Spielman AI, Vowels BR, Leyden JJ, Biemann K, Preti G: A human axillary odorant is carried by apolipoprotein D. Proc Natl Acad Sci USA 1996;93:6626–6630.
54 Natsch A, Gfeller H, Gygax P, Schmid J, Acuna G: A specific bacterial aminoacylase cleaves odorant precursors secreted in the human axilla. J Biol Chem 2003;278:5718–5727.
55 Stern K, McClintock MK: Regulation of ovulation by human pheromones. Nature 1998;392:177–179.
56 Cutler WB, Preti G, Krieger A, Huggins GR, Garcia CR and Lawley HJ: Human axillary secretions influence women's menstrual cycles: the role of donor extract from men. Horm Behav 1986;20:463–473.
57 Savic I, Berglund H, Gulyas B, Roland P: Smelling of odorous sex hormone-like compounds causes sex-differentiated hypothalamic activations in humans. Neuron 2001;31:661–668.
58 Gower DB, Ruparelia BA: Olfaction in humans with special reference to odorous 16-androstenes: their occurrence, perception and possible social, psychological and sexual impact. J Endocrinol 1993;137:167–187.
59 Wysocki CJ, Preti G: Facts, fallacies, fears, and frustrations with human pheromones. Anat Rec 2004;281A:1201–1211.
60 Porter RH, Cernoch JM, Balogh RD: Odor signatures and kin recognition. Physiol Behav 1985;34:445–448.
61 Jacob S, McClintock MK, Zelano B, Ober C: Paternally inherited HLA alleles are associated with women's choice of male odor. Nat Genet 2002;30:175–179.
62 Knecht M, Lundstrom JN, Witt M, Hüttenbrink KB, Heilmann S, Hummel T: Assessment of olfactory function and androstenone odor thresholds in humans with or without functional occlusion of the vomeronasal duct. Behav Neurosci 2003;117:1135–1141.
63 Zhang X, Rodriguez I, Mombaerts P, Firestein S: Odorant and vomeronasal receptor genes in two mouse genome assemblies. Genomics 2004;83:802–811.
64 Kouros-Mehr H, Pintchovski S, Melnyk J, Chen YJ, Friedman C, Trask B, Shizuya H: Identification of non-functional human VNO receptor genes provides evidence for vestigiality of the human VNO. Chem Senses 2001;26:1167–1174.
65 Rodriguez I, Feinstein P, Mombaerts P: Variable patterns of axonal projections of sensory neurons in the mouse vomeronasal system. Cell 1999;97:199–208.
66 Matsuoka M, Yoshida-Matsuoka J, Iwasaki N, Norita M, Costanzo RM, Ichikawa M: Immunocytochemical study of G(i)2alpha and G(o)alpha on the epithelium surface of the rat vomeronasal organ. Chem Senses 2001;26:161–166.
67 Mombaerts P: Genes and ligands for odorant, vomeronasal and taste receptors. Nat Rev Neurosci 2004;5:263–278.
68 Giorgi D, Friedman C, Trask BJ, Rouquier S: Characterization of nonfunctional VIR-like pheromone receptor sequences in human. Genome Res 2000;10:1979–1985.
69 Woodley SK, Cloe AL, Waters P, Baum MJ: Effects of vomeronasal organ removal on olfactory sex discrimination and odor preferences of female ferrets. Chem Senses 2004;29:659–669.
70 Wysocki C, Yamazaki K, Curran M, Wysocki L, Beauchamp G: Mice (*Mus musculus*) lacking a vomeronasal organ can discriminate MHC-determined odortypes. Horm Behav 2004;46:241–246.
71 Stensaas LJ, Lavker RM, Monti-Bloch L, Grosser BI, Berliner DL: Ultrastructure of the human vomeronasal organ. J Steroid Biochem Mol Biol 1991;39:553–560.

72 Moran DT, Jafek BW, Rowley JC: The vomeronasal (Jacobson's) organ in man: ultrastructure and frequency of occurrence. J Steroid Biochem Mol Biol 1991;39:545–552.
73 Garcia-Velasco J, Mondragon M: The incidence of the vomeronasal organ in 1,000 human subjects and its possible clinical significance. J Steroid Biochem Mol Biol 1991;39:561–563.
74 Gaafar HA, Tantawy AA, Melis AA, Hennawy DM, Shehata HM: The vomeronasal (Jacobson's) organ in adult humans: frequency of occurrence and enzymatic study. Acta Otolaryngol (Stockh) 1998;118:409–412.
75 Won J, Mair EA, Bolger WE, Conran RM: The vomeronasal organ: an objective anatomic analysis of its prevalence. Ear Nose Throat J 2000;79:600–605.

Martin Witt, MD, PhD
Smell and Taste Clinic
Dept. of Otorhinolaryngology, and
Department of Anatomy, University of Technology
Fetscherstrasse 74
DE–01307 Dresden (Germany)
Tel. +49 351 4586103, Fax +49 351 4586303, E-Mail martin.witt@mailbox.tu-dresden.de

Hummel T, Welge-Lüssen A (eds): Taste and Smell. An Update.
Adv Otorhinolaryngol. Basel, Karger, 2006, vol 63, pp 84–98

Assessment of Olfactory Function

Thomas Hummel[a], *Antje Welge-Lüessen*[b]

[a]Smell & Taste Clinic, Department of Otorhinolaryngology, University of Dresden Medical School ('Technische Universität Dresden'), Dresden, Germany; [b]Department of Otorhinolaryngology, University of Basel, Basel, Switzerland

Abstract

Numerous techniques are available for the investigation of chemosensory functions in humans. They include psychophysical measures of chemosensory function, e.g. odor identification, odor discrimination, odor thresholds, odor memory, and retronasal perception of odors. In order to assess changes related to the patients' quality of life or effects of qualitative olfactory dysfunction, questionnaires are being used. Measures relying to a lesser degree on the subjects' cooperation are e.g. chemosensory event-related potentials, odor-induced changes of the EEG, the electroolfactogram, imaging techniques, or measures of respiration. In a clinical context, however, psychophysical techniques are most frequently used, e.g. tests for odor identification, and odor thresholds. Interpretation of results from these measures is frequently supported by the assessment of chemosensory event-related potentials. Other techniques await further standardization before they will become useful in a clinical context.

Disturbances of the chemical senses are frequent. It is estimated that a complete loss of the sense of smell is found in at least 1% of the US population [1–3]. Recent epidemiological research indicates that at least 5% of the population exhibit a significant loss of olfactory function rendering them functionally anosmic [4, 5]. This appears to be largely due to aging as in individuals aged 53–97 years 24% were found to have impaired olfactory function [6]. Thus, based on a survey, each year over 70,000 patients are treated in German ENT clinics at least in part because of their olfactory loss [7].

Olfactory loss often appears to go unnoticed [8–10]. This seems to be especially true in older people or patients with Alzheimer's disease who are frequently unaware of their olfactory deficit [6, 11]. Similar findings have been reported in Parkinson's disease [12], diabetes mellitus [13], laryngectomy [14], or chronic renal failure [15]. While this highlights the limited attention the sense of smell receives in daily life [16], it also points towards the necessity to measure olfactory function in order to obtain reliable information on the patients' olfactory abilities [17, 18].

During the last centuries, standardized tests of olfactory function have been developed (e.g. The University of Pennsylvania Smell Identification Test, a 'scratch and sniff' odor identification test [19]; 'sniffin' sticks', pen-like odor-dispensing devices which include tests for odor identification, discrimination, and thresholds [20, 21], or the Connecticut Chemosensory Clinical Research Center Test, a combined odor identification and odor threshold test [22]). In addition, methods are currently being developed to quantify qualitative olfactory dysfunction, e.g. parosmia and phantosmia [23]. In terms of less biased measures of olfactory function, EEG-derived measures, such as the recording of olfactory event-related potentials (ERPs), are available [24] plus a large array of other techniques which are less well established in clinical routine.

Psychophysical Methods of Olfactory Testing

During psychophysical assessment of olfactory function, subjects/patients are confronted with an odor and their response is monitored. The critical advantage of this 'low-tech' approach is the speed of testing which allows psychophysical tests to serve as quick screening tools for olfactory dysfunction [25–28].

While screening tests only differentiate between normal and pathologic states, more extensive tests allow for the reliable discrimination between anosmic, hyposmic, and normosmic subjects, respectively. The ideal test should be based on normative data acquired and validated on large samples of healthy subjects and patients, respectively. This includes comparison of the results with other validated tests and a good test-retest reliability. These requirements are fulfilled only by a handful of olfactory tests [20, 21, 29–31]. The best-validated olfactory tests include the University of Pennsylvania Smell Identification Test [30], the Connecticut Chemosensory Clinical Research Center Test [29], and the 'sniffin' sticks' [20, 21].

Many tests are based on a forced-choice verbal identification of odors [20, 21, 29–33]. An odorant is presented at suprathreshold concentration and the subject has to identify the odor from a list of descriptors of odors (e.g. the subject gets rose odor to smell, and is asked whether the perceived odor was 'banana', 'anise', 'rose', or 'lilac'). This forced-choice procedure controls the subjects' response bias. Its strength lies in the fact that its concept is easily understood by both patient and investigator. In addition, the simplicity of its administration is a clear advantage over other olfactory tests. Odor identification tests based on forced choice also potentially allow to detect malingerers since anosmic subjects will produce a few correct answers provided random selection of items. Accordingly, if nonanosmic subjects willingly avoid selecting the correct items, they will have a very low score below random probability, which may indicate malingering. However, this method is insufficient for medicolegal investigations since well-read or hyposmic malingerers may overcome these pitfalls.

Most tests are based on the identification of 10–40 odors – the more items tested the more reliable the results. A major problem of odor identification is, however, that it strongly taps onto the verbal abilities of the subject which, on average, makes female subjects perform better than their male counterparts [34, 35]. In addition, odor identification tests have a strong cultural connotation as not all odors are known equally well around the world. Tests used in North America, for example, are composed of odors many of which are unfamiliar to Europeans or Asians (e.g. root beer, or wintergreen). The odors tested should therefore be adapted to the patients' cultural background [36] in order to minimize a familiarity bias in odor identification.

Other widely used test designs are threshold tests and tests of odor discrimination – tests for odor memory are only rarely used in a clinical context [37, 38]. In addition, tests for retronasal olfactory function are not in general use [39] despite the fact that the orthonasal and retronasal olfactory function can be dissociated in patients [40, 41].

The idea of threshold tests is to expose a subject repeatedly to ascending and descending concentrations of the same odorant and to identify the least detectable concentration for this individual odor [42]. Other techniques are based on logistic regression [43, 44]. Discrimination tasks are frequently based on a 3-alternative forced-choice technique. Two of the administered odors are identical, one is different. The subjects' task is to find out the different one. In principal, tests for odor threshold/odor discrimination are nonverbal. In addition, they can be used repetitively – which is more difficult with odor identification tests as subjects remember the answers they gave in previous tests of odor identification.

It seems intuitive to assume that different olfactory tests address different portions of olfactory processing, although there is no definitive proof for this idea [45]. Generally, identification and discrimination tests are believed to reflect central olfactory processing while thresholds are thought to reflect peripheral aspects of olfactory function to a stronger degree. Accordingly, it has been claimed by several authors [15, 46–51] that patients with diseases of the central nervous processing of odorous information exhibit selective disturbances of discrimination and identification while threshold results (or other measures of olfactory function, such as olfactory ERPs) may be normal. Although this idea of a certain pattern pathognomonic for 'central' olfactory dysfunction is highly attractive, the vast majority of studies have failed to confirm such typical pathology-associated patterns [52, 53]. One reliable and recurrent test pattern in olfactory disturbance, however, may be a low threshold and close to normal identification and discrimination in patients with sinunasal disease [54].

Besides the solid body of literature and their clinical convenience, psychophysical tests are limited insofar as they rely on the patients' cooperation. This becomes problematic in medicolegal investigations, or in subjects unable to respond appropriately, e.g. demented, unconscientious, or inexperienced patients.

Measures of Changes of Quality of Life Associated with Olfactory Loss

Loss of olfactory function affects the patients' quality of life [10, 16]. This impairment is especially severe in cases where patients develop qualitative olfactory dysfunction such as parosmia and phantosmia [55, 56]. Such deficits cannot be assessed with tests routinely administered to investigate quantitative olfactory loss [57].

Several questionnaires are available to measure mood states or general quality of life [58], e.g. the Beck Depression Inventory [59], 'mood inventories' [60], or the Short Form-36 Health Survey [61]. While these questionnaires allow the quantification of changes in the patients' quality of life in general, only recently have questionnaires been introduced which specifically address nasal dysfunction, e.g. the Sinonasal Outcome Test-16 [62]. Other questionnaires specifically address olfactory distortions, e.g. the Questionnaire for Olfactory Dysfunction [23] or other scales [63–65]. They have been developed in analogy to questionnaires used to quantify the degree of tinnitus [66]. Based upon these questionnaires, it was found that patients with smell loss indicated to be significantly impaired in areas of food, safety, personal hygiene, and additionally, in their sexual life [67, 68]. In addition, Varga et al. [69] presented a questionnaire to assess the impact of chemosensory dysfunction on everyday life which also includes utility-based or time trade-off scales, with a particular focus on the value placed by patients on chemosensory function. Interestingly, approximately half of the patients reported to be willing to spend more than 20% of their annual household income to successfully treat chemosensory dysfunction.

These developments seem to be of specific interest to studies on therapy of olfactory dysfunctions as they allow the assessment of a dimension related to olfactory loss which is not addressed by quantitative tests of olfactory function.

Chemosensory (Olfactory) Event-Related Potentials

In many clinics, chemosensory (olfactory) ERPs have become part of the routine investigation of patients with olfactory loss (for a review, see Hummel and Kobal [24]). According to the growing use of this technique, guidelines have been published for both recording and reporting of chemosensory ERPs [70, 71].

Chemosensory ERPs are derived from the EEG after intranasal chemical stimulation. As revealed by means of magnetoencephalographic techniques, the chemosensory ERPs' cortical sources are found in the temporal and insular cortices [72]. ERPs are recorded from the scalp; averaging is needed to increase the signal-to-noise ratio [73–75], which requires repeated stimulation that, in turn, makes the recording of olfactory ERPs a lengthy procedure.

Stimulators used to elicit chemosensory ERPs should allow for precise control of stimulus concentration and duration of stimuli, the rise time of stimuli should be in the range of 20 ms. Two major issues are of importance. (1) Chemical stimulation should not lead to the concomitant activation of mechano- or thermosensors of the nasal mucosa. Otherwise, the resulting ERPs will not reflect specific responses to odors [73, 76, 77]. (2) Stimulants should be characterized with regard to the degree to which they activate the trigeminal nerve [78–80]. Stimulants typically used to elicit chemosensory ERPs are vanillin, phenylethyl alcohol, and H_2S [73]; for relatively specific trigeminal stimulation, gaseous CO_2 is frequently employed [81].

It has been established that chemosensory ERPs are useful in the detection of malingering [82–85]. Typically, recordings are made after lateralized stimulation with two olfactory stimuli (e.g. phenylethyl alcohol and H_2S). In general, the interstimulus interval used lies between 30 and 40 s [73], the stimulus duration is in the range of 200 ms [86]. In anosmic patients, only responses to trigeminal stimulation can be elicited [82–84], and chemosensory ERPs discriminate between normosmic and hyposmic states [87]. In addition, transient changes of olfaction can be tracked by means of chemosensory ERPs [88–90].

The Human Electroolfactogram

Other than the cortically generated ERPs, electroolfactograms (EOGs) allow the assessment of the peripheral input signal to the olfactory system [73, 91, 92]. For recording, a tubular electrode is typically used (AgAgCl electrode, insulating tubing with an inner diameter of 0.3 mm, filled with 1% Ringer-agar, impedance $<10\,k\Omega$ at 1 kHz [93]); an electrode placed contralaterally on the bridge of the nose serves as reference. Placement of the electrode's tip on the mucosa is performed under endoscopical control [94]. After positioning, it is stabilized by means of adjustable clips on a frame similar to lensless glasses [95]. Recordings are performed with a high-resistance amplifier (DC, low pass 30 Hz, input resistance $>10\,M\Omega$).

Even under endoscopical control, EOGs cannot be recorded in all subjects [96]. This may be due to the topographical distribution of specific olfactory receptor neurons in combination with the relatively few number of odorants used, or the presence of metaplasias of the olfactory epithelium the extent of which may increase with the subjects' age [97, 98].

While the EOG has been proven to be useful in basic research [93], its use in the routine clinical diagnostic armamentarium of olfactory disorders is far from being established. As noted above, this may partly be due to (a) the difficulties associated with its recording, (b) the relatively poor reliability of the recordings that are only successful in approximately 60–80% of the subjects, which makes a negative result difficult to interpret, and (c) the strain imposed on the patient since local anesthesia cannot be performed. Although recent reports on 'EOG' recordings from the skin overlying the nose attracted much interest [99] based on previous research by Getchell [100, 101], it seems highly unlikely that the signal would be recordable from the cutaneous surface.

Contingent Negative Variation

The contingent negative variation (CNV) is a negative DC shift of the EEG that occurs in expectation of a stimulus. Recording procedures start usually with an initial warning stimulus (S1), which is followed by a second stimulus (S2). Subjects are asked to respond to S2, e.g. by pushing a button. During the interval between S1 and S2, the negativity slowly builds up, and breaks down after occurrence of S2. This EEG-related component was first described for visual and acoustic stimuli by Walter et al. [102]. Its first portion is related to S1 and is typically largest over frontal brain areas; the later part is related to the preparation to respond to S2 and is typically largest over the motor cortex [103]. The CNV is largely governed by psychological variables [104–106].

CNV-based paradigms are applied in olfactory tests in patients. Auffermann et al. [107] (see also Yamamoto [108]) used an odorant as S1. S2 was a tone that was only presented along with a certain odor. As the patients expect the tone after having perceived this certain odor, the CNV develops. Although this technique allows to test qualitative differences between odors, it cannot be performed without the patient paying attention to the paradigm, and even if subjects are cooperative, the CNV is not clearly present in all subjects. It is therefore of limited use, e.g., for medicolegal investigations.

EEG Changes in Response to Odors

In the 1950s, electroencephalographic changes were reported after presentation of odorous stimuli [109]. As a rule, an arousal reaction was observed in response to olfactory and trigeminal stimulation [110, 111]. Perbellini and Scolari [112] investigated 50 patients by means of three stimulants, i.e. pyridine, vanillin, and 'rose' odor. Their conclusion was that stimulus-induced EEG changes would be particularly useful in malingering patients. However, they also observed a number of cases where no arousal reaction could be recorded, although the subjects reported an olfactory sensation. Thus, when using this technique, only positive responses could be viewed as an unambiguous result. In line with this finding, in an investigation on 24 subjects, no electroencephalographic changes appeared when only weak odorous stimuli were perceived [113].

Apart from research oriented towards the diagnosis of olfactory disorders, EEG recordings have contributed to basic research, e.g. the study of subthreshold odor intensities where changes of the EEG have been observed even in the presence of undetected odors [114]. Others reported that individual odors can be discriminated by means of changes in the alpha-activity recorded at different sites [115]. Thus, although recording and analysis of stimulus-related EEG activity appear to be less complex than the recording of chemosensory ERP, it seems to be premature to recommend the use of these measurements in a medicolegal context.

Localization of Olfactory-Induced Activation in the Brain

Recent progress in the field of imaging opened the opportunity to study the functional topography of the human olfactory system in detail [72, 116, 117]. There are three major techniques being used: positron emission tomography [116, 118, 119], functional magnetic resonance imaging [120–122], and magnetic source imaging based on magnetoencephalography [123, 124]. While biomagnetic fields directly reflect neuronal activity, positron emission tomography and functional magnetic resonance imaging reflect either changes in blood flow or changes in metabolism which are epiphenomena of neuronal activity. Other major differences between these techniques relate to their temporal and spatial resolution [72]. All three techniques have been used extensively to perform basic research, e.g. on olfactory-induced emotions, odor memory, mechanisms of sniffing, or differences between orthonasal and retronasal olfactory perceptions [125], or age- and sex-related differences in terms of olfactory function

[126]. However, in order to become relevant for routine clinical investigations [127], these intriguing techniques await further standardization.

Other Measures of Olfactory Activation

Compared to EEG-related parameters, other measures of olfactory activation have never received the same degree of attention in a clinical context. For example, respiratory changes in response to odorous stimulation have been investigated in patients with olfactory loss [128–130]. Here, the perception of an odorant is followed by changes in respiratory frequency and both amplitude and pattern of the respiratory cycle. Investigations in respiratory changes following odorous stimulation have their place in research, e.g. in the investigation of the perception of subthreshold olfactory stimuli [131, 132]. In fact, devices for the clinical investigation of these respiratory changes are currently being developed [133].

Since the 1920s, the psychogalvanic skin response has been thought to be of use in the assessment of olfactory disorders [134–136]. In addition, pupillary reflexes in response to olfactory stimuli have been investigated [137, 138], but have not reached the level of routine clinical application. In addition, measurement of odor-induced eye blinks has been investigated as a means for the assessment of olfactory thresholds [139]. However, it appears that a protective reflex like blinking is rather evoked by chemical irritants than by excitation of the olfactory system. Yet other measures include changes of body posture following olfactory stimulation [140].

Methods Used for Assessment of Trigeminal Function

Most of the methods developed to quantify trigeminally mediated sensations such as stinging, burning, or tickling in humans are based on psychophysical and electrophysiological techniques. Psychophysical approaches include assessment of thresholds [141, 142], ratings of stimulus intensity of suprathreshold stimulus concentrations, or the assessment of the subject's ability to localize chemosensory stimuli [79, 143].

Electrophysiological measures appear to allow the assessment of sensory functions in a more detailed fashion. That is, the negative mucosal potential recorded from the surface of the respiratory epithelium is thought to reflect functional aspects of trigeminal chemosensors (= nociceptors) [144]. The ERPs recorded in response to trigeminal stimulants are of cortical origin [145] and reflect different stages of the cognitive processing of trigeminal function

[146, 147]. In a clinical context, only psychophysical measures of trigeminal function are used; very few centers also investigate trigeminal ERPs.

Conclusions

Numerous techniques are available for the investigation of chemosensory functions in humans. In a clinical context, psychophysical techniques are most frequently used, e.g. tests for odor identification, and odor thresholds. Interpretation of results from these measures is frequently supported by the assessment of chemosensory ERPs. Other techniques await further standardization before they will become useful in a clinical context.

References

1 Panel on Communicative Disorders to the National Advisory Neurological and Communicative Disorders and Stroke Council. NIH Publication No 79–1914. Washington, Public Health Service, 1979.

2 Gilbert AN, Wysocki CJ: The smell survey results. Natl Geogr Mag 1987;172:514.

3 Olsson P, Berglind N, Bellander T, Stjärne P: Prevalence of self-reported allergic and non-allergic rhinitis symptoms in Stockholm: relation to age, gender, olfactory sense, and smoking. Acta Otolaryngol 2003;123:75–80.

4 Landis BN, Konnerth CG, Hummel T: A study on the frequency of olfactory dysfunction. Laryngoscope 2004;114:1764–1769.

5 Brämerson A, Johansson L, Ek L, Nordin S, Bende M: Prevalence of olfactory dysfunction: the Skövde population-based study. Laryngoscope 2004;114:733–737.

6 Murphy C, Schubert CR, Cruickshanks KJ, Klein BE, Klein R, Nondahl DM: Prevalence of olfactory impairment in older adults. JAMA 2002;288:2307–2312.

7 Damm M, Temmel A, Welge-Lüssen A, Eckel HE, Kreft MP, Klussmann JP, et al: Epidemiologie und Therapie von Riechstörungen in Deutschland, Österreich und der Schweiz. HNO 2004;52:112–120.

8 Deems DA, Doty RL, Settle RG, Moore-Gillon V, Shaman P, Mester AF, et al: Smell and taste disorders: a study of 750 patients from the University of Pennsylvania Smell and Taste Center. Arch Otorhinolaryngol Head Neck Surg 1991;117:519–528.

9 Callahan CD, Hinkebein J: Neuropsychological significance of anosmia following traumatic brain injury. J Head Trauma Rehabil 1999;14:581–587.

10 Temmel AF, Quint C, Schickinger-Fischer B, Klimek L, Stoller E, Hummel T: Characteristics of olfactory disorders in relation to major causes of olfactory loss. Arch Otolaryngol Head Neck Surg 2002;128:635–641.

11 Nordin S, Monsch AU, Murphy C: Unawareness of smell loss in normal aging and Alzheimer's disease: discrepancy between self-reported and diagnosed smell sensitivity. J Gerontol 1995;50: P187–P192.

12 Muller A, Mungersdorf M, Reichmann H, Strehle G, Hummel T: Olfactory function in Parkinsonian syndromes. J Clin Neurosci 2002;9:521–524.

13 Jorgensen MB, Buch NH: Studies on the sense of smell and taste in diabetics. Arch Otolaryngol 1961;53:539–545.

14 van Dam FS, Hilgers FJ, Emsbroek G, Touw FI, van As CJ, de Jong N: Deterioration of olfaction and gustation as a consequence of total laryngectomy. Laryngoscope 1999;109:1150–1155.

15 Frasnelli JA, Temmel AF, Quint C, Oberbauer R, Hummel T: Olfactory function in chronic renal failure. Am J Rhinol 2002;16:275–279.
16 Miwa T, Furukawa M, Tsukatani T, Costanzo RM, DiNardo LJ, Reiter ER: Impact of olfactory impairment on quality of life and disability. Arch Otolaryngol Head Neck Surg 2001;127: 497–503.
17 Landis BN, Hummel T, Hugentobler M, Giger R, Lacroix JS: Ratings of overall olfactory function. Chem Senses 2003;28:691–694.
18 Welge-Lüssen A, Hummel T, Stojan T, Wolfensberger M: Subjective impairment due to olfactory dysfunction. Am J Rhinol, in press.
19 Doty RL, Shaman P, Dann M: Development of the University of Pennsylvania Smell Identification Test: a standardized microencapsulated test of olfactory function. Physiol Behav 1984;32: 489–502.
20 Hummel T, Sekinger B, Wolf S, Pauli E, Kobal G: 'Sniffin' sticks': olfactory performance assessed by the combined testing of odor identification, odor discrimination and olfactory threshold. Chem Senses 1997;22:39–52.
21 Kobal G, Klimek L, Wolfensberger M, Gudziol H, Temmel A, Owen CM, et al: Multicenter investigation of 1,036 subjects using a standardized method for the assessment of olfactory function combining tests of odor identification, odor discrimination, and olfactory thresholds. Eur Arch Otorhinolaryngol 2000;257:205–211.
22 Cain WS: Testing olfaction in a clinical setting. Ear Nose Throat J 1989;68:321–328.
23 Frasnelli J, Hummel T: Towards a quantitative measure of qualitative olfactory dysfunction. Laryngoscope, submitted.
24 Hummel T, Kobal G: Olfactory event-related potentials; in Simon SA, Nicolelis MAL (eds): Methods and Frontiers in Chemosensory Research. Boca Raton, CRC Press, 2001, pp 429–464.
25 Davidson TM, Freed C, Healy MP, Murphy C: Rapid clinical evaluation of anosmia in children: the Alcohol Sniff Test. Ann NY Acad Sci 1998;855:787–792.
26 Simmen D, Briner HR, Hess K: Screeningtest des Geruchssinnes mit Riechdisketten. Laryngorhinootologie 1999;78:125–130.
27 Kobal G, Palisch K, Wolf SR, Meyer ED, Huttenbrink KB, Roscher S, et al: A threshold-like measure for the assessment of olfactory sensitivity: the 'random' procedure. Eur Arch Otorhinolaryngol 2001;258:168–172.
28 Hummel T, Konnerth CG, Rosenheim K, Kobal G: Screening of olfactory function with a four-minute odor identification test: reliability, normative data, and investigations in patients with olfactory loss. Ann Otol Rhinol Laryngol 2001;110:976–981.
29 Cain WS, Gent JF, Goodspeed RB, Leonard G: Evaluation of olfactory dysfunction in the Connecticut Chemosensory Clinical Research Center (CCCRC). Laryngoscope 1988;98:83–88.
30 Doty RL, Shaman P, Kimmelman CP, Dann MS: University of Pennsylvania Smell Identification Test: a rapid quantitative olfactory function test for the clinic. Laryngoscope 1984;94:176–178.
31 Kondo H, Matsuda T, Hashiba M, Baba S: A study of the relationship between the T&T olfactometer and the University of Pennsylvania Smell Identification Test in a Japanese population. Am J Rhinol 1998;12:353–358.
32 Kobal G, Hummel T, Sekinger B, Barz S, Roscher S, Wolf S: 'Sniffin' sticks': screening of olfactory performance. Rhinology 1996;34:222–226.
33 Briner HR, Simmen D: Smell diskettes as screening test of olfaction. Rhinology 1999;37: 145–148.
34 Larsson M, Nilsson LG, Olofsson JK, Nordin S: Demographic and cognitive predictors of cued odor identification: evidence from a population-based study. Chem Senses 2004;29:547–554.
35 Larsson M, Bäckman L: Odor identification in old age: demographic, sensory, and cognitive correlates. Aging Neuropsychol Cog, in press.
36 Ho WK, Kwong DL, Wei WI, Sham JS: Change in olfaction after radiotherapy for nasopharyngeal cancer – A prospective study. Am J Otolaryngol 2002;23:209–214.
37 Larsson M, Oberg C, Backman L: Recollective experience in odor recognition: Influences of adult age and familiarity. Psychol Res 2006;70:68–75.
38 Doty RL, Kerr KL: Episodic odor memory: influences of handedness, sex, and side of nose. Neuropsychologia 2005;43:1749–1753.

39 Heilmann S, Strehle G, Rosenheim K, Damm M, Hummel T: Clinical assessment of retronasal olfactory function. Arch Otorhinolaryngol Head Neck Surg 2002;128:414–418.
40 Landis BN, Giger R, Ricchetti A, Leuchter I, Hugentobler M, Hummel T, et al: Retronasal olfactory function in nasal polyposis. Laryngoscope 2003;113:1993–1997.
41 Landis BN, Frasnelli J, Reden J, Lacroix JS, Hummel T: Differences between orthonasal and retronasal olfactory functions in patients with loss of the sense of smell. Arch Otolaryngol Head Neck Surg 2005;131:977–981.
42 Ehrenstein WH, Ehrenstein A: Psychophysical methods; in Windhorst U, Johansson H (eds): Modern Techniques in Neuroscience Research. Berlin, Springer, 1999, pp 1211–1241.
43 Lotsch J, Lange C, Hummel T: A simple and reliable method for clinical assessment of odor thresholds. Chem Senses 2004;29:11–317.
44 Linschoten MR, Harvey LO Jr, Eller PM, Jafek BW: Fast and accurate measurement of taste and smell thresholds using a maximum-likelihood adaptive staircase procedure. Percept Psychophys 2001;63:1330–1347.
45 Doty RL, Smith R, McKeown DA, Raj J: Tests of human olfactory function: principle component analysis suggests that most measure a common source of variance. Percept Psychophys 1994;56:701–707.
46 Hawkes CH, Shephard BC: Selective anosmia in Parkinson's disease? Lancet 1993;341:435–436.
47 Koss E, Weiffenbach JM, Haxby JV, Friedland RP: Olfactory detection and recognition in Alzheimer's disease. Lancet 1987;i:622.
48 Koss E, Weiffenbach JM, Haxby JV, Friedland RP: Olfactory detection and identification performance are dissociated in early Alzheimer's disease. Neurology 1988;38:1228–1232.
49 Hornung DE, Kurtz DB, Bradshaw CB, Seipel DM, Kent PF, Blair DC, et al: The olfactory loss that accompanies an HIV infection. Physiol Behav 1998;15:549–556.
50 Jones-Gotman M, Zatorre RJ: Olfactory identification deficits in patients with focal cerebral excision. Neuropsychologia 1988;26:387–400.
51 Peters JM, Hummel T, Kratzsch T, Lotsch J, Skarke C, Frolich L: Olfactory function in mild cognitive impairment and Alzheimer's disease: an investigation using psychophysical and electrophysiological techniques. Am J Psychiatry 2003;160:1995–2002.
52 Mesholam RI, Moberg PJ, Mahr RN, Doty RL: Olfaction in neurodegenerative disease: a meta-analysis of olfactory functioning in Alzheimer's and Parkinson's diseases. Arch Neurol 1998;55:84–90.
53 Daum RF, Sekinger B, Kobal G, Lang C: Riechprüfung mit 'Sniffin' Sticks' zur klinischen Diagnostik des Morbus Parkinson. Nervenarzt 2000;71:643–650.
54 Klimek L, Hummel T, Moll B, Kobal G, Mann WJ: Lateralized and bilateral olfactory function in patients with chronic sinusitis compared with healthy control subjects. Laryngoscope 1998;108:111–114.
55 Leopold D: Distorted olfactory perception; in Doty RL (ed): Handbook of Olfaction and Gustation. New York, Dekker, 1995, 441–454.
56 Frasnelli J, Landis BN, Heilmann S, Hauswald B, Huttenbrink KB, Lacroix JS, et al: Clinical presentation of qualitative olfactory dysfunction. Eur Arch Otorhinolaryngol 2003;11:11–13.
57 Nordin S, Murphy C, Davidson TM, Quinonez C, Jalowayski AA, Ellison DW: Prevalence and assessment of qualitative olfactory dysfunction in different age groups. Laryngoscope 1996;106: 739–744.
58 Bullinger M: Assessing health-related quality of life in medicine. An overview over concepts, methods and applications in international research. Restor Neurol Neurosci 2002;20:93–101.
59 Beck AT, Ward CM, Mendelson M, Mock JE, Erbaugh JK: An inventory for measuring depression. Arch Gen Psychiatry 1961;4:561–571.
60 von Zerssen D: Die Befindlichkeitsskala. Göttingen, Beltz Test, 1975.
61 Ware JE Jr: SF-36 health survey update. Spine 2000;25:3130–3139.
62 Anderson ER, Murphy MP, Weymuller EAJ: Clinimetric evaluation of the sinonasal outcome test-16. Otolaryngol Head Neck Surg 1999;121:702–707.
63 de Jong N, Mulder I, de Graaf C, van Staveren WA: Impaired sensory functioning in elders: the relation with its potential determinants and nutritional intake. J Gerontol A Biol Sci Med Sci 1999;54:B324–B331.

64 Hufnagl B, Lehrner J, Deecke L: Development of a questionnaire for the assessment of self reported olfactory functioning. Chem Senses 2003;28:E27.
65 Nordin S, Brämerson A, Murphy C, Bende M: A Scandinavian adaptation of the Multi-Clinic Smell and Taste Questionnaire: evaluation of questions about olfaction. Acta Otolaryngol 2003;123:536–542.
66 Goebel G, Hiller W: The tinnitus questionnaire. A standard instrument for grading the degree of tinnitus. Results of a multicenter study with the tinnitus questionnaire. HNO 1994;42:166–172.
67 Tennen H, Affleck G, Mendola R: Coping with smell and taste disorder; in Getchell TV, Doty RL, Bartoshuk LM, Snow JB (eds): Smell and Taste in Health and Disease. New York, Raven Press, 1991, pp 787–802.
68 Van Toller S: Assessing the impact of anosmia: review of a questionnaire's findings. Chem Senses 1999;24:705–712.
69 Varga EK, Breslin PA, Cowart BJ: The impact of chemosensory dysfunction on quality of life. Chem Senses 2000;25:654.
70 Evans WJ, Starr A: Stimulation parameters and temporal evolution of the olfactory evoked potential in rats. Chem Senses 1992;17:61–78.
71 Hummel T, Klimek L, Welge-Lussen A, Wolfensberger G, Gudziol H, Renner B, et al: Chemosensorisch evozierte Potentiale zur klinischen Diagnostik von Riechstörungen. HNO 2000;48:481–485.
72 Kettenmann B, Hummel T, Kobal G: Functional imaging of olfactory activation in the human brain; in Simon SA, Nicolelis MAL (eds): Methods and Frontiers in Chemosensory Research. Baco Raton, CRC Press, 2001, pp 477–506.
73 Kobal G: Elektrophysiologische Untersuchungen des menschlichen Geruchssinns. Stuttgart, Thieme, 1981.
74 Covington J, Geisler MW, Morgan CD, Ellison DW, Murphy C: Optimal number of trials to obtain a reliable olfactory event-related potential. Chem Senses 1996;21:489–490.
75 Covington JW, Polich J, Murphy C: Robust OERPs produced with fewer trials: effects of age and gender. Chem Senses, in press.
76 Bauer LO, Mott AE: Differential effects of cocaine, alcohol, and nicotine dependence on olfactory evoked potentials. Drug Alcohol Depend 1996;42:21–26.
77 Herberhold C: Typical results of computer-olfactometry. Rhinology 1976;14:109–116.
78 Doty RL, Brugger WPE, Jurs PC, Orndorff MA, Snyder PJ, Lowry LD: Intranasal trigeminal stimulation from odorous volatiles: psychometric responses from anosmic and normal humans. Physiol Behav 1978;20:175–185.
79 Kobal G, Van Toller S, Hummel T: Is there directional smelling? Experientia 1989;45:130–132.
80 Hummel T: Assessment of intranasal trigeminal function. Int J Psychophysiol 2000;36:147–155.
81 Handwerker H, Kobal G: Psychophysiology of experimentally induced pain. Physiol Rev 1993;73:639–671.
82 Hummel T, Pietsch H, Kobal G: Kallmann's syndrome and chemosensory evoked potentials. Eur Arch Otorhinolaryngol 1991;248:311–312.
83 Cui L, Evans WJ: Olfactory event-related potentials to amyl acetate in congenital anosmia. Electroencephalogr Clin Neurophysiol 1997;102:303–306.
84 Geisler MW, Schlotfeldt CR, Middleton CB, Dulay MF, Murphy C: Traumatic brain injury assessed with olfactory event-related brain potentials. J Clin Neurophysiol 1999;16:77–86.
85 Kobal G, Hummel T: Olfactory and intranasal trigeminal event-related potentials in anosmic patients. Laryngoscope 1998;108:1033–1035.
86 Frasnelli J, Wohlgemuth C, Hummel T: The influence of stimulus duration on odor perception. Int J Psychophysiol, in press.
87 Gudziol H, Wurschi A: Pre- and postoperative functional results with patients after ethmoidectomy. Chem Senses 1996;21:484.
88 Kobal G, Hummel T: Olfactory evoked potentials in humans; in Getchell TV, Doty RL, Bartoshuk LM, Snow JBJ (eds): Smell and Taste in Health and Disease. New York, Raven Press, 1991, pp 255–275.
89 Hummel T, Barz S, Pauli E, Kobal G: Chemosensory event-related potentials change as a function of age. Electroencephalogr Clin Neurophysiol 1998;108:208–217.

90 Hummel T, Rothbauer C, Barz S, Grosser K, Pauli E, Kobal G: Olfactory function in acute rhinitis. Ann NY Acad Sci 1998;855:616–624.
91 Hosoya Y, Yoshida H: Über die bioelektrischen Erscheinungen an der Riechschleimhaut. J Med Sci III Biophysics 1937;5:22.
92 Ottoson D: Sustained potentials evoked by olfactory stimulation. Acta Physiol Scand 1954;32: 384–386.
93 Knecht M, Hummel T: Recording of the human electro-olfactogram. Physiol Behav 2004;83:13–19.
94 Hummel T, Knecht M, Kobal G: Peripherally obtained electrophysiological responses to olfactory stimulation in man: electro-olfactograms exhibit a smaller degree of desensitization compared with subjective intensity estimates. Brain Res 1996;717:160–164.
95 Leopold DA, Hummel T, Schwob JE, Hong SC, Knecht M, Kobal G: Anterior distribution of human olfactory epithelium. Laryngoscope 2000;110:417–421.
96 Knecht M, Hummel T, Wolf S, Kobal G: Assessment of the peripheral input signal to the olfactory system in man: the electro-olfactogram. Eur J Physiol 1995:R47.
97 von Brunn A: Beiträge zur mikroskopischen Anatomie der menschlichen Nasenhöhle. Arch Mikrosk Anat 1892;39:632–651.
98 Feron F, Perry C, McGrath JJ, Mackay-Sim A: New techniques for biopsy and culture of human olfactory epithelial neurons. Arch Otolaryngol Head Neck Surg 1998;124:861–866.
99 Wang L, Hari C, Chen L, Jacob T: A new non-invasive method for recording the electro-olfactogram using external electrodes. Clin Neurophysiol 2004;115:1631–1640.
100 Getchell TV: Electrogenic sources of slow voltage transients recorded from frog olfactory epithelium. J Neurophysiol 1974;37:1115–1130.
101 Getchell TV: Analysis of intracellular recordings from salamander olfactory epithelium. Brain Res 1977;123:275–286.
102 Walter WG, Cooper R, Aldrige VJ, McCallum WC, Winter AL: Contingent negative variation: an electric sign of sensori-motor association and expectancy in the human brain. Nature 1964;203: 380–384.
103 Tecce JJ: Contingent negative variation and psychological processes in man. Psychol Bull 1972;77:73–108.
104 Lorig TS, Roberts M: Odor and cognitive alteration of the contingent negative variation. Chem Senses 1990;15:537–545.
105 Donchin E: Methodological issues in CNV research. Electroencephalogr Clin Neurophysiol 1973;33(Suppl):3–19.
106 Kanamura S, Kawasaki M, Indo M, Fukuda H, Torii S: Effects of odors on the contingent negative variation and the skin potential level. Chem Senses 1988;13:327–328.
107 Auffermann H, Gerull G, Mathe F, Mrowinski D: Olfactory evoked potentials and contingent negative variation simultaneously recorded for diagnosis of smell disorders. Ann Otol Rhinol Laryngol 1993;102:6–10.
108 Yamamoto T: Objective evaluation of human olfactory dysfunction by event-related brain potentials. Chem Senses 1996;21:83.
109 Archilei G, Moretti E: Olfattometria ed elettoencefalografia. Valsalva 1958;34:201–207.
110 Motokizawa F, Furuya N: Neural pathways associated with the EEG arousal response by olfactory stimulation. Electroencephalogr Clin Neurophysiol 1973;35:83–91.
111 Moncrieff RW: Effects of odours on EEG records. Perfum Essent Oil Rec 1962;53:757–760, 825–828.
112 Perbellini D, Scolari R: L'elettroencefalo-olfattometria. Ann Laringol Otol Rinol Faringol 1966;65:421–429.
113 Bartalena G, Romeo G: Olfattometria ematogena elettroencefalografica in sogetti normali. Boll Mal Orecch Gola Naso 1962;80:14–23.
114 Lorig TS, Huffman E, DeMartino A, DeMarco J: The effects of low concentration odors on EEG activity and behaviour. J Psychophysiol 1991;5:69–77.
115 Van Toller S, Behan J, Howells P, Kendall-Reed M, Richardson A: An analysis of spontaneous human cortical EEG activity to odours. Chem Senses 1993;18:1–16.
116 Savic I: Imaging of brain activation by odorants in humans. Curr Opin Neurobiol 2002;12: 455–461.

117 Zald DH, Pardo JV: Functional neuroimaging of the olfactory system in humans. Int J Psychophysiol 2000;36:165–181.
118 Small DM, Jones-Gotman M, Zatorre RJ, Petrides M, Evans AC: Flavor processing: more than the sum of its parts. Neuroreport 1997;8:3913–3917.
119 Kareken DA, Sabri M, Radnovich AJ, Claus E, Foresman B, Hector D, et al: Olfactory system activation from sniffing: effects in piriform and orbitofrontal cortex. Neuroimage 2004;22:456–465.
120 Sobel N, Prabhakaran V, Zhao Z, Desmond JE, Glover GH, Sullivan EV, et al: Time course of odorant-induced activation in the human primary olfactory cortex. J Neurophysiol 2000;83: 537–551.
121 Poellinger A, Thomas R, Lio P, Lee A, Makris N, Rosen BR, et al: Activation and habituation in olfaction – An fMRI study. Neuroimage 2001;13:547–560.
122 Anderson AK, Christoff K, Stappen I, Panitz D, Ghahremani DG, Glover G, et al: Dissociated neural representations of intensity and valence in human olfaction. Nat Neurosci 2003;6:196–202.
123 Kettenmann B, Hummel C, Stefan H, Kobal G: Magnetoencephalographical recordings: separation of cortical responses to different chemical stimulation in man. Funct Neurosci (EEG Suppl) 1996;46:287–290.
124 Ayabe-Kanamura S, Endo H, Kobayakawa T, Takeda T, Saito S: Measurement of olfactory evoked magnetic fields by a 64-channel whole-head SQUID system. Chem Senses 1997;22:214–215.
125 Small DM, Gerber JC, Mak YE, Hummel T: Differential neural responses evoked by orthonasal versus retronasal odorant perception in humans. Neuron 2005;47:593–605.
126 Yousem DM, Maldjian JA, Hummel T, Alsop DC, Geckle RJ, Kraut MA, et al: The effect of age on odor-stimulated functional MR imaging. Am J Neuroradiol 1999;20:600–608.
127 Henkin RI, Levy LM, Lin CS: Taste and smell phantoms revealed by brain functional MRI (fMRI). J Comput Assist Tomogr 2000;24:106–123.
128 Adema JM, Montserrat JM: Olfacto-rhinomanometry. Int Rhinol 1982;20:21–28.
129 Hoshino T, Usui N: Objective olfactometry by the method of recordings of respiratory resistances. Jibiinkoka 1987;90:516–522.
130 Gudziol H, Gramowski KH: Respirations-Olfaktometrie – eine objektivierende Methode zur quantitativen Bewertung einer Hyposmie. Laryngol Rhinol Otol 1987;66:570–572.
131 Kendall-Reed M, Walker JC: Human respiratory responses to odorants. Chem Senses 1996; 21:486.
132 Gudziol H, Wächter R: Gibt es olfaktorisch evozierte Atemänderungen? Laryngorhinootologie 2004;83:367–373.
133 Frank RA, Dulay MF, Niergarth KA, Gesteland RC: A comparison of the sniff magnitude test and the University of Pennsylvania Smell Identification Test in children and nonnative English speakers. Physiol Behav 2004;81:475–480.
134 Asaka H: The studies on the objective olfactory test by galvanic skin response. J Otorhinolaryngol Soc Jap1965;68:100–112.
135 Auld JSM: Psycho-galvanic measurement of smell. J Inst Petroleum Technol 1923;9:389–391.
136 Borsanyi SJ, Blanchard DL, Baker FJ: Psychogalvanic skin response olfactometry. Ann Otol 1962;74:213–221.
137 Nishida H, Kumagami H, Jinnouchi H: Pupillary reaction following olfactory stimulation – Use in objective olfactometry. Nippon Jibiinkoka Gakkai Kaiho 1973;76:1449–1458.
138 Sneppe R, Gonay P: Evaluation objective, quantitative et qualitative de l'olfaction. Electrodiagn Ther 1973;10:5–17.
139 Ichihara M, Komatsu A, Ichihara F, Asaga H, Hirayoshi K: Test of smell based on the wink response. Jibiinkoka 1967;39:947–953.
140 Delank KW: Subjektive und objektive Methoden zur Beurteilung der Riechfunktion. HNO 1998;46:182–190.
141 Cometto-Muniz JE, Cain WS: Trigeminal and olfactory sensitivity: comparison of modalities and methods of measurement. Int Arch Occup Environ Health 1998;71:105–110.
142 Lötsch J, Ahne G, Kunder J, Kobal G, Hummel T: Factors affecting pain intensity in a pain model based upon tonic intranasal stimulation in humans. Inflamm Res 1998;47:446–450.
143 Berg J, Hummel T, Huang G, Doty RL: Trigeminal impact of odorants assessed with lateralized stimulation. Chem Senses 1998;23:587.

144 Thürauf N, Friedel I, Hummel C, Kobal G: The mucosal potential elicited by noxious chemical stimuli: is it a peripheral nociceptive event. Neurosci Lett 1991;128:297–300.

145 Hari R, Portin K, Kettenmann B, Jousmaki V, Kobal G: Right-hemisphere preponderance of responses to painful CO_2 stimulation of the human nasal mucosa. Pain 1997;72:145–151.

146 Livermore A, Hummel T, Kobal G: Chemosensory evoked potentials in the investigation of interactions between the olfactory and the somatosensory (trigeminal) systems. Electroencephalogr Clin Neurophysiol 1992;83:201–210.

147 Pause BM, Sojka B, Krauel K, Ferstl R: The nature of the late positive complex within the olfactory event-related potential. Psychophysiology 1996;33:168–172.

Thomas Hummel, MD
Smell & Taste Clinic, Department of Otorhinolaryngology
University of Dresden Medical School ('Technische Universität Dresden')
Fetscherstrasse 74
DE–01307 Dresden (Germany)
Tel. +49 351 458 4189, Fax +49 351 458 4326
E-Mail thummel@rcs.urz.tu-dresden.de

Hummel T, Welge-Lüssen A (eds): Taste and Smell. An Update.
Adv Otorhinolaryngol. Basel, Karger, 2006, vol 63, pp 99–107

Posttraumatic Olfactory Loss

Richard M. Costanzo[a], *Takaki Miwa*[b]

[a]Virginia Commonwealth University, School of Medicine, Richmond, Va., USA; [b]Kanazawa University Graduate School of Medical Sciences, Kanazawa, Japan

Abstract

Head injury is the leading cause of posttraumatic anosmia. Complete or partial loss of olfactory function may occur when the nasal passages are blocked, olfactory nerves are injured or there are contusions or hemorrhages in olfactory centers of the brain. Evaluation of patients with posttraumatic olfactory loss should include a physical examination by the otolaryngologist. Nasal endoscopy and radiological studies should be performed as well as olfactory function tests to determine the degree and type of olfactory impairment. Although treatment options may be limited, physicians should provide information and counseling regarding the risks and hazards associated with loss of olfactory function. For some individuals such as cooks, firefighters, and research scientists, an assessment of vocational activities should be performed prior to reentry into the workplace. Individuals with impaired olfactory function may be unable to detect important warning signs such as gas leaks, volatile chemical fumes and fires and therefore place themselves and coworkers at an increased risk for serious injury or death.

Mechanisms of Injury

Head injury is one of the most common causes of posttraumatic olfactory loss. Mechanisms leading to a partial or complete loss of olfactory function include damage to the olfactory epithelium or nasal passages, injury to olfactory nerve fibers, and contusions and hemorrhage lesions in olfactory areas of the brain [1].

Damage to Nose or Nasal Passages

Direct injury, edema, mucosal hematoma, and scarring of the olfactory epithelium all contribute to impairment of olfactory function. Trauma to the nasal passages and conductive pathways can block airflow to the olfactory cleft region and impair olfactory function. Injury to the olfactory receptor cells in the nasal epithelium, including degeneration of the olfactory vesicles and cilia have been observed in biopsy specimens following traumatic anosmia [2, 3]. Fractures including nasozygomatic-Le Fort fractures, fronto-orbital fractures, and pure Le Fort fractures have been associated with posttraumatic smell disturbances [4]. The degree of impairment varies depending on the etiology of the fracture, the severity of the trauma, and the involvement of specific facial bone regions. Olfactory dysfunction is more likely to occur when there is a skull fracture, loss of consciousness for more than 1 h, or a severe head injury. Zusho [5], who studied 212 patients with posttraumatic anosmia, found that 44.8% had facial or skull fractures and 11.3% had facial contusions with fractures of the nasal bones. Identification of injuries to conductive pathways is important in assessing posttraumatic olfactory loss since corrective surgery may help to restore olfactory function.

Nerve Injury

Tearing or trauma to the delicate olfactory nerve fibers can occur with nasal and skull base fractures or secondary to the coup-contra-coup forces generated in blunt head injury. Studies have shown that anosmia is more likely to occur when there is a blow to the occipital region of the head in comparison to other sites [6]. Histopathological and immunocytochemical findings suggest that posttraumatic anosmia involves, at least in part, damage to the olfactory nerve fibers [7] with associated changes in the olfactory bulb. In more severe injuries, regeneration of olfactory nerve fibers and recovery of function may be blocked by the formation of scar tissue and gliosis [8].

Brain Injury

Olfactory disorders may occur following severe, moderate and even mild head injury [9]. Costanzo and Becker [10] found that 14–27% of patients with moderate to severe head injuries had smell disorders. Brain contusions, especially in the olfactory bulb and orbital frontal pole region, are a common cause of posttraumatic olfactory loss. Intraparenchymal hemorrhage or cortical contusions are often associated with disorders of odor discrimination [9]. Yousem et al. [11] investigated the primary sites of injury in patients with posttraumatic anosmia and hyposmia. Using magnetic resonance imaging (MRI), they found the highest incidence of posttraumatic encephalomalacia was in the olfactory bulbs and tracts, subfrontal lobes, and temporal lobes. Other studies suggest that impairment of olfactory recognition without anosmia may occur with local or diffuse injury to the orbitofrontal and temporal lobe regions of the brain [12].

In addition, behavioral and memory disorders frequently accompany impaired olfactory recognition.

Clinic Evaluation

History and Physical Exam

Patients with posttraumatic anosmia often present to the otolaryngologist well after the initial injury. In many cases, facial fractures, nasal intubation and neurosurgical procedures complicate the evaluation of patients with olfactory complaints. A detailed medical history should include the assessment of any previous incidents of olfactory loss, nasal obstruction, rhinosinusitis, head trauma, and upper respiratory infections. Details regarding the nature and degree of loss should be obtained. The patient should be asked if the onset was sudden or gradual, if the loss is complete or partial, and if there are associated dysosmias such as paraosmias or olfactory hallucinations. This information may be helpful in identifying and localizing the problem to a central or peripheral mechanism. A history of postnasal drip or clear rhinorrhea may indicate a cerebrospinal fluid leak, typically associated with fractures of the anterior cranial skull base or adjacent ethmoid sinuses [13]. Inpatient hospital records should be reviewed including any operative notes and radiological studies. MRI and computed tomography scans of the head should be carefully reexamined since they may have been initially obtained to evaluate areas of brain injury and not specific olfactory structures.

A physical examination by an otolaryngologist is essential in cases of posttraumatic anosmia. Nasal endoscopy should include inspection of the inferior, middle and superior meatuses, as well as the nasopharynx. Obstruction of nasal airflow by septal deviation, hypertrophy of the turbinates, neoplasia and polyposis should be determined. Visualization of the olfactory cleft is critical.

Sensory Testing

Patients with posttraumatic anosmia should undergo formal olfactory testing to determine the degree and nature of the loss. Standardized tests are available to assess detection thresholds and odor identification and recognition [14]. The University of Pennsylvania Smell Identification Test uses booklets containing microencapsulated odorants that are sampled using the 'scratch and sniff' technique [15]. This test has the advantage of being self-administered and can be used to detect malingerers. Another test developed at the University of Connecticut employs both odor detection and odor identification subtests [16]. Detection thresholds are useful in evaluation of receptor cell function, and identification test scores often help uncover cortical contusions and brain injuries.

a 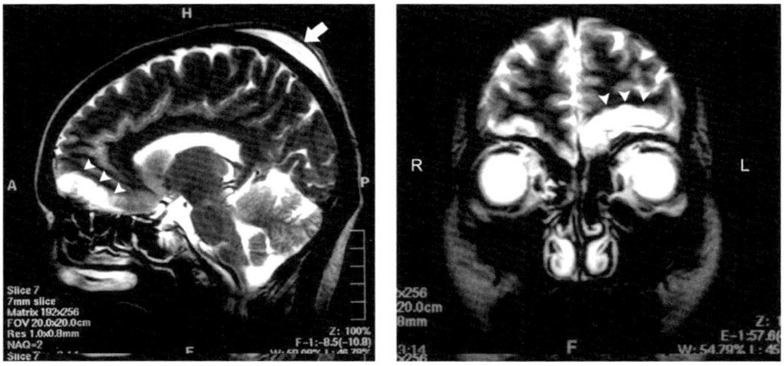 b

Fig. 1. Brain image (T_2) of a 42-year-old woman who fell and hit the back of her head while playing badminton. ***a*** Sagittal section showing the area of impact (arrow) and brain injury in the orbitofrontal lobe region (arrowheads). ***b*** Coronal section showing areas of increased intensity in the left orbitofrontal lobe (arrowheads).

In Japan, olfactory testing is often performed using a graded series of odorants presented on strips of blotting paper [17]. In Germany, 'sniffing sticks' are routinely used to test olfactory function [18] and olfactometers that deliver controlled odor stimuli have made it possible to measure olfactory evoked potentials [19]. A more comprehensive review of olfactory testing methods is presented in a separate chapter in this volume [chapter by Hummel et al., this vol, pp 84–98].

Radiology

Although plain-film radiographs may be of some diagnostic use in determining skull or facial fractures, high-resolution computed tomography and MRI scans are preferred in diagnosing posttraumatic anosmia [11, 20]. Hematomas and contusions are potential causes of posttraumatic anosmia and may be evaluated with MR images. When observed in the olfactory bulb and the orbitofrontal pole region, they represent clinical signs commonly associated with impaired olfactory function. Figure 1 shows the brain scan of a 42-year-old woman who suffered impaired olfactory function after hitting the back of her head. Olfactory function testing revealed that although she could detect some of the test odors presented, there was a complete loss in her ability to recognize or identify odors. A decrease in metabolism in the orbitofrontal cortex and medial prefrontal cortex areas has been demonstrated in posttraumatic anosmia using quantitative positron emission tomography scanning [21]. Functional MRI has also been used successfully to visualize olfactory function in different brain regions [22, 23].

Table 1. Compensatory strategies for individuals with posttraumatic anosmia

1	Install and maintain smoke and gas detectors in the home and workplace
2	Identify natural gas appliances and avoid use if possible
3	Date and label perishable foods to assure freshness
4	Identify and pay attention to personal hygiene issues such as body and breath odors
5	Determine appropriate measured amounts when applying fragrances and colognes
6	Use spices and colorful presentations to enhance the enjoyment of food
7	Avoid use of chemicals and cleaning agents that may release harmful vapors

Prognosis and Management

Factors That Determine Likelihood of Recovery

Recovery from a posttraumatic loss of olfactory function depends on the severity and type of injury. Sumner [6] suggested that recovery from posttraumatic anosmia may occur with repair of sinonasal obstruction, or the resolution of cerebral hemorrhage or contusion. Costanzo and Becker [10] found that among those patients with moderate to severe injuries, about 33% showed some improvement, while 27% worsened. Doty et al. [24] tested olfactory function in 66 patients with olfactory dysfunction following head injury and found that 36% improved, 45% showed no change and 18% worsened. In most cases of posttraumatic anosmia, treatment options are limited. Improvement over time may occur spontaneously without intervention due to the regenerative capacity of the olfactory system [25, 26]. Recovery is most likely to occur within the first 6 months to 1 year following injury. After 2 years, the chances of improvement decrease to less than 10% [10, 27]. Posttraumatic nasal obstruction and other conductive deficits may benefit from surgical procedures. The administration of steroids has also been used with some success and may help to resolve mucosal edema and possibly contribute to the promotion of neural regeneration [28, 29].

Compensatory Strategies

While medical treatment options of posttraumatic anosmia may be limited, patient counseling should be included in the management plan. There are a number of strategies that can help compensate for olfactory loss (table 1). Awareness of the loss and understanding detection limitations are a critical step. Precautions should be taken to reduce the risk of exposure to gas leaks, ingestion of spoiled food, and inability to detect smoke or fires [30]. Purchase of smoke detectors and gas alarm devices for the home and work environment should be considered. Perfumes and colognes should be carefully measured to avoid reaching offensive concentration levels. The use of seasonings and selection of colorful foods may help compensate for the loss of food flavor associated with olfactory impairment. Labeling and dating of refrigerated and frozen foods will help to prevent ingestion of spoiled foods. Special instructions should be given to patients with cardiovascular disease and hypertension so that they do not compensate for loss of flavor by adding excessive amounts of salt to their food.

Impact of Olfactory Loss

Risks and Hazards

The health and safety of patients with posttraumatic olfactory impairment are at significant risk due to their inability to detect gas leaks, smoke, spoiled foods and other olfactory-related warning signs that may be present in the home and at the workplace [31]. This inability may be further complicated by trauma to brain regions resulting in impaired cognitive and motor function. Instructions and counseling are essential for this patient population. A commonly cited activity of concern for patients with impaired olfactory function is the detection of spoiled food [32]. The detection of gas leaks, smoke, eating and cooking were also a concern of over half of the patients surveyed. Santos et al. [31] studied 445 patients and found that patients with impaired olfactory function are more likely to experience an olfactory-related hazardous event than those with normal olfactory function. Forty-five percent of these patients reported experiencing cooking-related incidents, 25% reported ingestion of spoiled food, and 23% the inability to detect a gas leak. Food hygiene is a major concern for those with olfactory impairment. Perishable foods should be stored in a refrigerator and eaten as soon as possible. Checking of expiration dates on food packages and discarding out-of-date foods should be carefully adhered to. If in doubt about the freshness of food, a family member or friend should be consulted prior to serving. Gas and smoke detectors are essential safety devices for those with impaired olfactory function. When possible, electric appliances should be used instead of gas appliances. Cooking times and temperatures should be monitored carefully to avoid burning food or igniting grease fires.

Quality of Life Issue

The patients who have olfactory impairment report significant changes in quality of life [30]. Major concerns were bad breath and body odor, worrying about the detection of gas leaks and smoke, and alterations in the taste of foods and loss of appetite. Additional concerns included exposure to cleaning solutions, pesticides and chemicals, and pet odors. Individuals with sustained olfactory impairment washed clothes and cleaned their house more frequently than those that had improvement in their olfactory function. Perfume and cologne usage decreased and deodorant increased for those with impaired olfactory function. They also reported a decrease in their enjoyment of hobbies and other activities. In a separate study of 278 individuals with hyposmia or anosmia, the predominant complaints were food related [33]. Quality of life issues seemed to be more of a concern with younger people than older individuals, and women seemed to be affected more strongly than men.

Overall, individuals with sustained olfactory impairment reported that they were more dissatisfied with their life than those with improved olfactory function. Medical or other health-related workers should be aware of and understand the effects of olfactory impairment on patients. Deems et al. [34] reported that the prevalence of depression in patients with chemosensory disturbances was higher than in controls. Counseling by either a psychologist or psychiatrist may be helpful for some patients.

Vocational Issues

Olfactory impairment may affect vocational reentry for patients undergoing rehabilitation from traumatic head injury. In a study of 40 patients with anosmia following closed head injury, major vocational problems were encountered during the 2 or more years after reentry into the workplace [35]. Most demonstrated psychosocial disorders associated with orbital frontal cortex damage, an important brain region for the processing of olfactory information. Perfumists, florists, and cooks are particularly impacted by olfactory impairment. Firefighters and chemists with impaired olfaction may put themselves and others at risk. Employees and employers should be educated regarding issues related to reentry into the workplace. Safety devices including smoke and gas detectors should be considered where appropriate and specific tasks evaluated relative to the degree of olfactory loss.

Conclusions

Loss of olfactory function is a common occurrence following traumatic head and brain injuries. Trauma can cause blockage of nasal passages, damage to the olfactory nerves and contusions and hemorrhaging in olfactory regions of the brain. Individuals with impaired olfactory function may be unable to detect important warning signs such as gas leaks, smoke fumes and spoiled foods and the quality of their life may be adversely affected. The evaluation and management of patients with olfactory impairment may lead to improved outcomes. Olfactory function and other diagnostic tests are useful in the assessment of patients with anosmia. Information and counseling regarding compensatory strategies and the likelihood of recovery are often helpful to patients coping with an olfactory loss.

Promising new research on the regenerative capacity of the olfactory system may lead to new treatment options for individuals with posttraumatic anosmia. The injection of harvested olfactory precursor cells may provide a new method for replacing injured or degenerated neurons within the olfactory epithelium. The identification and administration of growth factors may be of therapeutic value by enhancing the outgrowth of olfactory axons and restoring olfactory connections. Methods that enhance recovery by slowing down degenerative processes or promote cell growth and recovery are likely to play an important role in future treatment strategies for patients with posttraumatic anosmia.

References

1 Costanzo RM, DiNardo LJ, Reiter ER: Head injury and olfaction; in Doty RL (ed): Handbook of Olfaction and Gustation. New York, Dekker, 2003, pp 629–638.
2 Hasegawa S, Yamagishi M, Nakano Y: Microscopic studies of human olfactory epithelia following traumatic anosmia. Arch Otorhinolaryngol 1986;243:112–116.
3 Jafek BW, Eller PM, Esses BA, Moran DT: Post-traumatic anosmia. Ultrastructural correlates. Arch Neurol 1989;46:300–304.
4 Renzi G, Carboni A, Gasparini G, Perugini M, Becelli R: Taste and olfactory disturbances after upper and middle third facial fractures: a preliminary study. Ann Plast Surg 2002;48:355–358.
5 Zusho H: Posttraumatic anosmia. Arch Otolaryngol 1982;108:90–92.
6 Sumner D: Disturbance of the senses of smell and taste after head injuries; in Vinken PJ, Bruyn GW (eds): Handbook of Clinical Neurology. Amsterdam, North-Holland Publishing Co, 1975, pp 1–25.
7 Delank KW, Fechner G: Pathophysiology of post-traumatic anosmia. Laryngorhinootologie 1996;75:154–159.
8 Costanzo RM, Zasler ND: Epidemiology and pathophysiology of olfactory and gustatory dysfunction in head trauma. J Head Trauma Rehabil 1992;7:15–24.
9 Costanzo RM, Zasler ND: Head trauma; in Getchell TV, Doty RL, Bartoshuk LM, Snow JB Jr (eds): Smell and Taste in Health and Disease. New York, Raven Press, 1991, pp 711–730.
10 Costanzo RM, Becker DP: Smell and taste disorders in head injury and neurosurgery patients; in Meiselman HL, Rivlin RS (eds): Clinical Measurements of Taste and Smell. New York, MacMillian Publishing Company, 1986, pp 565–578.
11 Yousem DM, Geckle RJ, Bilker WB, Kroger H, Doty RL: Posttraumatic smell loss: relationship of psychophysical tests and volumes of the olfactory bulbs and tracts and the temporal lobes. Acad Radiol 1999;6:264–272.
12 Levin HS, High WM, Eisenberg HM: Impairment of olfactory recognition after closed head injury. Brain 1985;108:579–591.
13 DiNardo LJ, Costanzo RM: Post-traumatic olfactory loss; in Seiden A (ed): Taste and Smell Disorders. New York, Thieme Medical Publishers, 1997, pp 79–87.
14 Doty RL, Kobal G: Current trends in the measurement of olfactory function; in Doty RL (ed): Handbook of Olfaction and Gustation. New York, Dekker, 1995, pp 191–225.
15 Doty RL, Shaman P, Kimmelman CP, Dann MS: University of Pennsylvania Smell Identification Test: a rapid quantitative olfactory function test for the clinic. Laryngoscope 1984;94:176–178.
16 Cain WS, Gent JF, Goodspeed RB, Leonard G: Evaluation of olfactory dysfunction in the Connecticut Chemosensory Clinical Research Center. Laryngoscope 1988;98:83–88.

17 Takagi SF: A standardized olfactometer in Japan. A review over ten years. Ann NY Acad Sci 1987;510:113–118.
18 Kobal G, Klimek L, Wolfensberger M, Gudziol H, Temmel A, Owen CM, Seeber H, Pauli E, Hummel T: Multicenter investigation of 1,036 subjects using a standardized method for the assessment of olfactory function combining tests of odor identification, odor discrimination, and olfactory thresholds. Eur Arch Otorhinolaryngol 2000;257:205–211.
19 Hummel T, Kobal G: Differences in human evoked potentials related to olfactory or trigeminal chemosensory activation. Electroencephalogr Clin Neurophysiol 1992;84:84–89.
20 Yousem DM, Geckle RJ, Bilker WB, McKeown DA, Doty RL: Posttraumatic olfactory dysfunction: MR and clinical evaluation. AJNR Am J Neuroradiol 1996;17:1171–1179.
21 Varney NR, Pinkston JB, Wu JC: Quantitative PET findings in patients with posttraumatic anosmia. J Head Trauma Rehabil 2001;16:253–259.
22 Levy LM, Henkin RI, Hutter A, Lin CS, Martins D, Schellinger D: Functional MRI of human olfaction. J Comput Assist Tomogr 1997;21:849–856.
23 Smejkal V, Druga R, Tintera J: Olfactory activity in the human brain identified by fMRI. Bratisl Lek Listy 2003;104:184–188.
24 Doty RL, Yousem DM, Pham LT, Kreshak AA, Geckle R, Lee WW: Olfactory dysfunction in patients with head trauma. Arch Neurol 1997;54:1131–1140.
25 Graziadei PPC, Monti Graziadei GA: Neurogenesis and plasticity of the olfactory sensory neurons. Ann NY Acad Sci 1985;457:127–142.
26 Schwob JE: Neural regeneration and the peripheral olfactory system. Anat Rec 2002;269:33–49.
27 Sumner D: Post-traumatic anosmia. Brain 1964;87:107–120.
28 Fujii M, Fukazawa K, Takayasu S, Sakagami M: Olfactory dysfunction in patients with head trauma. Auris Nasus Larynx 2002;29:35–40.
29 Ikeda K, Sakurada T, Takasaka T, Okitsu T, Yoshida S: Anosmia following head trauma: preliminary study of steroid treatment. Tohoku J Exp Med 1995;177:343–351.
30 Reiter ER, Costanzo RM: The overlooked impact of olfactory loss: safety, quality of life and disability issues. Chem Senses 2003;6:1–4.
31 Santos DV, Reiter ER, DiNardo LJ, Costanzo RM: Hazardous events associated with impaired olfactory function. Arch Otolaryngol Head Neck Surg 2004;130:317–319.
32 Miwa T, Furukawa M, Tsukatani T, Costanzo RM, DiNardo LJ, Reiter ER: Impact of olfactory impairment on quality of life and disability. Arch Otolaryngol Head Neck Surg 2001;127: 497–503.
33 Temmel AF, Quint C, Schickinger-Fischer B, Klimek L, Stoller E, Hummel T: Characteristics of olfactory disorders in relation to major causes of olfactory loss. Arch Otolaryngol Head Neck Surg 2002;128:635–641.
34 Deems DA, Doty RL, Settle RG, Moore Gillon V, Shaman P, Mester AF, Kimmelman CP, Brightman VJ, Snow JB Jr: Smell and taste disorders, a study of 750 patients from the University of Pennsylvania Smell and Taste Center. Arch Otolaryngol Head Neck Surg 1991;117:519–528.
35 Varney NR: Prognostic significance of anosmia in patients with closed-head trauma. J Clin Exp Neuropsychol 1988;10:250–254.

Richard M. Costanzo PhD
Professor of Physiology and Otolaryngology
School of Medicine, Box 980551, Virginia Commonwealth University
Richmond, VA 23298–0551 (USA)
Tel. +1 804 828 4774, Fax +1 804 828 7382
E-Mail rcostanz@vcu.edu

Hummel T, Welge-Lüssen A (eds): Taste and Smell. An Update.
Adv Otorhinolaryngol. Basel, Karger, 2006, vol 63, pp 108–124

Chronic Rhinosinusitis and Olfactory Dysfunction

Joseph R. Raviv, Robert C. Kern

Department of Otolaryngology – Head and Neck Surgery, Northwestern University Feinberg School of Medicine, Chicago, Ill., USA

Abstract

Chronic rhinosinusitis encompasses a group of disorders characterized by inflammation of the mucosa of the nose and paranasal sinuses of at least 12 weeks' duration. In addition to nasal obstruction and discharge, chronic sinusitis is a common cause of olfactory dysfunction. However, smell loss is often overlooked in the clinical setting of sinusitis, with attention instead focused on the respiratory complaints of nasal obstruction, hypersecretion, and facial pressure and pain. Olfactory dysfunction can result in problems including safety concern, hygiene matters, appetite disorders, and changes in emotional and sexual behavior. Although smell loss related to sinonasal disease is probably the most treatable form of olfactory dysfunction, most studies show that improved olfactory sensation in this setting is usually transient and incomplete.

Chronic rhinosinusitis (CRS) encompasses a group of disorders characterized by inflammation of the mucosa of the nose and paranasal sinuses of at least 12 consecutive weeks' duration [1]. CRS is diagnosed using clinical criteria and typically confirmed with CT scan findings of mucosal inflammation [1]. This entity is often associated with the presence of allergic and nonallergic rhinitis, as well as nasal polyposis. In addition to the more common patient complaints of nasal obstruction and discharge, CRS is also a frequent cause of olfactory dysfunction particularly in the setting of nasal polyposis, and is believed to account for at least 25% of all cases of smell loss possibly affecting more than 10 million people [2–5]. However, olfaction is often overlooked in the clinical setting of CRS, with attention instead focused on the respiratory complaints of nasal obstruction, hypersecretion, and facial pressure and pain. Moreover, smell loss related to sinonasal disease is probably the most treatable form of olfactory dysfunction and, as a result, has engendered less interest from even those few clinicians interested in olfaction.

The etiology of CRS, whether allergic or nonallergic, with or without nasal polyps remains unclear. No single factor has been found in all patients and theories proposed for the pathogenesis of CRS include staphylococcal exotoxin exposure, T-cell disturbances, chronic infection of the underlying ethmoid bone and non-IgE-mediated hypersensitivity reactions directed against fungal antigens [1]. Although smell loss can be seen in all forms of sinonasal inflammatory disease, CRS with polyps is most commonly associated with olfactory dysfunction in general and anosmia in particular. While nasal polyps can be associated with systemic disorders such as cystic fibrosis, primary ciliary dyskinesia and aspirin sensitivity, it is most commonly seen in the scenario of typical CRS. No definite association between polyposis and atopy has been firmly established [6–8]. Furthermore, it remains a matter of controversy whether CRS with and without polyps represent distinct disease entities. The more commonly accepted hypothesis proposes that CRS with polyps represents the most advanced, end stage form of the disease. The alternative theory proposes that polyposis is a distinct entity resulting from separate pathologic processes [9]. Further work at the molecular level is necessary to resolve this question, but the common thread that links all forms of CRS is persistent mucosal inflammation.

Anatomy and Physiology

A basic understanding of olfactory anatomy and physiology is crucial to understanding olfactory dysfunction in the setting of CRS. The receptive surface of the human olfactory system is an approximately 1- to 2-cm^2 patch of pseudostratified columnar epithelium situated within each nasal vault on the cribiform plate and segments of both the superior and middle turbinates. The olfactory neuroepithelium harbors sensory receptors of the main olfactory (cranial nerve I) system and some trigeminal (cranial nerve V) somatosensory nerve endings. Access of odorants to the olfactory epithelium occurs via direct orthonasal airflow and retronasal flow through the nasopharynx. The receptive cell of the olfactory system is the olfactory sensory neuron (OSN). OSNs are true bipolar neurons present within the nasal epithelium, which synapse with second-order neurons in the olfactory bulb mediating the peripheral mechanisms of the sense of smell. The OSN cell bodies are in close proximity to the ambient environment and the dendritic processes project into the nasal lumen optimizing olfactory transduction, but rendering these neurons vulnerable to injury from inflammatory, infectious and chemical agents. This results in a baseline level of OSN death, which has the biochemical and morphologic characteristics of apoptosis [10]. As an adaptive mechanism, the mammalian olfactory epithelium maintains the ability to replace OSNs lost as a result of injury throughout life [11]. The link between OSN loss and regeneration has been termed *olfactory neuronal homeostasis*, wherein a balance is maintained between neuronal death and proliferation allowing the animal to maintain an adequate number of OSNs necessary for olfactory sensation [12]. It has been hypothesized that disease processes trigger uncompensated increases in the death rate, and a net loss of OSNs [13].

Classification

Clinical olfactory disorders have been classified as transport (conductive) disorders, sensory disorders, and neural disorders [14]. Neural disorders reflect injury to the olfactory bulb and central olfactory pathways. Transport olfactory loss reflects diminished access of odorants to the olfactory neuroepithelium and sensory disorders involve direct damage to the neuroepithelium itself. This classification is based on site of lesion and is analogous to that used for the evaluation of hearing loss. In the auditory system, various tests are used to identify the location and nature of the pathologic process, and these tests have been validated by histopathologic temporal bone studies. In the olfactory system, no site of lesion testing is currently available and histopathologic studies are scant. Current clinical olfactory testing evaluates only the degree of deficit, telling us nothing about the nature of the problem or the anatomical site of injury [15]. In this regard, functional radiological imaging may be capable of addressing this diagnostic limitation in the future, but not at present. For this reason, the pathologic process in hyposmic and anosmic patients has been categorized primarily on the basis of history, endoscopic examination and radiographic studies.

Pathology in Chronic Rhinosinusitis with Olfactory Dysfunction

Traditionally, olfactory deficits in CRS patients have been attributed to nasal respiratory inflammation with diminished airflow and odorant access to the olfactory cleft, classifying them as conductive or transport defects [16]. Specifically, less air with its volatilized odorants would enter the nose in this setting, and the amount of odorants delivered to the olfactory mucosa may fall below detection threshold [2]. In essence, sinonasal disease as a cause for smell loss was believed to be analogous to ear wax as a cause for hearing deficits, wherein the end organs are intact and completely normal. It was believed that the olfactory epithelium was spared the inflammation seen in the respiratory regions of the nose of CRS patients, apparently remaining intact and unaffected. Lacking definitive data, the conductive hypothesis of CRS and smell loss was supported by three main factors. First, clinical observations indicated that olfactory deficits often rapidly improve following a burst of oral steroids (see below). It was presumed that a rapid response to treatment was most consistent with the reversal of a mechanical process, as any direct inflammatory damage to the neuroepithelium would require a significant amount of time for full recovery. However, the clinical observation of rapid olfactory recovery in this setting was anecdotal and not based on clinical testing, so both the precise time course and full extent of recovery are uncertain and probably highly variable. Secondly, the olfactory mucosa was believed to be an 'immunologically privileged' site similar to the eye, incapable of mounting a normal immune response to foreign proteins, in order to spare the neuroepithelium any associated inflammatory damage. Studies from the Getchell laboratory, however, demonstrated that the olfactory mucosa at least in rodents possessed the requisite 'synthetic and secretory capacity to respond with immune and non-immune mechanisms to pathogenic challenge' [17]. Furthermore, biopsies of the human olfactory mucosa confirmed this assessment, demonstrating a marked inflammatory response in the olfactory epithelium in the setting of CRS, but the clinical significance was unclear [18]. The third point in support of the 'conductive hypothesis' was the demonstration by the Jafek laboratory (see below) of at least some normal-appearing OSNs on electron microscopic studies of biopsies obtained from patients with polyps and anosmia. Until relatively recently, the conductive hypothesis of CRS and smell loss had remained unchallenged.

Historically, the first attempt to evaluate the mechanism of smell loss in CRS was by Jafek et al. [19], who reported on 2 patients with nasal polyps and anosmia. Both patients remained anosmic following surgery despite obvious improvement in their nasal airways so they were started on oral steroid regimens. Long-term correction of the anosmia was achieved in both patients through this combination of corticosteroids and surgery. Biopsies of the olfactory mucosa at the time of surgery in both patients showed at least some normal OSNs, supporting the theory of mechanical obstruction as the cause for olfactory dysfunction. The need for steroids, however, caused Jafek et al. to speculate that, rather than being only an obstructive phenomenon from polyps, anosmia at least in part resulted from the direct effects of inflammatory processes on the olfactory epithelium, the surface of the olfactory receptors, or the olfactory mucus bathing the receptors.

Biopsies from the olfactory epithelium taken from patients with smell loss after head trauma or acute viral insult have demonstrated obvious histological abnormalities in the neuroepithelium [20, 21]. The demonstration of damage to the olfactory mucosa in these entities confirms that the nature of the problem is, at least in part, sensory. Pathologic changes in the olfactory mucosa of patients with CRS had not been documented, and these disorders were therefore classified as conductive disorders as mentioned above. In 2000, olfactory biopsies were performed on 30 patients undergoing nasal surgery [16]. Nineteen of the 30 had actual olfactory mucosa on the biopsy sample. Nine of 19 patients demonstrated normal olfactory mucosa and normal olfactory function. Ten patients demonstrated pathologic changes in the olfactory mucosa with an influx of lymphocytes, macrophages, and eosinophils. Seven of these 10 patients had olfactory deficits as confirmed by the University of Pennsylvania Smell Identification Test (UPSIT). This was the first study emphasizing inflammatory changes in the olfactory mucosa of patients with sinonasal disease, and indicated that the pathological process present in the respiratory regions of the nose involved the olfactory mucosa as well. In addition, grading of the intensity of the inflammatory response within the olfactory mucosa was performed and generally showed that severer inflammatory changes occurred in patients with decreased UPSIT scores, further suggesting that pathology within the olfactory epithelium contributed to hyposmia and anosmia in these patients.

Theoretically, inflammation within the olfactory neuroepithelium may trigger smell loss by a variety of potential mechanisms. Mediators released by lymphocytes and macrophages are known to trigger hypersecretion in respiratory and Bowman's glands [22]. Olfactory mucus, produced by Bowman's glands, is a highly specialized substance vastly different from nasal respiratory mucus, serving a function analogous to cochlear endolymph [23, 24]. Hypersecretion would likely alter the ion concentrations of olfactory mucus, affecting the microenvironment of olfactory neurons and possibly the transduction process [25]. Moreover, the presence of these inflammatory cells in the olfactory mucosa provides a direct mechanism for the action of corticosteroids on anosmia. Type II corticosteroid receptors are found in these inflammatory cells and activation with a systemic glucocorticoid would rapidly suppress the local cytokine response [23, 25]. In addition to secretory effects, these same cytokines and mediators may be toxic to neurons [12, 26]. In particular, inflammatory mediators released by lymphocytes, macrophages, and eosinophils may trigger caspase-3 activation in OSNs [27]. Caspase-3 is the primary executioner caspase in mammalian tissues, and detection of the active form within a cell designates it for apoptotic proteolysis. Recent studies from our laboratory demonstrated that OSN death, demonstrated by caspase-3 activation, appeared to be a significant component of olfactory dysfunction in chronic sinusitis. Olfactory tissue of patients with a normal sense of smell despite CRS demonstrated increased caspase-3 activity when compared with normal human olfactory mucosa. Furthermore, olfactory tissue from patients with CRS *and* smell loss demonstrated a severer inflammatory response and much more extensive caspase-3 activity in both the olfactory epithelium as well as the nerve bundles, suggesting that increased OSN apoptosis was at least partly responsible for the smell deficits in rhinosinusitis patients [28].

Overall, histopathologic studies suggest that anosmia secondary to sinonasal disease involves direct effects on the olfactory epithelium (sensory disorder) in addition to any gross changes in the airflow to the olfactory cleft (transport disorder). The sensory component may potentially involve both alteration of mucus ion concentration as well as direct loss of olfactory neurons. The frequent clinical observation that both steroids and surgery are required for optimal management supports the position that smell loss in CRS patients is most often a mixed disorder.

Clinical Studies of Chronic Sinonasal Disease and Anosmia

The prevalence of olfactory dysfunction among patients with sinonasal disease has been well documented. The first quantitative, large-scale empirical study of smell in allergic rhinitis was performed by Cowart et al. [29] in 1993. Olfactory thresholds for phenylethyl alcohol were measured in 91 patients with symptoms of allergic rhinitis and 80 nonatopic patients. Olfactory thresholds were significantly higher in allergic patients than in control subjects, with 23.1% of the patients demonstrating smell loss. Clinical or radiographic evidence of sinusitis and/or nasal polyps was significantly associated with hyposmia. In addition, nasal resistance measurements using anterior rhinometry were performed to determine the degree to which nasal congestion contributed to hyposmia in these patients. Interestingly, although nasal resistance was significantly higher among patients than among controls, it was not related to olfactory threshold in either group. This finding suggested that even substantial obstruction of the lower nasal airway was not sufficient to produce a significant reduction in olfactory sensitivity, and left open the possibility that (1) other factors specific to the allergic process other than nasal congestion play a role in allergy-related hyposmia, and (2) measures of airway resistance do not reflect small areas of focal inflammation within the nasal passages that can potentially disrupt transport to the olfactory epithelium.

In 1995, Apter et al. [30] reported on 62 patients with olfactory loss from a broad spectrum of sinonasal disease. Endoscopic rhinoscopy was performed in order to better assess the impact of mechanical obstruction in the superior-posterior portions of the nasal cavity. The mean olfactory score (based on a composite of odor identification and detection tests) of 34 patients with obstruction of the olfactory cleft (by polyps or severe CRS) was consistent with anosmia. Twenty-eight patients with sinonasal disease and no associated gross anatomic obstruction of the olfactory cleft (most commonly allergic rhinitis alone) had an average olfactory score consistent with hyposmia. Although smell dysfunction was worse in patients with nasal polyps, the presence of hyposmia in the patients without gross obstructive findings suggested that inflammatory processes within the neuroepithelium may play a role in olfactory dysfunction.

Five years later, Simola and Malmberg [31] provided the most extensive study of olfactory dysfunction in chronic sinonasal disease. This study compared 105 rhinitis patients with 104 healthy controls to analyze possible relationships between sense of smell and rhinitis, age, sex, smoking, prick test results, nasal resistance, and history of paranasal surgery. Age and rhinitis were the only variables with a significant effect on the olfactory threshold in the whole series. These results were interpreted to suggest that even when age-related changes are considered, chronic nasal inflammation impairs the sense of smell. Interestingly, the nonallergic rhinitis patients' sense of smell was found to be poorer than that of the patients with seasonal or perennial allergic rhinitis. More recent work by Olsson et al. [32] has supported these results. In a study surveying over 10,000 adults to estimate the prevalence of self-reported allergic and nonallergic symptoms, Olsson et al. found the pervasiveness of olfactory dysfunction to be significantly higher among individuals with nonallergic rhinits than among those with allergic rhinitis or nonrhinitic individuals. More objective studies by Mann et al. [33] also demonstrate greater olfactory dysfunction among patients with nonallergic rhinits when compared with allergic rhinits.

Medical Treatment of Olfactory Dysfunction in Chronic Rhinosinusitis

In 1956, Hotchkiss [34] performed one of the first studies describing recovery of olfaction using systemic corticosteroids. Hotchkiss described his findings in 30 patients who suffered from 'massive' nasal polyposis and subjective, self-reported anosmia. All patients were treated with a total dose of 70 mg of prednisone over a 6-day period and reexamined on the seventh day. A dramatic polyp shrinkage response was reported, and restoration of olfactory function was found to be proportional to the amount of polyp shrinkage and unrelated to previous duration of olfactory loss. In patients in whom prednisone was discontinued, a reversal to the original anosmic state was noted in about 10 days.

Ten years after Hotchkiss presented his findings, Fein et al. [35] reported on 18 patients with nasal allergy and anosmia. Of these patients, the 14 that had additional pathology such as nasal polyposis and/or sinusitis were also found to have severer olfactory dysfunction. The analysis of patient outcomes showed that combined therapy of hyposensitization, antibiotics, steroids, and surgery afforded the greatest relief of anosmia for the longest time. Again, no objective measurement of olfaction was performed and, similar to Hotchkiss' results, all

18 patients treated experienced only a temporary relief of anosmia before their loss of smell recurred.

In contrast to the largely subjective assessment of smell disturbance prior to the 1980s, the development of clinically practical tests to evaluate olfaction has given clinicians methods to quantitatively assess smell dysfunction and evaluate the efficacy of various treatment modalities [36, 37]. In 1996, Golding-Wood et al. [38] evaluated the efficacy of topical steroid treatment in patients with rhinitis. Twenty-five patients with perennial rhinitis were included in the study, 15 of which initially expressed a weak sense of smell as a significant symptom. Olfactory tests were administered once before and once after 6 weeks of topical betamethasone treatment. Scores of each of the 15 patients with symptoms of hyposmia significantly improved after the steroid treatment, whereas the other 10 patients showed no objective olfactory improvement despite a significant decrease in the sensation of nasal obstruction. Posttreatment testing in both groups was still indicative of mild hyposmia despite the therapy.

One year later, Mott et al. [39] sought to determine the efficacy of topical corticosteroid nasal spray treatment in the head-down-forward position for severe olfactory loss associated with nasal and sinus disease. Thirty-nine patients were treated with flunisolide for at least 8 weeks, with concurrent antibiotic administration for any bacterial infection. Olfactory scores significantly improved following treatment, signs of nasal and sinus disease decreased, and 66% of patients reported subjective improvement in their sense of smell. Objective scores significantly improved following treatment for the group as a whole, including patients with and without nasal polyposis. Nine patients with olfactory function that had initially improved chose to continue the topical corticosteroid treatment regimen and returned for a second follow-up more than 6 months after starting treatment. For this subgroup, olfactory function did not decrease significantly from the mean posttreatment value. More recently, a double-blind, placebo-controlled, randomized prospective study evaluated the effects of topical steroid spray on olfactory performance in 24 patients with seasonal allergic rhinitis. Odor threshold measurements significantly improved after 2 weeks of treatment with mometasone furoate when compared with placebo. These results appeared to be independent of improvement in nasal airflow, suggesting that olfactory dysfunction in allergic rhinitis is primarily due to allergic inflammation rather than reduced nasal airflow [40]. A separate study, however, failed to confirm the long-term necessity of topical steroids for maintenance of this olfactory improvement [41]. Selection criteria for inclusion in this study may have been flawed, as a significant percentage of patients may have suffered from postviral hyposmia rather than olfactory deficits secondary to chronic sinonasal inflammation.

Oral corticosteroids have long been used for the management of CRS and smell loss and the effects are generally more substantial than those seen with topical analogues. In 1984, Goodspeed et al. [42] tested olfactory function in 20 anosmic and hyposmic patients before and after a 1-week course of systemic steroids. Ten patients had sinonasal disease, 4 had olfactory loss after upper respiratory infection, and 6 were considered idiopathic. Only 6 patients responded, all of who had nasal and sinus disease. In 1995, Ikeda et al. [43] documented olfactory function before and after systemic corticosteroid therapy in 12 patients with anosmia refractory to topical steroid treatment. Significant efficacy was achieved with a short course of high-dose oral corticosteroids in nonallergic sinus disease. On the other hand, anosmia induced by upper respiratory infection failed to respond to systemic steroid treatment. The authors speculated that the lack of improvement in patients after upper respiratory infection was due to permanent damage to olfactory receptor cells, while the effectiveness observed in patients with sinus disease was explained by improvement of mucosal thickening in the area of the olfactory cleft, leading to increased access of odorants to the olfactory epithelium. More recently, in 2001, Seiden and Duncan [44] presented a retrospective study of consecutive patients presenting with a primary complaint of olfactory loss. A 'pulse' dose of systemic steroids was used to support the diagnosis of olfactory dysfunction in a subset of the patients reviewed. Thirty-six patients received a tapering course of systemic steroids, and 30 (83%) experienced an improvement in their sense of smell. In contrast, only 13 of 52 patients (25%) who were given topical steroids noted any improvement. In their discussion, the authors conclude that while long-term systemic steroid treatment for anosmia may not be appropriate, a short course of high-dose therapy can be helpful for diagnosing a reversible 'conductive loss'.

Finally, recent studies have suggested a role for leukotriene receptor antagonist (LRA) therapy in the management of anosmia in CRS. In 2001, Wilson et al. [45] reported on 32 patients with CRS who were treated with an LRA (montelukast, 10 mg/day). Significant improvement in subjective scoring of sense of smell was reported at a median follow-up duration of 14 weeks. More recently, Otto et al. [46] presented their work supporting the role of LRAs in the medical management of anosmia related to CRS. In their study, 12 patients were treated with a combination of topical and systemic steroids, LRAs, and sinus surgery in necessary cases. The average UPSIT score increased by 17.5 points with a minimum follow-up of 1 year.

Surgical Treatment of Olfactory Dysfunction in Chronic Rhinosinusitis

Overall, studies suggest that the degree of olfactory loss is usually associated with the severity of sinonasal disease, with the greatest loss occurring in patients who have concurrent rhinosinusitis and polyposis. While smell loss improved in some patients treated with topical steroids, normal olfactory function was only rarely restored, implying either a permanent olfactory loss or failure to completely reverse the underlying cause. The most common operative procedures impacting on olfaction are performed in patients with CRS. The work by Jafek et al. [19] in 1987, discussed earlier, was one of the first reports of the influences of nasal surgery on smell function. In 1988, Seiden and Smith [47] continued to evaluate the role of surgery in sinusitis and olfactory dysfunction. Olfaction was tested in 5 patients before and after surgery. Preoperatively, patients ranged from anosmic to moderately hyposmic (average UPSIT score = 16.5). Four to eight weeks following surgery, all 5 patients showed significant improvement (average UPSIT score = 33.5). None of the patients in this study were treated with systemic steroids and long-term results were not available. Also in 1988, Leonard et al. [48] administered pre- and postoperative olfactory function testing to 25 patients with known olfactory dysfunction. Patients underwent unilateral or bilateral ethmoidectomy. Nine patients achieved normal status in one or both nostrils postsurgically, whereas 4 remained with mild hyposmia, and 5 with moderate to severe hyposmia. Seven patients showed no improvement. Surgery on one side of the nose appeared to significantly improve function in the contralateral nostril in some cases. The authors mentioned, however, that whether the contralateral improvement derived from bilateral release of obstruction or some more 'obscure mechanism' remained unclear. In 1989, Yamagishi et al. [49] provided one of the first reports of a relatively large number of patients undergoing sinus surgery, with evaluation of the sense of smell to assess outcome. Twenty patients who had olfactory dysfunction caused by localized inflammation of the ethmoid sinuses were studied before and after bilateral ethmoidectomy. At 6 months after surgery, the improvement rate was 70% subjectively and 80% on olfactometry (T&T olfactometry and Alinamin test). Yamagishi et al. [49] found that when inflammation was localized in the ethmoid sinuses, there may be no severe nasal symptoms despite significant olfactory disturbance. They theorized that chronic ethmoidal sinusitis triggered obstruction of the olfactory cleft from local inflammatory changes in the respiratory epithelium. As in the study by Seiden and Smith [47], no component of steroid therapy was included in this study population. Several years later, Hosemann et al. [50] described the preoperative and postoperative results of a 'qualitative and semi-quantitative olfactory function test' on 111 patients with chronic polypoid ethmoiditis, 78% of whom required complete ethmoidectomy. Before surgery, 39 patients (35%) had normal olfactory function, 34 patients (31%) were hyposmic, and 38 patients (34%) suffered from anosmia. Postoperatively, 89 patients (80%) had normal smell, 13 patients (12%) showed hyposmia, and 9 patients (8%) experienced anosmia. No patients, whose sense of smell before surgery was normal, deteriorated. The authors concluded that the 'mechanical viability of the olfactory cleft played the main causative role' in olfactory disturbance. In addition, the immediate relief after surgery was interpreted to mean 'that an inflammatory affection of the sense organ itself could not be responsible' for olfactory dysfunction. In 1994, Eichel [51] tested olfaction preoperatively and then at 6-month intervals in the postoperative period in 10 anosmic patients with advanced obstructive bilateral nasal polyposis and pansinusitis. The procedures performed included bilateral nasal polypectomies, sphenoethmoidectomies, and nasal antral windows. On the seventh postoperative day, topical nasal corticosteroid sprays were initiated and continued indefinitely. Olfactory improvement was recorded in 7 of the 10 patients, although postoperative scores were still indicative of olfactory dysfunction.

In 1994, Lund and Scadding [52] reported on 50 hyposmic patients in a series of 200 patients with long-term follow-up. All patients had symptoms of CRS and had failed conservative medical management, which included intranasal steroids, antibiotics, antihistamines, and allergen avoidance. The endoscopic procedure included uncinectomy, anterior ethmoidectomy, and perforation of the ground lamella of the middle turbinate in all cases, with posterior ethmoidectomy, sphenoidectomy, clearance of the frontal recess and enlargement of the maxillary ostium as required. Patients continued with intranasal steroid therapy up to the time of surgery and for at least 3 months postoperatively. After surgery, with a mean follow-up of 2.3 years for the 200 patients, a significant olfactory improvement was detected in the 50 hyposmics. Again, however, the average postoperative results were still indicative of marked hyposmia. One year later, Min et al. [53] assessed changes in olfaction using butanol thresholds before and after sinus surgery in 80 patients with CRS. Overall, the percentage of patients with impaired olfactory function decreased from 78 to 64% 12 months following endoscopic sinus surgery. Although preoperative butanol threshold scores were significantly lower as the severity of sinusitis increased (graded by CT scan results), the degree of postoperative improvement showed no significant correlation with the severity of sinusitis. The study did not mention to what extent medical treatment was employed following surgery. That same year, El Nagger et al. [54] attempted to quantitatively assess the effect of steroid nasal spray on olfaction after nasal polypectomy. Twenty-nine patients with bilateral nasal polyps received a 6-week course of beclomethasone nasal spray following intranasal polypectomy. Pre- and postoperative UPSIT scores were obtained for each nostril separately, and no significant difference between treated and untreated nostrils was found. The postoperative smell function for either nostril was in the anosmic to severely hyposmic range. Downey et al. [55] provided further evidence that surgical treatment of patients with CRS and anosmia had only incomplete effects on olfactory sensation. In their study, 50 patients with subjective anosmia and varying degrees of rhinosinusitis underwent surgical treatment. Postoperatively,

24 (48%) patients had an 'unimproved' olfactory status despite satisfactory resolution of other complaints. In this study, the extent of mucosal disease (stage) [56] was a reliable prognostic indicator for improvement in olfaction. Increasingly widespread mucosal disease was associated with significantly lower success rates in alleviating anosmia. Postoperative endoscopic findings of persistent polypoid mucosa strongly correlated with unresolved olfactory disturbances. No attempt was made in this series to treat unresolved postoperative anosmia with corticosteroids.

In 1997, Klimek et al. [57] provided further evidence that olfactory function after sinus surgery in patients with nasal polyposis is only transiently improved. Olfactory function testing was performed in 31 patients with nasal polyposis 1–3 days before endoscopic sinus surgery and at 6 postoperative times. The study demonstrated severe hyposmic changes preoperatively, best olfactory recovery (mild hyposmia) occurring at approximately 3 months after surgery, and a decrease in olfactory function to the preoperative hyposmic state between months 3 and 6. Additionally, wound healing and mucosal status were evaluated endoscopically with particular attention to signs of inflammation, crust formation, and secretion in the area of the olfactory cleft. Although mechanical obstruction appeared to explain early postoperative decrease in olfactory function, the mucosal status seemed unlikely to be the only reason for the late decrease between months 3 and 6 following surgery. The authors speculated that 'other mechanisms like changes in composition and function of the olfactory mucus… and dysfunction of the olfactory receptor cells caused by toxic inflammatory mediators' might partially explain postoperative hyposmia.

Also in 1997, Rowe-Jones and Mackay [58] prospectively collected data on 115 patients before and 6 weeks after endoscopic sinus surgery with adjuvant medical treatment for CRS. All patients received a postoperative 3-week prednisolone taper (30 mg per day for 1 week, 20 mg per day for 1 week and 10 mg per day for 1 week) and 2 weeks of co-amoxiclav. Ninety patients (87%) with decreased olfaction preoperatively had subjective improvement. A visual analogue, patient-rated symptom score improved in 94 (82%) patients. Acoustic rhinometry was performed pre- and postoperatively in 96 patients and improvement in olfactory symptom scores was found to correlate with increase in nasal volume.

In 1998, Delank and Stoll [59] evaluated odor detection thresholds in 115 patients suffering from CRS before and after endoscopic sinus surgery. Preoperatively, only 58% of the patients complained of subjective olfactory deficits, however, olfactory threshold testing found 83% to be either hyposmic (52%) or anosmic (31%). Despite improvements in 70% of patients after surgery, normosmia was achieved in only 25% of the hyposmics and 5% of the anosmics. There was no mention to what extent medical treatment was employed following surgery. As in prior studies, the authors noted that the extent of sinus disease as measured by the degree of nasal polyposis correlated with levels of preoperative olfactory dysfunction, and that the rate of improvement following surgery was generally lower than assumed.

Summary of Current Therapy

As described, a large number of clinical studies of variable quality have sought to determine the efficacy of standard therapy in the management of the olfactory component of sinonasal disease. Although improvement in olfaction is often possible, it is frequently transient and incomplete. In addition to antibiotics and surgery, both systemic and topical steroids are helpful in attempting to alleviate olfactory dysfunction in this setting. While systemic steroids are usually more effective than topically administered steroids, prescription of systemic steroids over an extended period is usually unwarranted and places the patient at risk for side effects including gastric ulceration, diabetes, and osteoporosis [60]. Instead, it has been suggested that systemic steroids can be used as an effective diagnostic tool to help determine if a patient has any functioning olfactory mucosa, at which point therapy is continued with locally administered steroids. Repeated administration of short courses of systemic steroids with a long enough interval between courses to avoid untoward side effects may also be effective [61].

The mechanism of olfactory dysfunction in CRS remains controversial. As mentioned above, some investigators believe that obstruction of the olfactory cleft via polyps or edema is the sole significant cause of smell loss in this setting. Furthermore, the often rapid response to treatment described with both corticosteroids and surgery supports this hypothesis. This rapid response is unlikely to result in normal olfactory sensation, however, and most reports of 'immediate' return are subjective or anecdotal. The histopathologic data, on the other hand, support the concept that direct injury to the neuroepithelium is a component of the problem in addition to any superimposed obstruction. The common clinical observation of persistent smell loss despite adequate medical or surgical treatment of other sinonasal complaints also supports this alternative hypothesis. Overall, it is the authors' opinion that the weight of current evidence supports the theory that olfactory dysfunction in the setting of sinonasal disease is a mixed problem, with varying degrees of conductive and sensory losses in individual patients.

Future Therapy

The fundamental limitation of current therapy for CRS and smell loss is that while surgery and corticosteroids can effectively treat the mechanical or obstructive components of sinonasal disease in most cases, it is often impossible to alter the underlying process of mucosal inflammation. While this residual inflammation in the respiratory regions of the nose is more often minimally symptomatic after surgery, persistent inflammation in the olfactory cleft may result in smell loss. The reason(s) for the mucosal inflammation is/are unclear, as the fundamental etiologies of CRS remain obscure. Recent reports have implicated fungi or bacterial superantigens as primary agents in CRS, but definitive evidence and subsequent therapeutic options are lacking. Antibiotics and antifungals have some apparent efficacy but the inflammatory triggers are likely multiple, and may reflect defects of the innate mucosal immune system. Progress in this area will likely go a long way toward more effective treatment of the smell loss component of CRS. Therapy directed at the olfactory mucosa in cases of sinusitis and smell loss also holds some promise. As discussed earlier, increased OSN apoptosis has been implicated as a contributing mechanism responsible for the smell deficits in rhinosinusitis patients. Furthermore, OSN apoptosis may be important in a wide array of olfactory disorders including age-related and postviral anosmia [28]. Antiapoptotic drugs are the subject of a number of current investigational trials in acute and chronic neurodegenerative disorders including Parkinson's disease, stroke and spinal cord trauma. The established capacity of the olfactory epithelium for regeneration makes it a particularly attractive target for antiapoptotic therapy. One drug in particular, the tetracycline analogue minocycline, has both antibiotic and antiapoptotic properties making it an intriguing choice as a drug for the treatment of rhinosinusitis and smell loss [62]. Minocycline is well tolerated in the chronic treatment of acne but it is currently unknown whether this drug will improve smell in sinusitis patients. Histologic studies from our laboratory have demonstrated inhibition of experimentally induced OSN apoptosis (axotomy and bulbectomy) in mice treated with minocycline [63]. Electrical olfactory recovery in the same experimental population occurred more rapidly in minocycline-treated mice, suggesting that these OSNs remain viable and capable of participating in the recovery process [64]. It remains unclear whether this will impact OSN loss in CRS. Nevertheless, antiapoptotic drugs are likely to play a future role in the treatment of neurologic diseases in general, including possibly disorders of smell.

References

1 Benninger MS, Ferguson BJ, Hadley JA, Hamilos DL, Jacobs M, Kennedy DW, Lanza DC, Marple BF, Osguthorpe JD, Stankiewicz JA, Anon J, Denneny J, Emanuel I, Levine H: Adult chronic rhinosinusitis: definitions, diagnosis, epidemiology, and pathophysiology. Otolaryngol Head Neck Surg 2003;129(3 suppl):S1–S32.

2 Doty RL: Olfaction; in Lilholdt T (ed): Nasal Polyposis: An Inflammatory Disease and Its Treatment. Copenhagen, Munksgaard, 1997, pp 153–159.

3 Cullen MM, Leopold DA: Disorders of smell and taste. Med Clin North Am 1999;83:57–74.

4 Seiden AM, Duncan HJ: The diagnosis of a conductive olfactory loss. Laryngoscope 2001;111: 9–14.
5 Loury MC, Kennedy DW: Chronic sinusitis and nasal polyposis; in Getchell TV, Bartoshuk LM, Doty RL, Snow J (eds): Smell and Taste in Health and Disease. New York, Raven Press, 1991, pp 711–730.
6 Caplin I, Haynes JT, Spahn J: Are nasal polyps an allergic phenomenon? Ann Allergy 1971;29: 631–634.
7 Patriarca G, Romano A, Schiavino D, Venuti A, Di Rienzo V, Fais G, Nucera E: ASA disease: the clinical relationship of nasal polyposis to ASA intolerance. Arch Otorhinolaryngol 1986;243: 16–19.
8 Delaney JC: Aspirin idiosyncracy in patients admitted for nasal polypectomy. Clin Otolaryngol 1976;1:27–30.
9 Rudach C, Sachse F, Alberty J: Chronic rhinosinusitis – Need for further classification? Inflamm Res 2004;53:111–117.
10 Magrassi L, Graziadei PP: Cell death in the olfactory epithelium. Anat Embryol (Berl) 1995;192: 77–87.
11 Farbman AI: Cell Biology of Olfaction. Developmental and Cell Biology Series No 27. New York, Cambridge University Press, 1992.
12 Holcomb JD, Graham S, Calof AL: Neuronal homeostasis in mammalian olfactory epithelium: a review. Am J Rhinol 1996;10:125–134.
13 Robinson AM, Conley DB, Shinners MJ, Kern RC: Apoptosis in the aging olfactory epithelium. Laryngoscope 2002;112:1431–1435.
14 Snow JB: Causes of olfactory and gustatory disorders; in Getchell TV, Bartoshuk LM, Doty RL, Snow J (eds): Smell and Taste in Health and Disease. New York, Raven Press, 1991, pp 445–449.
15 Doty RL, Shaman P, Kimmelman CP, Dann MS: University of Pennsylvania Smell Identification Test: a rapid quantitative olfactory function test for the clinic. Laryngoscope 1984;94:176–178.
16 Baroody F, Naclerio R: Allergic rhinitis; in Getchell TV, Bartoshuk LM, Doty RL, Snow J (eds): Smell and Taste in Health and Disease. New York, Raven Press, 1991, pp 731–740.
17 Mellert TK, Getchell ML, Sparks L, Getchell TV: Characterization of the immune barrier in human olfactory mucosa. Otolaryngol Head Neck Surg 1992;106:181–188.
18 Kern RC: Chronic sinusitis and anosmia: pathologic changes in the olfactory mucosa. Laryngoscope 2000;110:1071–1077.
19 Jafek BW, Moran DT, Eller PM, Rowley JC, Jafek TB: Steroid-dependent anosmia. Arch Otolaryngol 1987;113:547–549.
20 Jafek BW, Hartman D, Eller P, Johnson EW, Strahan RC, Moran DT: Post-viral olfactory dysfunction. Am J Rhinol 1990;4:91–100.
21 Yamagashi M, Okazoe R, Ishizuka Y: Olfactory mucosa of patients with olfactory disturbance following head trauma. Ann Otol Rhinol Otolaryngol 1994;103:218–226.
22 Getchell M, Mellert T: Olfactory mucus secretion; in Getchell TV, Bartoshuk LM, Doty RL, Snow J (eds): Smell and Taste in Health and Disease. New York, Raven Press, 1991, pp 445–449.
23 Robinson AM, Kern RC, Foster JD, Fong KJ, Pitovski DZ: Expression of glucocorticoid receptor mRNA and protein in the olfactory mucosa. Laryngoscope 1999;109:383–388.
24 Joshi H, Getchell ML, Zielinski B, Getchell TV: Spectrophotometric determination of cation concentrations in olfactory mucus. Neurosci Lett 1987;82:321–326.
25 Kern RC, Foster JD, Pitovski DZ: Glucocorticoid (type II) receptors in the olfactory mucosa of the guinea pig: RU 28362. Chem Senses 1997;22:313–319.
26 Nakashima T, Kimmelman C, Snow JB: Immunohistopathology of human olfactory epithelium, nerve, and bulb. Laryngoscope 1985;95:391–398.
27 Ge Y, Tsukatani T, Nishimura T, Furukawa M, Miwa T: Cell death of olfactory receptor neurons in a rat with nasosinusitis infected artificially with staphylococcus. Chem Senses 2002;27:521–527.
28 Kern RC, Conley DB, Haines GK 3rd, Robinson AM: Pathology of the olfactory mucosa: implications for the treatment of olfactory dysfunction. Laryngoscope 2004;114:279–285.
29 Cowart BJ, Flynn-Rodden K, McGeady SJ, Lowry LD: Hyposmia in allergic rhinitis. J Allergy Clin Immunol 1993;91:747–751.

30 Apter AJ, Mott AE, Frank ME, Clive JM: Allergic rhinitis and olfactory loss. Ann Allergy Asthma Immunol 1995;75:311–316.
31 Simola M, Malmberg H: Sense of smell in allergic and nonallergic rhinitis. Allergy 1998;53: 190–194.
32 Olsson P, Berglind N, Bellander T, Stjarne P: Prevalence of self-reported allergic and non-allergic rhinits symptoms in Stockholm: relation to age, gender, olfactory sense, and smoking. Acta Otolaryngol 2003;123:75–80.
33 Mann SS, Maini S, Nageswari KS, Mohan H, Handa A: Assessment of olfactory status in allergic and non-allergic rhinitis patients. Indian J Physiol Pharmacol 2002;46:186–194.
34 Hotchkiss WT: Influence of prednisone on nasal polyposis with anosmia. Arch Otolaryngol 1956;64:478–479.
35 Fein B, Kamin P, Fein NS: The loss of smell in nasal allergy. Ann Allergy 1966;24:278–283.
36 Doty RL, Smith R, McKeown DA, Raj J: Development of the University of Pennsylvania Smell Identification Test: a standardized microencapsulated test of olfactory function. Physiol Behav 1984;32:489–502.
37 Kobal G, Hummel T, Sekinger B, Barz S, Roscher S, Wolf S: 'Sniffin' sticks': screening of olfactory performance. Rhinology 1996;34:222–226.
38 Golding-Wood DG, Holmstrom M, Darby Y, Scadding GK, Lund VJ: The treatment of hyposmia with intranasal steroids. J Laryngol Otol 1996;110:132–135.
39 Mott AE, Cain WS, Lafreniere D, Leonard G, Gent JF, Frank ME: Topical corticosteroid treatment of anosmia associated with nasal and sinus disease. Arch Otolaryngol Head Neck Surg 1997;123: 367–372.
40 Stuck BA, Blum A, Hagner AE, Hummel T, Klimek L, Homann K: Mometasone furoate nasal spray improves olfactory performance in seasonal allergic rhinitis. Allergy 2003;58:1195.
41 Blomqvist EH, Lundblad L, Bergstedt H, Stjarne P: Placebo-controlled, randomized, double-blind study evaluating the efficacy of fluticasone proprionate nasal spray for the treatment of patients with hyposmia/anosmia. Acta Otolaryngol 2003;123:862–868.
42 Goodspeed RB, Gent JF, Catalanotto FA, Cain WS, Zagraniski RT: Corticosteroids in olfactory dysfunction; in Meiselma HL, Rivlin RS (eds): Clinical Measurement of Taste and Smell. New York, Macmillan, 1986, pp 514–518.
43 Ikeda K, Sakurada T, Takasaka T: Efficacy of systemic corticosteroid treatment for anosmia with nasal and paranasal sinus disease. Rhinology 1995;33:162–165.
44 Seiden AM, Duncan HJ: The diagnosis of a conductive olfactory loss. Laryngoscope 2001;111: 9–14.
45 Wilson A, White P, Gardiner Q, Nassif R, Lipworth B: Effects of leukotriene receptor antagonist therapy in patients with chronic rhinosinusitis in a real life rhinology clinical setting. Am J Rhinol 2001;39:142–146.
46 Otto K, Lang E, DelGaudio J, Venkatraman G: Long-term medical management of anosmia related to chronic sinusits. Am Rhinol Soc Meet, New York, 2004.
47 Seiden AM, Smith DV: Endoscopic intranasal surgery as an approach to restoring olfactory function. Chem Senses 1988;13:736.
48 Leonard G, Cain W, Clavet G: Surgical correction of olfactory disorders. Chem Senses 1988;13:708.
49 Yamagishi M, Hasegawa S, Suzuki S, Nakamura H, Nakano Y: Effects of surgical treatment of olfactory disturbance caused by localized ethmoiditis. Clin Otolaryngol 1989;14:405–409.
50 Hosemann W, Goertzen W, Wohlleben R, Wolf S, Wigand ME: Olfaction after endoscopic endonasal ethmoidectomy. Am J Rhinol 1993;7:11–15.
51 Eichel BS: Improvement of olfaction following pansinus surgery. Ear Nose Throat J 1994;73: 248–250.
52 Lund VJ, Scadding GK: Objective assessment of endoscopic sinus surgery in the management of chronic rinosinusitis: an update. J Laryngol Otol 1995;109:941–944.
53 Min Y-G, Yun Y-S, Song BH, Cho YS, Lee KS: Recovery of nasal physiology after functional endoscopic sinus surgery: olfaction and mucociliary transport. ORL J Otorhinolaryngol Relat Spec 1995;57:264–268.

54 el Naggar M, Kale S, Aldren C, Martin F: Effects of Beconase nasal spray on olfactory function in post-nasal polypectomy patients: a prospective controlled trial. J Laryngol Otol 1995;109: 941–944.
55 Downey LL, Jacobs JB, Lebowitz RA: Anosmia and chronic sinus disease. Otolaryngol Head Neck Surg 1996;115:24–28.
56 Kennedy D: Prognostic factors, outcomes and staging in ethmoid sinus surgery. Laryngoscope 1992;102(Suppl 51):1–18.
57 Klimek L, Moll B, Amedee RG, Mann WJ: Olfactory function after microscopic endonasal surgery in patients with nasal polyps. Am J Rhinol 1997;11:251–255.
58 Rowe-Jones J, Mackay I: A prospective study of olfaction following endoscopic sinus surgery with adjuvant medical treatment. Clin Otolaryngol 1997;22:377–381.
59 Delank KW, Stoll W: Olfactory function after functional endoscopic sinus surgery for chronic sinusitis. Rhinology 1998;36:15–19.
60 Scott AE: Caution urged in treating 'steroid-dependent anosmia'. Arch Otolaryngol Head Neck Surg 1989:115;109–110.
61 Wolfensberger M, Hummel T: Anti-inflammatory and surgical therapy of olfactory disorders related to sino-nasal disease. Chem Senses 2002:27;617–622.
62 Friedlander R: Apoptosis and caspases in neurodegenerative diseases. N Engl J Med 2003:348; 1365–1375.
63 Kern RC, Conley DB, Haines GK, Robinson AM: Treatment of olfactory dysfunction. 2. Studies with minocycline. Laryngoscope 2004:114;2200–2204.
64 Kern RC, Raviv JR, Conley DB, Robinson AM: Recovery after olfactory axotomy in mice: pharmacologic and genetic effects. 27th Annu Meet Assoc Chemorecept Sci, Sarasota, 2005.

Robert C. Kern, MD
Department of Otolaryngology – Head and Neck Surgery, Suite 15–200
657 N. St. Clair Street
Chicago, IL 60611 (USA)
Tel. +1 312 695 0805, Fax +1 312 695 7851, E-Mail r-kern@northwestern.edu

Hummel T, Welge-Lüssen A (eds): Taste and Smell. An Update.
Adv Otorhinolaryngol. Basel, Karger, 2006, vol 63, pp 125–132

Olfactory Disorders following Upper Respiratory Tract Infections

Antje Welge-Lüssen, Markus Wolfensberger

Department of Otorhinolaryngology, University of Basel, Basel, Switzerland

Abstract

Postviral olfactory disorders usually occur after an upper respiratory tract infection (URTI) associated with a common cold or influenza. With a prevalence between 11 and 40% they are among the common causes of olfactory disorders. Women are more often affected than men and post-URTI disorders usually occur between the fourth and eighth decade of life. The exact location of the damage in post-URTI is not yet known even though from biopsies a direct damage of the olfactory receptor cells is very likely. Nevertheless, central mechanisms cannot completely be ruled out. The diagnosis is made according to the history, clinical examination and olfactory testing. Affected patients usually recall the acute URTI and a close temporal connection should be present to establish the diagnosis. Spontaneous recovery might occur within 2 years. So far, no effective therapy exists even though specific olfactory training might be promising.

Definition of Postviral Olfactory Disorders

Postviral olfactory disorders were described for the first time more than 20 years ago [1, 2]. The most frequent cause of postviral smell loss is an upper respiratory tract infection (URTI) typically associated with the common cold or influenza. In such cases, there is a close temporal connection between the subsiding of the URTI and the development of olfactory disorders [3, 4].

Postviral olfactory disorders tend to be transient, but certain symptoms may be irreversible resulting in permanent parosmia, phantosmia, hyposmia, or anosmia [5]. Often, the respiratory infection is described by the affected patient as 'severe' [3] or at least as severer than usual [6]. After subsiding of the URTI symptoms, the olfactory disorder persists [7].

Although onset of the olfactory disorder is typically sudden, many patients postpone medical consultation assuming that their sense of smell will return. Nevertheless, patients tend to remember the causative URTI episode and ensuing loss of smell, which greatly facilitates the diagnosis of postviral olfactory disorder [6]. On the other hand, the correct diagnosis may be complicated by the presence of rhinitis or sinusitis in addition to URTI, since these conditions may also cause olfactory disorders. If the cause of an olfactory disorder is URTI, there will be no persisting nasal symptoms other than the loss of smell. Nevertheless, patients suffering from sinusitis may also have suffered a viral episode causing their olfactory impairment. Therefore, careful nasal examination including endoscopy is mandatory for the correct diagnosis of postviral olfactory disorder.

Epidemiology

Epidemiological data in the literature vary considerably. According to a survey conducted by Damm et al. [8], approximately 11% of olfactory disorders treated at German, Austrian, and Swiss university hospitals are caused by URTI. In other studies, the prevalence of olfactory disorders caused by URTI is stated to be as high as 20–40% [3, 9–11]. These discrepant data are thought to be due to the variable patient populations studied. While the survey by Damm et al. [8] focused on general ear, nose, and throat clinics, other studies were conducted in specialized smell and taste centers that would be expected to see a different patient population. Women are more often affected than men [12–14], and this disorder typically occurs between the fourth and eighth decade of life [14, 15].

Pathogenesis

The exact location and nature of the damage in olfactory disorders caused by URTI are not fully understood. Possible mechanisms are direct damage to the olfactory epithelium or to central pathways leading to retrograde degeneration [6]. Moreover, the virus itself may damage the olfactory receptor cells, or the damage may result from immune responses to the virus. Although the causative virus has not yet been identified, it is known that responsible viruses give rise to the common cold and/or neural symptoms. Viruses suspected to cause olfactory impairment are influenza virus, parainfluenza virus, respiratory syncytial virus, coxsackievirus, adenovirus, poliovirus, enterovirus, and herpesvirus.

Sugiura et al. [14] analyzed the monthly incidence of postviral olfactory disorders and the monthly incidence of virus isolation. Based on the antibody titers of affected patients, the authors suggested parainfluenza virus type 3 as the virus responsible for olfactory disorders. However, since this virus is widespread in the general public and often causes recurrent infections, it remains speculative whether this is the only causative virus.

A retrospective study by Konstantinidis et al. [16] showed that the incidence of olfactory disorders after URTI exhibits seasonal fluctuations, with the months March and May showing the highest incidences. The first peak in March appears to correlate with the peak occurrence of influenza, while the second peak in May could be due to climate conditions, such as low humidity and high temperatures, rendering the nasal epithelium particularly susceptible towards certain viral infections.

Biopsies have shown that direct damage to the peripheral epithelium is the most likely cause of olfactory disorders after URTI. Nevertheless, a central mechanism cannot be ruled out as the olfactory epithelium is directly connected with the brain thus providing a route for viruses to penetrate the brain [17]. In animals, experimental intranasal infection with influenza A virus led to apoptosis of the olfactory epithelium but did not result in death, while injection of the virus into the olfactory bulb led to a spreading of the anatomically connected brain areas with subsequent death of all treated animals [18]. It was therefore suggested that virus-induced apoptosis may be a protective response of the host [19]. Moreover, certain viruses (e.g. herpes simplex virus type 1, corona mouse hepatitis virus, rabies virus) affect central olfactory pathways after intranasal inoculation [20, 21].

In a study in 9 patients with sporadic Creutzfeldt-Jakob disease, an infectious prion protein was found after death in the olfactory cilia and olfactory pathway, but not in the respiratory epithelium [22]. The authors proposed that the olfactory pathway constitutes a route of infection and spreading of prions. The same group identified a protease-resistant prion protein in an olfactory biopsy taken only 45 days after disease onset, suggesting that involvement of the olfactory epithelium is an early event in sporadic Creutzfeldt-Jakob disease [23].

Histopathology

Based on histological findings from 17 patients with postviral olfactory disorders, Jafek et al. [24] established that anosmic patients had very few ciliated olfactory receptor cells. Moreover, the dendrites of the few ciliated olfactory receptor cells usually did not reach the epithelial surface. In contrast, biopsies of hyposmic patients contained a larger number of olfactory receptors, and some of the dendrites possessed sensory cilia while others did not. The authors concluded that the peripheral receptor damage in postviral olfactory disorders directly correlates with the degree of olfactory acuity.

However, Yamagishi et al. [25] were not able to correlate the number of olfactory receptor cells in biopsies from 13 patients with postviral olfactory disorders with the olfactory acuity measured by T&T olfactometry. Instead, they found a correlation between subsequent olfactory improvement and the number of olfactory receptor cells and intact nerve bundles. The authors suggested that in patients with reversible olfactory impairment, cells were only partially injured, e.g. showing damaged olfactory vesicles or olfactory cilia.

In later studies, the histological findings of a markedly disorganized epithelium, very few dendrites often not reaching the surface, and frequent junctions of the olfactory and respiratory epithelium were confirmed [26]. The authors described the excessively patchy distribution of the olfactory epithelium as 'checkerboard-like'.

Overall, replacement of the olfactory epithelium with the respiratory epithelium might take place, and olfactory receptors are reduced in number [27].

Clinical Examination

History

Affected patients tend to present at the clinic after their common cold has disappeared. While impaired smelling ability is commonly experienced during an acute cold [28, 29], postviral olfactory disorder typically persists after the acute symptoms have disappeared. Patients present with either quantitative disorders (hyposmia or anosmia) or qualitative disorders (phantosmia, parosmia). The incidence of qualitative disorders ranges from 10% [13] to 50% [1], 65% [30], and up to 70% [31]. Yamagishi et al. [25] observed dysosmia in 12.9% of patients whose biopsies showed slightly or moderately impaired olfactory mucosa, but not in those with completely destroyed olfactory mucosa. The authors suggested that dysosmia can be attributed to impaired residual receptor cells that are still connected to higher olfactory centers.

Patients usually recall the acute URTI, and a close temporal connection between the infection and first manifestation of smell loss should be established to diagnose postviral olfactory disorder [4]. When recording the medical history of the patient, it is important to elicit any concomitant medications, especially those taken to treat the cold, since a number of medications (including antibiotics) may cause olfactory dysfunction themselves [32, 33]. Sometimes, patients self-diagnose an olfactory disorder after URTI, but careful inquiry may bring to light that there has been a period of normal smell ability after the URTI. These cases should be considered as sinunasal olfactory disorders and might be treated with systemic steroids. Therefore, the exact medical history is of primary importance to the correct diagnosis.

Nasal Examination

In addition to establishing the exact medical history, a nasal examination including nasal endoscopy should be performed. It is important to rule out intranasal pathology such as polyps or tumors. Endoscopy should be performed with vasoconstriction and, if possible, the olfactory cleft should be viewed. In most cases, nasal examination is unremarkable showing no typical signs.

Neurological evaluation, including examination of other cranial nerves, should be performed to detect abnormalities suggestive of intracranial disease. In case of any doubts or if other neurological symptoms exist and the onset of the disorder cannot clearly be related to an URTI, we recommend to perform a magnetic resonance tomogram to rule out any intracranial pathology.

Olfactory Testing

Olfactory function is usually measured by threshold and/or odor identification tasks and odor discrimination tasks using validated olfactory tests, such as the University of Pennsylvania Smell Identification Test [34] or the 'Sniffin' Sticks' test battery [35]. In patients with postviral olfactory disorders, the degree of smell loss is usually less severe than in patients with head trauma [36], and is mostly partial rather than complete [13]. Nevertheless, typical patterns of olfactory test results that would unambiguously identify URTI-related olfactory disorders are still missing.

Prognosis

Spontaneous recovery of olfactory performance occurs in about one third of patients with postviral olfactory disorders [37]. Out of 262 patients, 32% improved within the first 13 months [38]. Recovery usually starts within the first 6 months after the infection and occurs more often in younger patients than in the elderly [37]. However, prognosis for the individual patient is difficult to make, although the highest chance of recovery is within the first 2 years. The longer the disorder has been persisting, the less likely is a recovery [38].

To predict the outcome of postviral olfactory disorders, Yamagishi et al. [25] proposed the use of biopsy results combined with intravenous Alinamin injection. Patients whose mucosal biopsies contained many receptor cells and intact nerve bundles had the highest chance of recovery. Duncan and Seiden [36] monitored 21 patients with URTI-related smell loss for 3 years. After this period, 19 patients had a significantly improved score in the University of

Pennsylvania Smell Identification Test, and 13 patients reported subjective improvement of olfactory performance. However, most patients were not anosmic initially, and improvement was modest only [7, 36]. Moreover, the authors pointed out that the percentage improvement quoted depends on the time point of assessing olfactory function. Mott and Leopold [12] reported improved olfactory function in 15% of patients suffering from postviral olfactory disorders who were reevaluated after 26 months, while Duncan and Seiden [36] reevaluated their patients after at least 36 months.

Therapy

At present, no effective therapy exists, but several medications and supplements have been used. Although zinc was thought to be an effective agent, a well-designed, double-blind, placebo-controlled study showed no benefit at all [39, 40]. Treatment with α-lipoid acid seemed to be promising initially. In an open, prospective study, patients (n = 23, 19 hyposmic patients, 4 functionally anosmic patients) received α-lipoid acid (600 mg/day) for 4.5 months on average [41]. Six patients experienced mild improvement and 8 patients showed clear improvement of olfactory performance. However, a subsequent double-blind study in approximately 140 patients did not confirm these results [Hummel T., pers. commun.].

Spontaneous recovery and regeneration are common in postviral olfactory disorders and may occur up to 2 years after viral exposure [36]. In parosmic patients, intranasal injection of hydrocortisone was used in the past [42], but newer data are missing. Specific olfactory training, applied twice a day over a period of 3 months, appears to be promising in promoting regeneration of olfactory function [43].

References

1 Henkin RI, Larson AL, Powell RD: Hypogeusia, dysgeusia, hyposmia and dysosmia following influenza-like infection. Ann Otol 1975;84:672–685.
2 Bendar M: Anosmie und Influenza. Med Klin 1930;48:1787–1789.
3 Duncan HJ: Postviral olfactory loss; in Seiden AM (ed): Taste and Smell Disorders. New York, Thieme, 1997, pp 72–78.
4 Riechstörungen – Leitlinie zur Epidemiologie, Pathophysiologie, Klassifikation, Diagnose und Therapie. http://www.uni-duesseldorf.de/AWMF/II/hno_II50.htm. 2004.
5 Doty RL: A review of olfactory dysfunction in man. Am J Otolaryngol 1979;1:57–79.
6 Seiden AM: Postviral olfactory loss. Otolaryngol Clin North Am 2004;37:1159–1166.
7 Murphy C, Doty RL, Duncan HJ: Clinical disorders of olfaction; in Doty RL (ed): Handbook of Olfaction and Gustation. New York, Dekker, 2003, pp 461–478.
8 Damm M, Temmel A, Welge-Luessen A, Eckel HE, Kreft M-P, Klussmann JP, et al: Epidemiologie und Therapie von Riechstörungen in Deutschland, Österreich und der Schweiz. HNO 2004;52: 112–120.
9 Seiden AM, Duncan HJ: The diagnosis of a conductive olfactory loss. Laryngoscope 2001;111:9–14.

10 Deems DA, Doty RL, Settle G, Moore-Gillon V, Shaman P, Mester AF, et al: Smell and taste disorders, a study of 750 patients from the university of Pennsylvania smell and taste center. Arch Otolaryngol Head Neck Surg 1991;117:519–528.
11 Quint C, Temmel AF, Schickinger B, Pabinger S, Ramberger P, Hummel T: Patterns of non-conductive olfactory disorders in eastern Austria: a study of 120 patients from the Department of Otorhinolaryngology at the University of Vienna. Wien Klin Wochenschr 2001;113:52–57.
12 Mott AE, Leopold DA: Disorders in taste and smell. Med Clin North Am 1991;75:1321–1353.
13 Leopold DA, Hornung DE, Youngentob SL: Olfactory loss after upper respiratory infection; in Getchell ML, Doty RL, Bartoshuk LM, Snow JBJ (eds): Smell and Taste in Health and Disease. New York, Raven Press, 1991, pp 731–734.
14 Sugiura M, Aiba T, Mori J, Nakai Y: An epidemiological study of postviral olfactory disorder. Acta Otolaryngol Suppl 1998;538:191–196.
15 Goodspeed RB, Gent JF, Catalanotto FA: Chemosensory dysfunction. Clinical evaluation results from a taste and smell clinic. Postgrad Med 1987;81:251–260.
16 Konstantinidis I, Müller A, Frasnelli J, Reden J, Hummel T: Seasonality of post-infectious olfactory dysfunction: retrospective study of 461 patients. Chem Senses, in press.
17 Baker H, Genter MB: The olfactory system and the nasal mucosa as portals of entry of viruses, drugs, and other exogenous agents into the brain; in Doty RL (ed): Handbook of Olfaction and Gustation. New York, Dekker, 2003, pp 549–573.
18 Mori I, Goshima F, Imai Y, Kohsaka S, Sugiyama T, Yoshida T, et al: Olfactory receptor neurons prevent dissemination of neurovirulent influenza A virus into the brain by undergoing virus-induced apoptosis. J Gen Virol 2002;83:2109–2116.
19 Mori I, Nishiyama Y, Yokochi T, Kimura Y: Virus-induced neuronal apoptosis as pathological and protective responses of the host. Rev Med Virol 2004;14:209–216.
20 Tomlinson AH, Esiri MM: Herpes simplex encephalitis. J Neurol Sci 1983;60:473–484.
21 Lavi E, Gilden DH, Highkin MK, Weiss SR: The organ tropism of mouse hepatitis virus A59 in mice is dependent on dose and route of inoculation. Lab Anim Sci 1986;36:130–135.
22 Zanusso G, Ferrari S, Cardone F, Zampieri P, Gelati M, Fiorini M, et al: Detection of pathologic prion protein in the olfactory epithelium in sporadic Creutzfeldt-Jakob disease. N Engl J Med 2003;348:711–719.
23 Tabaton M, Monaco S, Cordone MP, Colucci M, Giaccone G, Tagliavini F, et al: Prion deposition in olfactory biopsy of sporadic Creutzfeldt-Jakob disease. Ann Neurol 2004;55:294–296.
24 Jafek BW, Hartman D, Eller PM, Johnson EW, Strahan RC, Moran DT: Postviral olfactory dysfunction. Am J Rhinol 1990;4:91–100.
25 Yamagishi M, Fujiwara M, Nakamura H: Olfactory mucosal findings and clinical course in patients with olfactory disorders following upper respiratory viral infection. Rhinology 1994;32:113–118.
26 Jafek BW, Murrow B, Michaels R, Restrepo D, Linschoten M: Biopsies of human olfactory epithelium. Chem Senses 2002;27:623–628.
27 Douek E, Bannister LH, Dodson HC: Recent advances in the pathology of olfaction. Proc R Soc Med 1975;68:467–470.
28 Akerlund A, Bende M, Murphy C: Olfactory threshold and nasal mucosal changes in experimentally induced common cold. Acta Otolaryngol 1995;115:88–92.
29 Hummel T, Rothbauer C, Barz S, Grosser K, Pauli E, Kobal G: Olfactory function in acute rhinitis. Ann NY Acad Sci 1998;855:616–624.
30 Duncan HJ, Seiden AM, Paik SI, Smith DV: Differences among patients with smell impairment resulting from head trauma, nasal disease or prior upper respiratory infection. Chem Senses 1991;16:517.
31 Hummel T, Maroldt H, Frasnelli J, Landis BN, Hüttenbrink K-B, Heilmann S: Qualitative olfactory dysfunction: frequency and prognostic significance. Chem Senses 2005;30:A97.
32 Ackerman BH, Kasbekar N: Disturbances of taste and smell induced by drugs. Pharmacotherapy 1997;17:482–496.
33 Henkin RI: Drug-induced taste and smell disorders. Incidence, mechanisms and management related primarily to treatment of sensory receptor dysfunction. Drug Saf 1994;11:318–377.

34 Doty RL, Shaman P, Dann M: Development of the University of Pennsylvania Smell Identification Test: a standardized microencapsulated test of olfactory function (UPSIT). Physiol Behav 1984;32:489–502.
35 Hummel T, Sekinger B, Wolf SR, Pauli E, Kobal G: 'Sniffin' Sticks': olfactory performance assessed by the combined testing of odor identification, odor discrimination, and olfactory thresholds. Chem Senses 1997;22:39–52.
36 Duncan HJ, Seiden AM: Long-term follow-up of olfactory loss secondary to head trauma and upper respiratory tract infection. Arch Otolaryngol Head Neck Surg 1995;123:367–372.
37 Hummel T: Perspectives in olfactory loss following viral infections of the upper respiratory tract. Arch Otolaryngol Head Neck Surg 2000;126:802–803.
38 Reden J, Müller A, Konstantinidis I, Landis BN, Hummel T: Recovery of olfactory function following closed head injury or infections of the upper respiratory tract. Chem Senses, in press.
39 Henkin RI, Schecter PJ, Friedewald WT, Demets DL, Raff M: A double blind study of the effects of zinc sulfate on taste and smell dysfunction. Am J Med Sci 1976;272:285–299.
40 Quint C, Temmel AFP, Hummel T, Ehrenberger K: The quinoxaline derivative caroverine in the treatment of sensorineural smell disorders: a proof-of-concept study. Acta Otolaryngol 2002;122: 877–881.
41 Hummel T, Heilmann S, Hüttenbrink K-B: Lipoid acid in the treatment of smell dysfunction following viral infection of the upper respiratory tract. Laryngoscope 2002;112:2076–2080.
42 Szmeja Z, Obrebowski A: Die Behandlung der nach Influenza auftretenden Kakosmien vermittels lokaler Injektionen von Hydrokortison. HNO 1969;17:53–54.
43 Hummel T, Rissom K, Müller A, Reden J, Weidenbecher M, Hüttenbrink K-B: 'Olfactory training' in patients with olfactory loss. Chem Senses, in press.

PD Dr. Antje Welge-Lüssen
Department of Otorhinolaryngology
University Hospital Basel
Petersgraben 4
CH–4031 Basel (Switzerland)
Tel. +41 61 2654 109, Fax +41 61 2654 029, E-Mail awelge@uhbs.ch

Hummel T, Welge-Lüssen A (eds): Taste and Smell. An Update.
Adv Otorhinolaryngol. Basel, Karger, 2006, vol 63, pp 133–151

Olfaction in Neurodegenerative Disorder

Christopher Hawkes

Essex Centre for Neuroscience, Oldchurch Hospital, Romford, UK

Abstract

There has been gradual increase of interest in olfactory dysfunction since it was realised that anosmia was a common feature of idiopathic Parkinson's disease (IPD) and Alzheimer-type dementia. It is an intriguing observation that a premonitory sign of a disorder hitherto regarded as one of movement or cognition may be that of disturbed sense of smell. In this review of aging, IPD, parkinsonian syndromes, tremor, Alzheimer's disease (AD), motor neuron disease (MND), Huntington's chorea (HC) and inherited ataxia, the following observations are made: (1) olfactory senescence starts at about the age of 36 years in both sexes and accelerates with advancing years, involving pleasant odours preferentially; (2) olfactory dysfunction is near-universal, early and often severe in IPD and AD developing before any movement or cognitive disorder; (3) normal smell identification in IPD is rare and should prompt review of diagnosis unless the patient is female with tremor-dominant disease; (4) anosmia in suspected progressive supranuclear palsy and corticobasal degeneration is atypical and should likewise provoke diagnostic review; (5) subjects with hyposmia and one ApoE4 allele have an approximate 5-fold increased risk of later AD; (6) impaired sense of smell may be seen in some patients at 50% risk of parkinsonism, and possibly in patients with unexplained hyposmia; (7) smell testing in HC and MND where abnormality may be found is not likely to be of clinical value, and (8) biopsy of olfactory nasal neurons reveals non-specific changes in IPD and AD and at present will not aid diagnosis.

The term 'neurodegeneration' is a bad one without clear definition, but will be taken here to mean disorders where the cause is not known to be infectious, autoimmune, neoplastic nor inflammatory. The commonest form of neurodegeneration is aging itself and assessment of olfaction without allowance for changes with age is not possible. The following diseases will be discussed: idiopathic Parkinson's disease (IPD), parkinsonian syndromes, essential tremor

(ET), motor neuron disease (MND), Alzheimer's disease (AD), Huntington's chorea (HC) and inherited ataxias.

Aging

It is essential to be aware of the effect of aging itself as it is recognised to be a major variable affecting olfaction. We studied the effect of age on the University of Pennsylvania Smell Identification Test (UPSIT) score in 211 healthy controls by multiple regression against gender, age and age squared [1] and showed that identification ability begins a significant decline after the age of 36 years and thereafter more steeply. Females scored 1–2 points higher at all ages although they declined at exactly the same rate as their male counterparts. Aging may not affect all varieties of odours in the same way. We therefore studied the effect of age on olfactory identification and its principle components – pleasantness, irritation and intensity – in the same 211 controls using the 40 individual odours comprising the UPSIT. Older subjects scored less well than younger on odours with high hedonic or low intensity/irritation rating. Although intensity correlated better with age than pleasantness, the two dimensions predicted age discrepancy independently. In the over-50-year age group, we were able to select 7 odours with moderately high hedonic properties and low intensity score (chocolate, liquorice, grass, coconut, strawberry, rose and melon) and suggest these odours are more resistant to the effects of age and may be appropriate for testing elderly subjects. At present, there is no geriatric smell kit available commercially, but such a product should be more discriminatory for testing diseases of the elderly. This makes the assumption that degenerative diseases are not themselves a simple acceleration of 'normal' aging.

Sniffing and the Role of the Cerebellum

It has been shown that sniffing enhances smell detection and apart from redirection of airflow to the olfactory neuroepithelium, functional MRI studies have shown that sniffing activates the pyriform and orbitofrontal cortices [2]. In a meticulous study of IPD, Sobel et al. [3] showed that sniffing was impaired in IPD and this caused slight reduction in the performance on identification and detection threshold tests. This equates to a mean reduction of around 2–3 points on the UPSIT-40 test (see below). Increasing sniff vigour improved olfactory scores. Studies that have not allowed for this effect (which includes the majority) may tend to slightly exaggerate the severity of any defect especially where the disease is known to involve bulbar function (e.g. motor neurone disease).

To complicate matters further, it is suggested that the cerebellum is concerned with smell identification [2] and if correct this has clear relevance to disorders such as inherited ataxias and ET where the cerebellum is known to be abnormal.

Idiopathic Parkinson's Disease

Impairment of the sense of smell in IPD was first documented in 1975 by Ansari and Johnson [4]. Before that, many noted anecdotally that olfactory loss may precede IPD by many years, but at present there is just one prospective population-based study to be described below [5]. The majority of olfactory studies in IPD have used clinical diagnostic criteria and none have correlated changes in life with those diagnostic criteria found after death. This is of considerable relevance as the diagnostic error rate of neurologists contrasted with autopsy diagnosis is 10–26% [6, 7]. Despite this, it is reasonable to propose that patients with IPD have profound disorder of olfactory function [8, 9]. This observation is based on pathological abnormality, psychophysical tests and evoked potential studies.

Pathology

The rhinencephalon has only recently been investigated systematically in PD without dementia. Dystrophic neurites but no Lewy bodies were found in two of three autopsy-derived olfactory epithelia of patients with IPD, but several patients displayed accumulation of amyloid precursor protein fragments which would not allow distinction from AD [10]. All three varieties of synuclein (α, β, γ) are expressed in olfactory neuroepithelium, particularly α-synuclein, the hallmark of IPD. Unfortunately, the expression of α-synuclein was found to be no different from Lewy body disease (LBD), AD, multiple system atrophy (MSA) and seemingly healthy controls [11]. Nonetheless, it is possible that those patients coincidentally found to have α-synuclein-containing neurites may be in the preclinical stage of IPD and that the changes actually represent a disease-specific finding.

In a preliminary study, we examined, blinded to clinical information, the olfactory bulbs and tracts from formalin-fixed brains of 8 controls and 8 patients with a clinical and pathological diagnosis of IPD taken from the UK Parkinson's Disease Brain Bank [12]. By inspecting the olfactory bulb and tract all 8 cases were correctly diagnosed 'probable PD'. Lewy bodies were most numerous in the anterior olfactory nucleus but they were also found in mitral cells. The morphology of Lewy bodies at this site resembled their cortical counterparts but inclusions showing a classical trilaminar structure were rare. It was subsequently shown that loss of anterior olfactory neurons correlated with disease duration [13].

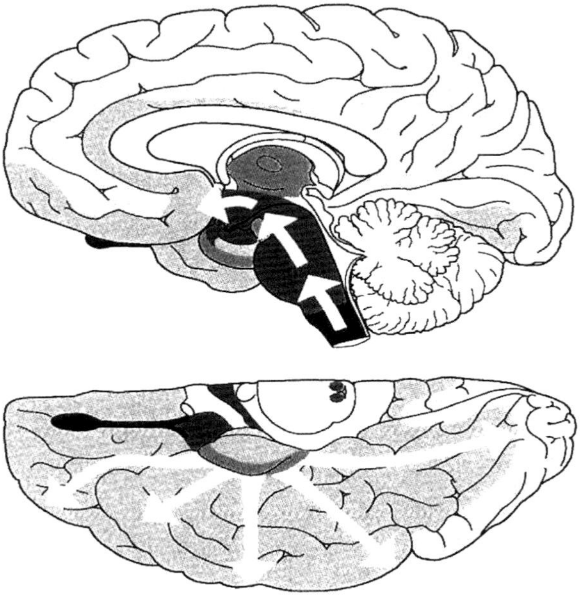

Fig. 1. The sequential development of PD starting in the olfactory bulb and dorsal nuclear complex of cranial nerves IX and X (shown in black). The process then ascends the brain stem to the cerebral cortex. Reproduced with permission from Braak et al. [14].

Braak et al. [14] performed a detailed analysis of pathology in IPD in 125 cases by immunoreaction with α-synuclein, the protein specific to PD, which is found in Lewy neurites and Lewy bodies. They demonstrated that the pathology process advances in a predictable sequence, but the earliest changes, even before the motor components appeared in life, were found in the dorsal motor nuclei of the glossopharyngeal and vagus nerves, the olfactory bulb and associated anterior olfactory nucleus (stage I; fig. 1). This is a pivotal study as it clearly identifies the dorsal medulla and olfactory bulb as starting points for IPD. Given that at least 40% of substantia nigra cells have to die before there are clinical symptoms [15], it is clear that the clinical motor manifestations of IPD represent the terminal stage of a process that probably started several decades previously. This point is borne out by the anecdotal clinical observation that patients regularly report smell impairment 10–20 years before their first motor symptoms. Involvement of central olfactory areas such as the entorhinal cortex takes place much later in Braak stage 3 [14].

There have been few studies of the olfactory bulb beyond anatomical description, but it is clear that there is considerable cell loss in the bulb. The mitral cells seem to vanish in IPD [Braak, H., pers. commun.]. One report [16] suggested that expression of tyrosine hydroxylase in the olfactory bulb is increased 100-fold and that this might explain hyposmia of IPD. In the mouse methylphenyltetrahydropyridine (MPTP) models of PD, there is a 4-fold increase of dopamine neurogenesis in the olfactory bulb [17] that probably relates to the migration of dopamine-secreting cells from the subventricular zone – the so-called rostral migratory stream. This experiment suggests that ongoing compensation is taking place in the adult brain and infers that only when the process fails do the symptoms of the disease become apparent.

The studies of Braak et al. [14] make it clear that the first olfactory abnormalities are peripheral, that is in the olfactory bulb, which led these authors to propose [18] that IPD starts in the stomach (Auerbach's Plexus) and the agent (virus or chemical) spread in a retrograde fashion up the motor vagal fibres into the dorsal medulla. This theory, whilst appealing, does not take account of the olfactory bulb changes.

Psychophysical Tests

The first case-control assessment [4] involved 22 patients with a clinical diagnosis of IPD by detection threshold to amyl acetate. They found a correlation between average olfactory threshold and more rapid disease progression. There appeared to be no influence from medication (levodopa, anticholinergic drugs) or smoking habit. A subsequent larger study of IPD [19] also used detection threshold tests to various concentrations of amyl acetate in 78 patients and 40 controls. Thresholds were reduced but no correlation was found with age, sex or use of levodopa. Unlike the first study, there was no association with disease duration. The next sizeable olfactory investigations [8, 20] using the UPSIT-40 showed that age-matched olfactory dysfunction did not relate to odour type; it was independent of disease duration and did not correspond with motor function, tremor or cognition. They also demonstrated that the deficit was of the same magnitude in both nostrils and not influenced by antiparkinsonian medication. Further evaluation in subtypes of presumed IPD showed that females with mild disability and tremor-dominant disease had a significantly higher age-matched UPSIT-40 score than males with moderate to severe disability and little or no tremor; age at disease onset was not relevant [21]. A comparable survey [9] was undertaken using the UPSIT-40 in 155 cognitively normal, depression-free IPD patients aged 34–84 years and 156 age-matched controls. The age-matched UPSIT scores for PD patients were dramatically lower than for controls. Only 19% (30/155) of the PD patients had a score within the level expected for 95% age-matched healthy controls. There were 65 (42%) who were graded anosmic, i.e. scoring less than 17. There was no correlation between disease duration and UPSIT score ($r = 0.074$) [9]. Analysis of the 40 individual odours in the UPSIT showed that pizza was the single most difficult odour to identify for patients and that the combination of pizza and wintergreen was the best discriminator with a sensitivity of 90% and specificity of 86% [22]. Impairment of the sense of smell has been documented in IPD patients using 'sniffin' sticks' [23]. There was a significant negative correlation between odour discrimination and disease severity suggesting as far as psychophysical tests go that there is in fact some correlation between olfaction and disease severity.

Neurophysiological Tests

One of the most informative and validated objective measurements of the sense of smell is the olfactory (chemosensory) evoked potential (OEP) pioneered by Kobal and Plattig [24]. We tested 73 patients with IPD by OEP recording [9] and compared them to 47 controls of similar age and sex. None were depressed, all were cognitively normal and had a clinical diagnosis of IPD. In 36 patients (49%), responses were either absent or unsatisfactory for technical reasons. Regression analysis on the 37 with a measurable trace showed that for hydrogen sulphide (H_2S) a highly significant latency difference existed between diagnostic groups (i.e. control or PD). Assuming the 36/73 who had no detectable OEP were anosmic and combining these with the abnormal 12/37 (32%) then 81% have abnormality on OEP which is the same as for UPSIT measurements. In 10 patients with normal UPSIT-40 scores, there was 1 with absent H_2S responses and 3 with significantly prolonged latency to H_2S, suggesting that the prevalence of olfactory disorder may be higher still. We used only one odour, whereas the UPSIT implements 40. If a large number of different gases were used, the sensitivity of OEP might well increase. Similar results were obtained in 31 patients with clinically labelled IPD tested by OEP to vanillin and H_2S [25]. Responses were found to both stimulants in all patients, which is remarkable given that many would be anosmic. Prolonged latencies were seen in the IPD patients whether they were taking medication for the disease or not. More marked changes were seen in those receiving medication, possibly because they were more disabled. The same group demonstrated a correlation between disability (as measured by Webster score) and latency to the H_2S OEP, complementing the psychophysical findings mentioned above [23].

Familial and Presymptomatic Parkinson's Disease

In the Michigan study [26] of familial parkinsonism, the UPSIT-40 was applied to 6 kindreds of which 3 had typical PD and 3 had a 'parkinsonism-plus' syndrome. In the typical families, there were 4 apparently healthy individuals at 50% risk of whom 3 were microsmic. In the PD-plus families, there were 8 at risk and 2 had abnormal UPSIT scores. It is not known as yet whether these subjects at risk will develop clinical evidence of parkinsonism. In PARK2, which is a dominant form of parkinsonism, the sense of smell is relatively preserved, whilst in PARK1, subjects are mildly hyposmic [27]. Others [28] have implemented a test battery to first-degree relatives of IPD patients which comprised motor function, olfaction (UPSIT-40) and mood. There were significant differences in first-degree relatives (both sons and daughters) particularly where the affected parent was the father. Another group [29] evaluated asymptomatic but hyposmic relatives of patients with IPD. Dopamine transporter binding was abnormal in 4/25 hyposmic relatives, 2 of whom subsequently developed IPD. None of the 23 normosmic relatives have so far developed IPD.

Sommer et al. [30] tested 30 patients with unexplained smell impairment to determine whether any might be in the premotor phase of IPD. Apart from detailed olfactory testing, subjects were evaluated by dopamine transporter imaging (DATScan) and transcranial sonography of the substantia nigra. Eleven displayed increased (abnormal) echogenicity on transcranial sonography and 10 volunteered for DATScan. Of these 10, 5 had abnormal DATScans and a further 2 were borderline, suggesting they might be in a presymptomatic phase of parkinsonism. This study now awaits long-term follow-up.

One long-term community-based prospective study has now been published [5]. The authors used the cross-cultural UPSIT-12 test in 2,263 healthy elderly men aged 71–95 years who participated in the Honolulu-Asia Aging Study. They were followed up for 7 years and during this period, 19 men developed PD at an average latency of 2.7 years from baseline assessment. After adjustment for multiple confounders, the relative odds for PD in the lowest tertile of the UPSIT-12 score was 4.3 (95% CI 1.1–16.1; $p = 0.02$).

The above evidence, while still provisional, suggests that isolated olfactory impairment is indeed an early warning sign of pending parkinsonism. It can be argued that even if olfactory impairment is an early sign of subsequent IPD, it may simply reflect ease of measurement and that although movement-related pathology is more difficult to assess in its earliest phase, it may still be of greater aetiological importance. This view is probably incorrect considering the strong neuropathological evidence [14].

Parkinsonism

This term refers to those diseases which phenotypically resemble IPD but differ on pathological and genetic grounds. This section will include LBD, MSA, progressive supranuclear palsy (PSP), corticobasal degeneration (CBD), drug-induced PD (DIPD), Guam PD-dementia complex (PDC), X-linked dystonia-parkinsonism ('Lubag') and vascular parkinsonism (VP). All data have been obtained by psychophysical measurement, rarely with pathological verification, and in most cases the number of patients studied has been small. Consequently, most observations should be regarded as provisional.

Lewy Body Disease

In comparison to IPD, LBD is characterised by rapid course, early onset of confusion, hallucinations, drug sensitivity and subsequent dementia. Neuropathologically, it is indistinguishable from IPD [14] but it is probable there are genetic differences to account for the different rate of progression. In a study of clinically defined LBD, severe impairment of olfactory identification and detection threshold was observed and test scores were independent of disease stage and duration [31, 32]. In another investigation [33], simple smell perception of one odour (lavender water) was examined in 92 patients with autopsy-confirmed dementia of whom 22 had LBD and 43 had AD. They were compared with 94 age-matched controls. The main finding was impaired smell perception in the LBD group but little or no defect in the AD patients. Although just one odourant was used for perception tests, this study confirms at the clinical and pathological level the clinically based conclusion [32] that impairment of smell is significant in LBD.

Multiple System Atrophy

There are two major varieties of MSA: a common predominantly parkinsonian variety (MSA-P; Shy-Drager syndrome), which comprises 80% of the total, where there is predominance of akinesia and rigidity. In the remaining 20% cerebellar ataxia predominates (MSA-C), but both display rapidly evolving parkinsonism with dysautonomia affecting bladder and orthostatic blood pressure control. Pathological changes of MSA may be seen in olfactory bulbs and characterised by cytoplasmic inclusions in oligodendrocytes sometimes called Papp-Lantos filaments [34]. In an early study [35] of smell identification in 29 patients with a clinical diagnosis of MSA-P, mild impairment of the UPSIT-40 score was demonstrated with a mean UPSIT-40 score of 26.7 compared to the control mean of 33.5. There were no differences between the parkinsonian and cerebellar types for smell identification. A study of 8 MSA-P patients using ‘sniffin’ sticks’ showed hyposmia in 7 of 8 patients [36]. A further investigation [37] focussed particularly on MSA-C in comparison to other ataxias of unknown aetiology and found no useful difference between their two categories. In conclusion, mild olfactory impairment would be expected in MSA-P, but the severity of this defect appears much less than in IPD.

Corticobasal Degeneration

Here parkinsonian features are compounded by limb dystonia, ideomotor apraxia and myoclonus. In the one small study [38] of 7 patients with clinically suspected CBD, smell identification scores (UPSIT-40) were in the low normal range with a mean of 27, a value not significantly different from their age-matched controls.

Progressive Supranuclear Palsy (Steele-Richardson-Olszewski Syndrome)

In this variety, there is failure of voluntary vertical gaze, rapid course, marked imbalance and dementia. The brain shows widespread deposits of tau protein in degenerating neurons. In a large study of olfactory bulbs [39], tau and α-synuclein pathology was found in only 9 of 27 bulbs. The bulbs which showed tau pathology also had coexisting AD or LBD implying that pure PSP is not a tauopathy. Normal identification values have been found in two surveys [38, 39]. In another [40], there was likewise no difference in age-matched UPSIT-40 scores and threshold tests to phenylethyl alcohol were not significantly different from control values ($p = 0.085$), but there was a trend to higher threshold values which may have failed to reach significance because of the relatively small number of patients. In all instances, the diagnosis once more has been clinically, not autopsy based. A more complex picture was found in relatives of patients with PSP [41]. A test battery that measured motor function, olfaction and mood was administered to 27 first-degree relatives of whom 9 scored in the abnormal range. The authors suggested that this may help the detection of asymptomatic carriers, but it is likely the familial prevalence in this survey is overstated due to selection bias. This confused picture emphasises the major need for pathologically confirmed studies, but present evidence suggests that olfaction is normal or nearly so in PSP.

Vascular Parkinsonism

Some patients with extensive cerebrovascular disease that involves the basal ganglia, particularly the putamen, may develop a syndrome that mimics IPD, but the response to levodopa is variable. If parkinsonism develops acutely, it is usually one-sided and affords little diagnostic difficulty, but a problem may arise when the onset is insidious or stepwise. Although brain MRI may show extensive vascular disease, it can be difficult to know if this is coincidental. A recently presented study of the UPSIT-40 in 14 patients fulfilling strictly defined criteria for VP showed no significant difference compared to age-matched controls (25.5 in VP and 27.5 in controls), suggesting that identification tests may aid differentiation from IPD [42].

Drug-Induced Parkinson's Disease

DIPD can be clinically indistinguishable from IPD and was common when broad-spectrum dopamine antagonists were widely used. With the advent of selective D_2 blockers, the prevalence has subsided. We undertook a small study of neuroleptic-induced PD in 10 patients [43] all of whom scored 27 or more on the Mini Mental State Examination test. Parkinsonism had developed in response to a variety of phenothiazine drugs that had been administered for at least 2 weeks. Of the 10 patients, 5 had abnormal age-matched UPSIT-40 scores and none made a complete recovery, whereas all but 1 of those who did recover had a normal UPSIT score. The interpretation is difficult, especially as some of these patients had a psychotic disorder which itself may be associated with olfactory impairment, but it implies that some patients with DIPD are predisposed to develop IPD and that taking a dopamine-depleting drug unmasks the disease. In parkinsonism induced by MPTP, 6 subjects were found to be normal for UPSIT-40 and detection threshold [44]. Although this is a small series, it implies that MPTP-PD is an unrepresentative model of its idiopathic counterpart.

Guam Parkinson's Disease-Dementia Complex

PDC is typified by the coexistence of Alzheimer-type dementia and sometimes MND. Pathologically, the presence of neurofibrillary tangles and absence of Lewy bodies place this disorder well apart from IPD. Initial findings of olfactory impairment using a culturally adapted form of the UPSIT [45] were confirmed by Doty et al. [46]. They administered the Picture Identification Test and the UPSIT-40 to 24 patients with PDC and found severe impairment of olfactory function of the magnitude comparable to that seen in IPD although a few had additional cognitive impairment [46].

X-Linked Dystonia-Parkinsonism (Lubag)

Lubag is an X-linked disorder affecting Filipino male adults with maternal roots from the Philippine Island of Panay. A single study of 20 affected males using the UPSIT-40 showed that olfaction is moderately impaired in Lubag even early on in the disorder and that it is independent of the degree of dystonia, rigidity, severity or duration of the disease [47].

Essential Tremor

Classical ET is usually diagnosed without difficulty but there are problems when the tremor appears to be dystonic or there is coexisting rigidity. There have been no pathological studies of the olfactory pathways. The first small study of identification ability using the UPSIT-40 in 15 subjects with benign ET found all to be normal [48]. Subsequently, Louis et al. [49] found that a significant proportion of ET patients have mild impairment of smell identification and that this may relate to the postulated olfactory function of the cerebellum. In a subsequent study of ET patients with isolated rest tremor, the UPSIT-40 score was no different from typical ET patients and from this it was proposed that involvement of the basal ganglia is part of the ET syndrome [50]. The most recent olfactory findings using the UPSIT-40 in ET [51, 52] are normal overall but errors may creep in if a patient with apparent ET is misdiagnosed as having benign tremulous PD – where the sense of smell may be impaired. Clearly it would help diagnostically if all ET subjects had normal olfaction as this would allow distinction from PD subjects with tremor. To resolve this dilemma completely will need detailed imaging; autopsy-based studies or better still the characterisation of candidate genes for ET.

Cerebellar Ataxia

If the cerebellum is concerned with olfaction (other than control of sniffing) then abnormalities would be expected in various ataxias. In a recent study [53], there were mild abnormalities of the UPSIT-40 score in Friedreich's ataxia but the changes did not correlate with trinucleotide repeat length, disease duration or walking disability. Other patients with a variety of ataxic disorder as a whole did not differ from the Friedreich's patients. Another group [54] examined a variety of ataxic subjects by using the UPSIT-40. Mild olfactory impairment was found in autosomal dominant spinocerebellar ataxia type 2 (SCA2) but not Machado-Joseph syndrome (SCA3). This is of relevance as patients with SCA2 or SCA3 may have parkinsonian features, hence the finding of normal olfaction in suspected IPD might alert the clinician to the presence of a cerebellar syndrome. All observations here have been made on small patient numbers and should be interpreted with caution. Furthermore, it would be premature to suggest on the basis of the above that the cerebellum is responsible for the smell defect in ataxic disorder until there has been neuropathological examination of the olfactory pathways.

Motor Neuron Disease and Amyotrophic Lateral Sclerosis

Nomenclature varies worldwide, but in this article the term MND will be used as a generic term for all varieties affecting the motor neuron of which amyotrophic lateral sclerosis (ALS) is the commonest, followed by bulbar forms and the milder form – progressive muscular atrophy (PMA). ALS in the American literature corresponds to MND in the UK. Pathologically, there has been just one study of the olfactory bulb in 8 cases of MND [55]. There was marked accumulation of lipofuscin in olfactory neurons compared to age-matched controls, suggesting defective lipid peroxidation. An initially clinically based pilot study [56] examined 15 patients with MND of whom 8 had moderate or severe bulbar involvement and 8 were chair bound. No test for dementia or sniffing was administered but significant lowering of the UPSIT-40 score was documented. In another study of 37 patients with ALS [57], 28 (75.7%) had significantly lower scores on the UPSIT-40 compared to age-matched controls. There were 4 (11%) with near or total anosmia. We examined 58 cognitively normal patients with an established diagnosis of MND [55]. Seven had PMA, 34 typical ALS and 17 had bulbar onset. Overall 9/58 (16%) scored abnormally on age-matched UPSIT-40. The effect of group status overall (i.e. MND or control) was statistically significant (p = 0.02). Within-group analysis, i.e. whether ALS, bulbar or PMA, showed that only bulbar patients were significantly different. OEPs were performed in 15 patients; in 9 the responses were normal for latency and amplitude measurements, 1 was delayed, in 2 the response was absent and in 3 the tracing could not be obtained. Our UPSIT-40 findings differ from previous publications and the seemingly conflicting results are probably an effect of case selection and diagnostic bias. Also there is a theoretical worsening of identification score from sniffing, i.e. if a patient has respiratory weakness which may well be present in bulbar MND, then spurious olfactory impairment may occur and this was shown to have a modest impact at least in patients with IPD [3]. The more objective findings from OEP suggest that olfaction is affected albeit to a mild degree.

Alzheimer's Disease

There was initial excitement when it was suggested AD could be identified by autopsy samples of nasal olfactory neuroepithelium [58]. Changes in morphology, distribution and immunoreactivity of neuronal structures were typical of AD. Subsequent studies cast doubt on this and although changes characteristic of AD have been confirmed they are not specific: similar changes are seen in IPD and even some healthy elderly controls [59]. Despite this, it could be argued that the apparently healthy controls were in the preclinical phase of the disease. There were subsequent attempts at diagnosis by olfactory mucosal biopsy. Further difficulties have been found: apart from the lack of specific changes, it is difficult to identify olfactory neurons as the neuroepithelium tends to be replaced progressively by respiratory epithelium with aging and this process may be more rapid in AD. In one study [60], only 6/13 samples contained olfactory neurons. Studies by Braak and Braak [61] suggest that one of the first areas of damage is in the transentorhinal cortex. This is a bottleneck entry zone for cortical sensory afferents to the hippocampus. Abnormalities in this region are followed by changes in the entorhinal cortex – an area concerned with memory, emotion and olfaction. Olfactory bulb changes in AD are well recognised [62] but the pathological time course of these events is not yet established, so it is not known whether AD starts in the peripheral or central rhinencephalon [Braak, H., pers. commun.].

Numerous psychophysical studies of olfaction in presumed AD have shown abnormalities and in some, a correlation of dementia severity with anosmia [63, 64]. The majority have used clinical criteria for diagnosis and rarely have autopsy data been available. Severe abnormalities have been documented in most cases for identification, recognition and threshold detection. In a metanalysis [64], defects in olfaction shown by patients with AD and PD were relatively uniform although there was a trend toward better performance on threshold tests than recognition and identification tests. Unfortunately, no measure could distinguish the two conditions.

In a post-mortem confirmed prospective study of the sense of smell in AD and LBD, anosmia was found to correlate better with the presence of Lewy bodies than Alzheimer pathology [33] and this group found no significant impairment of olfactory perception at all. The authors measured perception of just one uncalibrated odour (lavender water) which is grossly inadequate. The finding of normal olfaction for AD, whilst persuasive because there was autopsy confirmation, was probably an artefact of the small sample size and reliance on a single crude test of threshold. It is at variance with nearly every other study. Furthermore, AD patients have relative preservation of threshold in the early stages [65], thus measurement of threshold might well be within normal limits.

It has been suggested that hyposmia is an early and consistent change in AD. In a prospective population-based study [66], 1,836 healthy people were tested at baseline by the international UPSIT-12 test and a cognitive screening procedure ('CASI'). They found hyposmia and particularly anosmia significantly increased the risk of subsequent cognitive failure. Anosmics at baseline who had at least one ApoE4 allele had nearly 5 times the risk of subsequent cognitive decline. Another group [67] examined prospectively the olfactory identification score in patients with mild cognitive impairment. Those scoring 34 or less on the UPSIT-40 who were also unaware of their defect were more at risk of developing AD within 2 years. In theory, unawareness might have been a manifestation of their cognitive impairment but insight is usually well preserved in the early stages of AD.

We studied 8 non-depressed patients with AD [68] all of whom had mild or moderately severe forms of the disease. Age-matched UPSIT-40 scores were abnormal in all, but the H_2S OEP was normal in the 4 subjects who could be tested. Provisionally, this infers an olfactory defect that is severer centrally than peripherally, i.e. signals are reaching the brain but are incorrectly interpreted due to cognitive change. This concurs with two studies [69, 70] that demonstrated abnormal identification but no change in threshold tests, also implying a central more than peripheral defect. Others documented a significant delay in amyl acetate OEP and impairment of the UPSIT-40 score in 12 patients with a clinical diagnosis of AD [71]. It has been suggested that healthy individuals who are positive for ApoE4 have a significant delay in OEP in comparison to ApoE4-negative persons [72] – just as those who were hyposmic on the UPSIT-12 at baseline [67]. In conclusion, there is abundant evidence of olfactory impairment in AD which is probably severer and commences centrally. Decline of smell identification could therefore act as a biomarker of future cognitive impairment.

Down's Syndrome

Most subjects with Down's syndrome (DS) eventually develop Alzheimer-type dementia. In adolescent DS subjects [73], low identification and discrimination scores were found that were similar to other children of comparable age and cognitive function. They concluded that patients with DS first evidence loss of olfactory function at a time when Alzheimer-type pathology is just commencing and inferred that smell testing could not be used in DS to predict the onset of later AD. However, the UPSIT scores of their DS group were of similar magnitude to those seen in older DS patients and it could be argued that olfactory impairment is an early change predating AD-like cognitive impairment. In a study of DS by OEP to amyl acetate, a significant increase of latency was found [74] providing an objective basis for hyposmia in DS. While it may be true that hyposmia precedes the Alzheimer-type dementia of DS, it might simply reflect the ease of detecting smell impairment compared to very early cognitive decline. Characteristic AD pathology may be present in relatively silent cortical areas which are difficult to probe by current techniques.

Huntington's Chorea

HC is an autosomal dominant disorder of basal ganglia function typified by choreic movement, dementia and rarely muscular rigidity similar to PD (Westphal variant). Initial studies have documented early defective odour memory sometimes prior to cognitive defect or the onset of marked involuntary movement [75]. Subsequent studies using identification and detection tests have confirmed the presence of moderate olfactory impairment, affecting identification in particular and less than that seen in PD [76]. Olfactory testing of presymptomatic relatives at 50% risk has not shown abnormalities, implying that olfaction is impaired at the onset of motor or cognitive disorder [76]. In another study [77], however, odour detection presented good classification of sensitivity and specificity between the patients and controls, suggesting that olfactory testing may provide a sensitive measure of early disease process in

Table 1. Relative degree of perceptive olfactory dysfunction in neurodegeneration on an arbitrary scale

Disease	Relative severity of smell loss
IPD, LBD, PDC	++++
MSA, HC, DIPD, Lubag, AD, DS	++
MND, SCA2, Friedreich's ataxia	+
PSP, ET	0/+?
CBD, VP, PD, MPTP parkinsonism, idiopathic dystonia, SCA3	0?

++++ = Marked damage; + = mild; 0 = normal. IPD = Idiopathic Parkinson's disease, LBD = Lewy Body Disease, PDC = Guam PD-dementia complex, MSA = Multiple System Atrophy, HC = Huntington's chorea, DIPD = Drug induced PD, AD = Alzheimer's disease, DS = Down syndrome, MND = Motor neurone disease, SCA2 = spinocerebellar ataxia type 2, PSP = progressive supranuclear palsy, ET = essential tremor, CBD = Cortico-basal degeneration, VP = Vascular parkinsonism, PD = Parkin disease. Note that most of the above values are provisional and based on relatively small patient numbers.

HC patients. The utility of this observation is offset by the widely available and specific DNA test for HC.

Conclusions

The olfactory system is damaged to a varying degree in the presence of clinically evident parkinsonism (table 1). Severest changes are seen in the idiopathic, Guamanian and LBD varieties. Least involvement would be expected in CBD, PSP and intermediate damage in MSA. These differences could aid diagnosis. For example, if a patient is suspected to have IPD, the presence of normal olfaction on psychophysical tests should prompt review of the diagnosis especially in the akinetic rigid variety. Anosmia in CBD or PSP would also be unexpected. In a patient with predominant tremor, it may be difficult to know whether this is tremor-dominant PD, benign ET or inherited ataxia. Normal olfaction would favour ET or SCA3 with the proviso that females with tremor-dominant IPD might also have a normal result and that some ET patients according to one source are abnormal. Olfactory testing in HC and MND shows abnormalities but it is likely to prove less rewarding both for diagnosis and for presymptomatic testing. In AD, the majority view would be that the sense of smell is impaired probably at the central more than peripheral level but there is a need for large studies particularly with OEP recording to circumvent confounding by cognitive impairment. Preliminary evidence suggests that olfactory disorder may be an early precognitive feature of AD, although it might be argued that olfaction is just easier to measure than early cognitive decline. A parallel argument may be applied to IPD. There is a distinct possibility that olfactory testing in unaffected relatives of those with IPD or AD may allow identification of those at risk of subsequently expressed disease, i.e. olfactory testing could act as a useful biomarker. Prospective studies in families with AD and IPD containing substantial members at 50% risk will help solve this problem.

Finally, there is the question of the future value of smell testing in neurological disease. Hyposmia might be a readily detectable epiphenomenon of no real diagnostic value. Imaging dopamine distribution by SPECT (DATScan) or fluorodopa PET is more definitive both for confirming a diagnosis of suspected parkinsonism and for study of at-risk subjects, while increasing knowledge of genetic defects in parkinsonian syndromes make obsolete other diagnostic tests. At present, these non-olfactory procedures are expensive and olfactory testing may have a complementary role in PD and AD. We were surprised to discover that 50% of patients with DIPD had an impaired identification score. If this is confirmed in a larger series, it raises the possibility of genetically determined susceptibility to various classes of organic chemicals and if so neuroprotective measures could be undertaken to stop or slow down the onset of the disease.

References

1 Hawkes CH, Fogo A, Shah M: Smell identification declines from age 36 years and mainly affects pleasant odours. Mov Disord 2005;20(Suppl 10):160.

2 Sobel N, Prabhakaran V, Desmond JE, Glover GH, Goode RL, Sullivan EV, Gabrieli JD: Sniffing and smelling: separate subsystems in the human olfactory cortex. Nature 1998;392:282–286.

3 Sobel N, Thomason ME, Stappen I, Tanner CM, Tetrud JW, Bower JM, Sullivan EV, Gabrieli JD: An impairment in sniffing contributes to the olfactory impairment in Parkinson's disease. Proc Natl Acad Sci USA 2001;98:4154–4159.

4 Ansari KA, Johnson A: Olfactory function in patients with Parkinson's disease. J Chronic Dis 1975;28:493–497.

5 Ross W, Petrovitch H, Abbott RD, Tanner CM, Popper, Masaki K: Association of olfactory dysfunction with risk of future Parkinson's disease. Mov Disord 2005;20(Suppl 10):439.

6 Hughes AJ, Daniel SE, Lees AJ: Improved accuracy of clinical diagnosis of Lewy body Parkinson's disease. Neurology 2001;57:1497–1499.

7 Hughes AJ, Daniel SE, Kilford L, Lees AJ: Accuracy of clinical diagnosis of idiopathic Parkinson's disease: a clinico-pathological study of 100 cases. J Neurol Neurosurg Psychiatry 1992;55: 181–184.

8 Doty RL, Deems DA, Stellar S: Olfactory dysfunction in parkinsonism: a general deficit unrelated to neurologic signs, disease stage or disease duration. Neurology 1988;38:1237–1244.

9 Hawkes CH, Shephard BC, Daniel SE: Olfactory dysfunction in Parkinson's disease. J Neurol Neurosurg Psychiatry 1997;62:436–446.

10 Crino PB, Martin JA, Hill WD, Greenberg B, Lee VM, Trojanowski JQ: Beta-amyloid peptide and amyloid precursor proteins in olfactory mucosa of patients with Alzheimer's disease, Parkinson's disease and Down syndrome. Ann Otol Rhinol Laryngol 1995;104:655–661.

11 Duda JE, Shah U, Arnold SE, Lee VM, Trojanowski JQ: The expression of alpha-, beta-, and gamma-synucleins in olfactory mucosa from patients with and without neurodegenerative diseases. Exp Neurol 1999;160:515–522.
12 Daniel SE, Hawkes CH: Preliminary diagnosis of Parkinson's disease using olfactory bulb pathology. Lancet 1992;340:186.
13 Pearce RKB, Hawkes CH, Daniel SE: The anterior olfactory nucleus in Parkinson's disease. Mov Disord 1995;10:283–287.
14 Braak H, Del Tredici K, Rub U, de Vos RA, Jansen Steur EN, Braak E: Staging of brain pathology related to sporadic Parkinson's disease. Neurobiol Aging 2003;24:197–211.
15 Ross GW, Petrovitch H, Abbott RD, Nelson J, Markesbery W, Davis D, Hardman J, Launer L, Masaki K, Tanner CM, White LR: Parkinsonian signs and substantia nigra neuron density in descendents elders without PD. Ann Neurol 2004;56:532–539.
16 Huisman E, Uylings HB, Hoogland PV: A 100% increase of dopaminergic cells in the olfactory bulb may explain hyposmia in Parkinson's disease. Mov Disord 2004;19:687–692.
17 Yamada M, Onodera M, Mizuno Y, Mochizuki H: Neurogenesis in olfactory bulb identified by retroviral labeling in normal and MPTP-treated adult mice. Neuroscience 2004;124:173–181.
18 Braak H, Rub U, Gai WP, Del Tredici K: Idiopathic Parkinson's disease: possible routes by which vulnerable neuronal types may be subject to neuroinvasion by an unknown pathogen. J Neural Transm 2003;110:517–536.
19 Quinn NP, Rossor MN, Marsden CD: Olfactory threshold in Parkinson's disease. J Neurol Neurosurg Psychiatry 1987;50:88–89.
20 Doty RL, Stern MB, Pfeiffer C, Gollomp SM, Hurtig HI: Bilateral olfactory dysfunction in early stage treated and untreated idiopathic Parkinson's disease. J Neurol Neurosurg Psychiatry 1992;55:138–142.
21 Stern MB, Doty RL, Dotti M, et al: Olfactory function in Parkinson's disease subtypes. Neurology 1994;44:266–268.
22 Hawkes CH, Shephard BC: Selective anosmia in Parkinson's disease? Lancet 1993;341: 435–436.
23 Daum RF, Sekinger B, Kobal G, Lang CJ: Olfactory testing with 'sniffin' sticks' for clinical diagnosis of Parkinson disease. Nervenarzt 2000;71:643–650.
24 Kobal G, Plattig KH: Objective olfactometry: methodological annotations for recording olfactory EEG-responses from the awake human. EEG EMG Z Elektroenzephalogr Elektromyogr Verwandte Geb 1978;9:135–145.
25 Barz S, Hummel T, Pauli E, Majer M, Lang CJ, Kobal G: Chemosensory event-related potentials in response to trigeminal and olfactory stimulation in Parkinson's disease. Neurology 1997;49: 1424–1431.
26 Markoupoulou K, Larsen KW, Wszolek EK, Denson MS, Lang AE, Pfeiffer RF, Wszolek ZK: Olfactory dysfunction in familial parkinsonism. Neurology 1997;49:1262–1267.
27 Katzenschlager R, Lees AJ: Olfaction and Parkinson's syndromes: its role in differential diagnosis. Curr Opin Neurol 2004;17:417–423.
28 Montgomery EB Jr, Baker KB, Lyons K, Koller WC: Abnormal performance on the PD test battery by asymptomatic first-degree relatives. Neurology 1999;52:757–762.
29 Berendse HW, Booij J, Francot CM, Bergmans PL, Hijman R, Stoof JC, Wolters EC: Subclinical dopaminergic dysfunction in asymptomatic Parkinson's disease patients' relatives with a decreased sense of smell. Ann Neurol 2001;50:34–41.
30 Sommer U, Hummel T, Cormann K, Mueller A, Frasnelli J, Kropp J, Reichmann H: Detection of presymptomatic Parkinson's disease: combining smell tests, transcranial sonography, and SPECT. Mov Disord 2004;19:1196–1202.
31 Liberini P, Parola S, Spano PF, Antonini L: Olfaction in Parkinson's disease: methods of assessment and clinical relevance. J Neurol 2000;247:88–96.
32 Liberini P, Parola S, Spano PF, Antonini L: Olfactory dysfunction in dementia associated with Lewy bodies. Parkinsonism Relat Disord 1999;5:30.
33 McShane RH, Nagy Z, Esiri MM, King E, Joachim C, Sullivan N, Smith AD: Anosmia in dementia is associated with Lewy bodies rather than Alzheimer's pathology. J Neurol Neurosurg Psychiatry 2001;70:739–743.

34 Kovacs T, Papp MI, Cairns NJ, Khan MN, Lantos PL: Olfactory bulb in multiple system atrophy. Mov Disord 2003;18:938–942.
35 Wenning GK, Shephard B, Magalhaes M, Hawkes CH, Quinn NP: Olfactory function in multiple system atrophy. Neurodegeneration 1993;2:169–171.
36 Muller A, Mungersdorf M, Reichmann H, Strehle G, Hummel T: Olfactory function in Parkinsonian syndromes. J Clin Neurosci 2002;9:521–524.
37 Abele M, Riet A, Hummel T, Klockgether T, Wullner U: Olfactory dysfunction in cerebellar ataxia and multiple system atrophy. J Neurol 2003;250:1453–1455.
38 Wenning GK, Shephard BC, Hawkes CH, Lees A, Quinn N: Olfactory function in progressive supranuclear palsy and corticobasal degeneration. J Neurol Neurosurg Psychiatry 1995;57: 251–252.
39 Tsuboi Y, Wszolek ZK, Graff-Radford NR, Cookson N, Dickson DW: Tau pathology in the olfactory bulb correlates with Braak stage, Lewy body pathology and apolipoprotein epsilon4. Neuropathol Appl Neurobiol 2003;29:503–510.
40 Doty RL, Golbe LI, McKeown DA, Stern MB, Lehrach CM, Crawford D: Olfactory testing differentiates between progressive supranuclear palsy and idiopathic Parkinson's disease. Neurology 1993;43:962–965.
41 Baker KB, Montgomery EB Jr: Performance on the PD test battery by relatives of patients with progressive supranuclear palsy. Neurology 2001;56:25–30.
42 Katzenschlager R, Zijlmans J, Evans A, Watt H, Lees AJ: Olfactory function distinguishes vascular parkinsonism from Parkinson's disease. J Neurol Neurosurg Psychiatry 2004;75:1749–1752.
43 Hensiek AE, Bhatia K, Hawkes CH: Olfactory function in drug induced parkinsonism. J Neurol 2000;247(Suppl 3):P303.
44 Doty RL, Singh A, Tetrud J, Langston JW: Lack of major olfactory dysfunction in MPTP-induced parkinsonism. Ann Neurol 1992;32:87–100.
45 Ahlskog JE, Waring SC, Petersen RC, Esteban-Santillan C, Craig UK, et al: Olfactory dysfunction on Guamanian ALS, parkinsonism and dementia. Neurology 1988;51:1672–1677.
46 Doty RL, Perl DP, Steele JC, Chen KM, Pierce JD, Reyes P, Kurland LT: Odor identification deficit of the parkinsonism-dementia complex of Guam: equivalence to that of Alzheimer's disease and idiopathic Parkinson's disease. Neurology 1991;41(5 suppl 2):77–81.
47 Evidente VG, Esteban RP, Hernandez JL, Natividad FF, Advincula J, Gwinn-Hardy K, Hardy J, Singleton A, Singleton A: Smell testing is abnormal in 'lubag' or X-linked dystonia-parkinsonism: a pilot study. Parkinsonism Relat Disord 2004;10:407–410.
48 Busenbark KL, Huber SJ, Greer G, Pahwa R, Koller WC: Olfactory function in essential tremor. Neurology 1992;42:1631–1632.
49 Louis ED, Bromley SM, Jurewicz EC, Watner D: Olfactory dysfunction in essential tremor: a deficit unrelated to disease duration or severity. Neurology 2002;59:1631–1633.
50 Louis ED, Jurewicz EC: Olfaction in essential tremor patients with and without isolated rest tremor. Mov Disord 2003;18:1387–1389.
51 Shah M, Findley LJ, Muhammed N, Hawkes CH: Olfaction is normal in essential tremor and can be used to distinguish it from Parkinson's disease. Mov Disord 2005;20(Suppl 10):563.
52 Adler CH, Caviness JN, Sabbagh M, Hentz JG, Evidente VGH, Shill H: Olfactory testing in Parkinson's disease and other movement disorders: correlation with Parkinsonian severity. Mov Disord 2005;20(Suppl 10):231.
53 Connelly T, Farmer JM, Lynch DR, Doty RL: Olfactory dysfunction in degenerative ataxias. J Neurol Neurosurg Psychiatry 2003;74:1435–1437.
54 Fernandez-Ruiz J, Diaz R, Hall-Haro C, Vergara P, Fiorentini A, Nunez L, Drucker-Colin R, Ochoa A, Yescas P, Rasmussen A, Alonso ME: Olfactory dysfunction in hereditary ataxia and basal ganglia disorders. Neuroreport 2003;14:1339–1341.
55 Hawkes CH, Shephard BC, Geddes JF, Body GD, Martin JE: Olfactory disorder in motor neuron disease. Exp Neurol 1998;50:248–253.
56 Elian M: Olfactory impairment in motor neuron disease: a pilot study. J Neurol Neurosurg Psychiatry 1991;54:927–928.
57 Sajjadian A, Doty RL, Gutnick D, Chirurgi RJ, Sivak M, Perl D: Olfactory dysfunction in amyotrophic lateral sclerosis. Neurodegeneration 1994;3:153–157.

58 Talamo BR, Rudel R, Kosik KS, Lee VM, Neff S, Adelman L, Kauer JS: Pathological changes in olfactory neurons in patients with Alzheimer's disease. Nature 1989;23;337:736–739.
59 Trojanowski JQ, Newman PD, Hill WD, Lee VM: Human olfactory epithelium in normal aging, Alzheimer's disease, and other neurodegenerative disorders. J Comp Neurol 1991;310:365–376.
60 Yamagishi M, Ishizuka Y, Seki K: Pathology of olfactory mucosa in patients with Alzheimer's disease. Ann Otol Rhinol Laryngol 1994;103:421–427.
61 Braak H, Braak E: Evolution of neuronal changes in the course of Alzheimer's disease. J Neural Transm Suppl 1998;53:127–140.
62 Kovacs T, Cairns NJ, Lantos PL: Olfactory centres in Alzheimer's disease: olfactory bulb is involved in early Braak's stages. Neuroreport 2001;12:285–288.
63 Doty RL, Reyes PF, Gregor T: Presence of both odor identification and detection deficits in Alzheimer's disease. Brain Res Bull 1987;18:597–600.
64 Mesholam RI, Moberg PJ, Mahr RN, Doty RL: Olfaction in neurodegenerative disease: a meta-analysis of olfactory functioning in Alzheimer's and Parkinson's diseases. Arch Neurol 1998;55: 84–90.
65 Serby M, Larson P, Kalkstein D: The nature and course of olfactory deficits in Alzheimer's disease. Am J Psychiatry 1991;148:357–360.
66 Graves AB, Bowen JD, Rajaram L, McCormick WC, McCurry SM, Schellenberg GD, Larson EB: Impaired olfaction as a marker for cognitive decline: interaction with apolipoprotein E epsilon4 status. Neurology 1999;22:3:1480–1487.
67 Devanand DP, Michaels-Marston KS, Liu X, Pelton GH, Padilla M, Marder K, Bell K, Stern Y, Mayeux R: Olfactory deficits in patients with mild cognitive impairment predict Alzheimer's disease at follow-up. Am J Psychiatry 2000;157:1399–405.
68 Hawkes CH, Shephard BC: Olfactory evoked responses and identification tests in neurological disease. Ann NY Acad Sci 1998;855:608–615.
69 Koss E, Weiffenbach JM, Haxby JV, Friedland RP: Olfactory detection and identification performance are dissociated in early Alzheimer's disease. Lancet 1987;i:622.
70 Bacon AW, Bondi MW, Salmon DP, Murphy C: Very early changes in olfactory functioning due to Alzheimer's disease and the role of apolipoprotein E in olfaction. Ann NY Acad Sci 1998;855: 723–731.
71 Morgan CD, Murphy C: Olfactory event-related potential in Alzheimer's disease. J Int Neuropsychol Soc 2002;8:753–763.
72 Wetter S, Murphy C: Apolipoprotein E epsilon4 positive individuals demonstrate delayed olfactory event-related potentials. Neurobiol Aging 2001;22:439–447.
73 McKeown DA, Doty RL, Perl DP, Frye RE, Simms I, Mester A: Olfactory function in young adolescents with Down's syndrome. J Neurol Neurosurg Psychiatry 1996;61:412–414.
74 Wetter S, Murphy C: Individuals with Down's syndrome demonstrate abnormal olfactory event-related potentials. Clin Neurophysiol 1999;110:1563–1569.
75 Moberg PJ, Pearlson GD, Speedie LJ, Lipsey JR, Strauss ME, Folstein SE: Olfactory recognition: differential impairments in early and late Huntington's and Alzheimer's disease. J Clin Exp Neuropsychol 1987;9:650–664.
76 Nordin S, Paulsen JS, Murphy C: Sensory- and memory-mediated olfactory dysfunction in Huntington's disease. J Int Neuropsychol Soc 1995;1:281–290.
77 Hamilton JM, Murphy C, Paulsen JS: Odor detection, learning, and memory in Huntington's disease. J Int Neuropsychol Soc 1999;5:609–615.

C.H. Hawkes
Essex Neuroscience Centre, Oldchurch Hospital
Romford, Essex RM7 0BE (UK)
Tel./Fax +44 1708 708055, E-Mail chrishawkes@msn.com

Hummel T, Welge-Lüssen A (eds): Taste and Smell. An Update.
Adv Otorhinolaryngol. Basel, Karger, 2006, vol 63, pp 152–190

Human Taste: Peripheral Anatomy, Taste Transduction, and Coding

Paul A.S. Breslin, Liquan Huang

Monell Chemical Senses Center, Philadelphia, Pa., USA

Abstract

The anatomy, physiology and psychology of taste provide a glimpse into a uniquely heterogeneous sensory world; a world that is robust in its importance to flavor, redundant in its transductive heterogeneity and complexity, requisite in that feeding and hence life usually depend upon taste input, regenerative in that taste cells constantly turn over and regrow after tissue damage, and resistant to disease, loss of neural innervation and epithelial destruction. This chapter considers our current state of knowledge in anatomy, taste bud physiology, molecular biology of bitter, sweet, sour, savory and salty tastes, afferent signaling and quality coding, human perception, and pathophysiology and senescence of taste. We highlight some of the advances made in molecular biology of taste and point out areas where further research is needed ranging from taste bud development and regeneration, to within-taste bud processing, to central/perceptual coding networks for taste. Our hope is that this chapter will provide a background for greater understanding of taste physiology, perception, disease, and future sensory research.

Human Taste

Taste is the main sensory modality by which we evaluate whether a potential food is friend or foe. While we typically appreciate and evaluate a food's overall flavor as a gestalt, comprised of taste, odor, somatosensation and pain, it is taste that makes a substance seem like food. For example, while potential ingesta that are sweet, slightly sour, and odorless may pass as food depending on the context, items with an odor but that are tasteless are likely to appear as odorized Styrofoam. Everything we eat normally passes through the oral cavity, and so this portal provides a universal location for sensing and evaluating what should be digested and absorbed into the blood and what should not. Thus, the ultimate final motor outputs that result from taste stimulation involve the swallowing of food (acceptance) or the expectorating of food (rejection). If acceptable, there are other taste-cued reflexes, which anticipate and facilitate digestion of the forthcoming nutrients, involving both exocrine (e.g. gastric acid) and endocrine (e.g. insulin) secretions. These early taste-triggered secretions, labeled cephalic phase responses, are necessary for normal digestion and were the focus of Pavlov's famous digestion research [1, 2].

The acceptance of sweet tastes, signaling calories, and the rejection of strongly bitter tastes, signaling toxins, are brain stem reflexes apparent in humans prenatally [3, 4]. Our adult food preferences are built on top of these reflexes, which can be modified by experience but never eliminated. Thus, it is rare to find commercial foods that are strongly bitter, and among global markets for foods with requisite bitterness, such as coffee and beer, the less bitter the product the higher are global sales. Regardless of whether taste-guided behaviors are reflexive or are part of a more sophisticated developed appetite, the taste system must detect what is present in the mouth and enable discrimination and recognition of chemical components and levels there. These processes ultimately lead to the recognition of food as familiar or, when novel, as safe. When taste is severely disturbed, then feeding is almost always disturbed [5, 6].

Taste is arguably the only external sensory system required for life. It is well known that people live in our societies without sight, hearing, or smell. Even some somatosensory and proprioceptive deficits are overcome with visual feedback. But people without taste often do not eat and without medical intervention would die. This is most frequently seen in head and neck cancer patients who receive radiotherapy [7]. These patients typically experience a radiation-induced loss of taste that may be complete and always interferes with eating [8, 9]. They commonly require the placement of a chronic nasogastric tube for feeding, as they are an already nutritionally compromised population. In addition to its requisite role in feeding, taste in other species plays a role in detecting and identifying hydrocarbons that serve as a pheromonal social communication signal such as in courting and mating [10]. While olfaction is known to play a social communication role in humans, it is unknown whether taste in humans plays a similar social role [11].

Taste sensations may be divided into several psychological attributes: quality, intensity, oral location, and timing. All of these attributes are, in turn, rapidly evaluated and, within a given context, imbued with some degree of positive or negative hedonic value – yummy or yucky. The qualities of taste are the basic subdivisions of the modality labeled sweet, sour, salty, bitter and savory (or umami). Prototypical exemplars of these tastes are honey, lemon juice, table salt, strong black coffee, and chicken broth (or more specifically free glutamate). To evaluate the taste of these solutions without the influence of their intrinsic odors, simply taste them while pinching the nares of the nose closed. Without airflow, there is no sense of olfaction, which is why congestion can cause temporary loss of smell. This type of test is also clinically useful in patients who complain of taste loss. It is presently an area of active research whether identified fatty acid detection systems (fat perception) are part of the taste sensory system and, if so, whether these transduction systems generate an identifiable 'fat taste' quality [12–14]. One theory posits that fat receptors on taste receptor cells may be gustatory modulators or perhaps help trigger cephalic reflexes for fat metabolism.

The taste intensity is the magnitude of the qualitative sensations, such as weakly salty or strongly bitter. The location is the perceived region of the oral cavity from which a taste sensation arises. Taste unlike smell has a spatial dimension that enables humans to localize specific qualities and even to manipulate items in the mouth based on taste localization [15]. The timing of taste refers to whether taste sensations arise quickly and whether they linger, i.e. aftertastes. Timing and spatial cues are inherently used to evaluate stimuli and are largely responsible for why artificial sweeteners rarely taste like sucrose; most artificial sweeteners are localized more to the posterior oral cavity and linger longer than sucrose does.

Peripheral Taste Anatomy

Humans have taste receptors in several fields within the oral cavity including: all edges of the tongue and on the anterior dorsal surface of the tongue, on the soft palate, and in the pharyngeal and the laryngeal regions of the throat (fig. 1). Taste receptor cells predominantly reside within multicellular rosette clusters labeled 'taste buds' [16, 17]. There is presently a discussion as to whether single chemosensitive cells may also exist in humans; but there is evidence for solitary taste receptor cells in the larynx of rodents [18]. The taste buds on the dorsal surface and edges of the tongue occur in papillae: fungiform papillae pepper the anterior two thirds of the tongue, several continuous foliate papillae (folds) appear on the posterior lateral edges of the tongue, and circumvallate papillae (towers with motes) appear in an arc of approximately nine papillae on the posterior tongue just anterior to the lingual genu [17]. Other taste fields in the soft palate and pharynx reside in the epithelium but not within papillae. A lingual fungiform papilla can contain between 0 and 15 taste buds and averages approximately five taste buds [19]. Foliate and vallate papillae always contain taste buds and typically have many more than do fungiform papillae, often dozens. The taste buds live within the folds of foliate and in the 'mote' of the vallate papillae where von Ebner's glands secrete saliva and proteins into these recesses [20, 21].

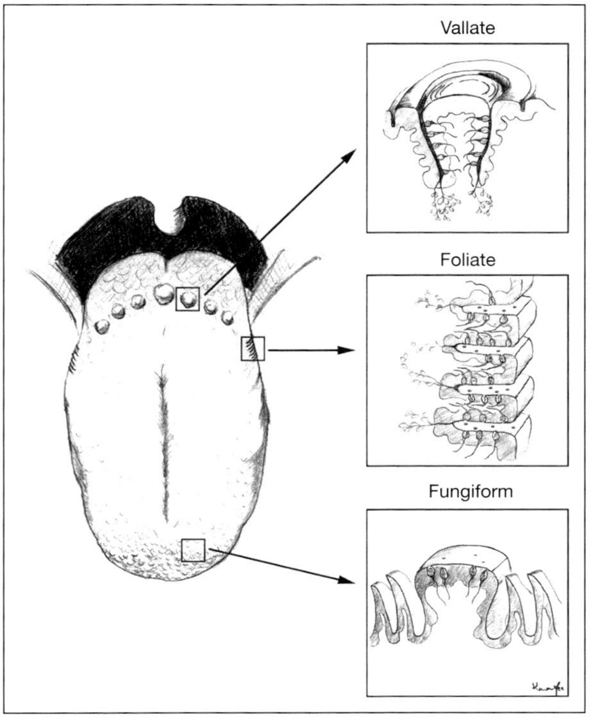

Fig. 1. The human tongue contains three types of taste papillae. Vallate and foliate papillae reside on the middle and sides of the posterior 1/3 of the tongue, respectively, and contain hundreds of taste buds. Fungiform papillae are scattered in the anterior 2/3 of the tongue, each harboring 0–15 taste buds. Taste buds are also located in the soft palate and pharynx but are in the flat epithelium rather than in papillae in these locations.

The receptor cells within a taste bud are not neural cells. Rather, they are specialized epithelial cells that share almost all of the same properties as neural cells except that they lack an axon [17, 22]. The afferent taste information is transmitted to neural fibers within the taste bud. The cell bodies of these taste neural fibers occur within the sensory ganglia of cranial nerves (CN) VII, IX, and X [6]. The fibers project into the central nervous system at the level of the brain stem and synapse initially onto the nucleus of the solitary tract [23, 24]. From there, afferent information projects cortically via the thalamus to the primary opercular and insular taste cortex, as well as to the orbitofrontal cortex, cingulate gyrus and to other multimodality integrative projection areas [25]. There is also a more ventral anterior pathway for taste signals that includes: the bed nucleus of the stria terminalis, hypothalamus, mammillary bodies, amygdala, hippocampus and other areas of the limbic system [25].

The facial nerve (CN VII) has its cell bodies in the geniculate ganglion and its taste information is carried by two nerve branches [6]. The chorda tympani nerve, so named because it passes through the middle ear behind the tympanum, innervates the whole anterior two thirds of the tongue including all fungiform and the most anterior foliate papillae [6]. The greater superficial petrosal nerve branch innervates the soft palate, which may contain as many taste buds as does the anterior tongue [6]. The glossopharyngeal nerve (CN IX) has its cell bodies in the petrosal ganglion and innervates most of the foliate and all of the vallate papillae in the posterior tongue via the lingual-tonsillar nerve branch [6]. The vagus nerve (CN X) has its cell bodies in the nodose ganglion and innervates taste buds in the pharynx and larynx via the superior laryngeal nerve branch [6]. A single axonal fiber of any of these nerves may innervate multiple cells within a taste bud and may also innervate multiple buds. Thus, the primary neural fibers are a potential site of taste signal integration. These fibers rarely have branches that cross the midline of the tongue, leaving the left and right sides of the tongue under independent peripheral control. Specific taste field loss that impacts both sides of the tongue is usually associated with peripheral damage to the receptor systems and in the tongue via physical or perhaps chemical trauma. Peripheral neural damage or central neural involvement is usually associated with asymmetrical effects. In addition to taste compound sensitivity, most, but not all, of these peripheral afferent fibers are touch and thermally sensitive in the anterior tongue and are additionally noxious stimulus sensitive in the posterior tongue.

The Taste Bud

The taste bud is a microscopic 'rosebud'-shaped structure that contains between 60–120 cells [26–28] (fig. 1). The receptor cells involved in primary taste signal transduction are in direct contact with the solutions of the oral cavity via microvilli at the apical end of the cells [17]. The microvilli contact the oral solutions via a small (approx. 20 μm) opening in the epithelium called the taste pore that lies at the tip of each bud [29]. Chemical stimuli are restricted in their flow past taste cells at the taste pore by tight junctions linking the cells in contact with the pore; generally, only small ions may pass tight junctions [29–31]. Adult human fungiform papillae contain approximately 4 or 5 taste buds on average although many contain zero buds [32, 33]. The foliate papillae contain several buds on either side of the epithelial walls that comprise each foliate groove [17]. Similarly many taste buds line the papillary sidewall of the 'mote' around each vallate papillae [17]. In contrast, the vast majority of papillae on the tongue are filiform, small conical-shaped papillae, which never contain taste buds [34]. These seem to serve the function of making the lingual surface mechanically rough, which facilitates food and beverage manipulation and may also enhance lingual somatosensory function. The larger and less dense fungiform papillae may be observed among the smaller more plentiful filiform papillae by simple direct visual examination of the anterior dorsal surface of the tongue.

There are four principal types of cells within a taste bud, but these may be further subdivided histochemically. The cell types were historically labeled dark, light, intermediate and basal cells based upon their electron-dense appearance, shape, and position in an electron microscope image of a taste bud [28, 35]. The basal cells were small round cells at the base of the taste bud. The other three cell types were elongated cells stretching from the basal to the apical end of the taste bud and appearing dark, light, and intermediate [36, 37]. Today these cell types are referred to as type I, II, and III cells, respectively [26–28, 38, 39]. Both type I and II cells possess microvilli with those of type II cells shorter than those of type I [35, 40]. Most primary signal transduction components such as receptors and effector enzymes are found only in type II cells [41]. This observation has led some to conclude that type II cells are the canonical taste receptor cell. However, most synapses with primary afferent axons are on type III cells, many of which are identified as being serotonergic [42–44]. In mice, synapses have also been occasionally identified in type I and II cells [27, 28, 44]. Rats, mice and rabbits appear to possess significant species differences in taste bud organization. The precise configuration of human taste buds remains to be determined [36, 37, 45].

An important question is how information passes from type II cells, where the transduction elements reside, to type III cells, where most synapses with primary afferents occur [44]. Bud cells can be electrically coupled via gap junctions, but more likely they communicate with one another neurochemically. Taste bud cells are known to possess serotonin and serotonin receptors, ATP, ADP and their P2X and P2Y receptors, glutamate and its mGluR1, mGluR4 and ionotropic receptors, as well as nitric oxide (NO) synthase and NO, which is a gas that freely enables cells to communicate with each other. These are all possible candidates for cell-to-cell chemical communication within the taste bud, so that primary receptor cells without neural synapses could communicate with taste bud cells with neural synapses [46–50]. Since the taste bud contains receptor cells specialized for different classes of chemicals, the problem arises as to how to coordinate the cells within a bud of a similar chemical sensitivity. This could be accomplished intragemmally pan-bud by the coordination of the chemical identity of what is released with the appropriate receptors on target cells within the bud. For example, glutamate might be a bitter compound intercellular signal within the bud, while serotonin might be a sweet compound intercellular signaler [47].

While taste bud cells differ from olfactory receptor cells in that they are not neurons, they are, however, similar to olfactory receptor cells in that they have a short life span and are replaced continuously throughout the life of the bud. The life of a taste bud cell is approximately 10 days [51]. Basal cells were believed to be the sole progenitors of the elongated cells within the taste bud and to give rise to the three elongate cell types, but this is not necessarily the case [26, 52, 53]. The stem cells that give rise to taste bud cells reside outside the bud, near its base in the stratum germinativum, and continuously migrate into the bud to generate new cells [52, 53]. Moreover, the cell types are not all different stages along a single cell's life cycle. Rather, the stem cells that give rise to different elongate bud cells are of different origins; that is, the elongate cells seem to have different lineages [54]. The exact role that each bud cell plays remains uncertain. For example, despite the fact that type I cells have many long microvilli extending into the taste pore and may occasionally have synapses with neurons in some species, it is unclear whether they are involved directly in signal transduction. Some have hypothesized that they play a secretory role for the taste pore and/or may even serve a glia-like function for the bud [55].

Bitterness and the TAS2R Receptors

Insensitivity to bitter-tasting compounds was serendipitously discovered in 1931 when it was found that some people could not identify the taste of a compound called phenylthiocarbamide (PTC) as strongly bitter, although the majority of people could [56]. Subsequent studies demonstrated that humans displayed this type of taster and nontaster bimodality in many structurally related compounds containing the N–C=S chemical moiety. This trait was found to be heritable. Since then genetic studies have identified several loci on both human and mouse chromosomes that control the sensitivity to several bitter compounds. For example, the PTC loci on human chromosomes 5p15 7q31 control the response to PTC and PROP [57, 58]; on the distal end of mouse chromosome 6, loci SOA [59–61], RUA [62], CYX [63] and QUI [64] are tightly linked to the response to sucrose octaacetate, raffinose undecaacetate, cycloheximide and quinine, respectively. Recently, progress in human and mouse genome sequencing projects has made it possible to identify a novel subfamily of G-protein-coupled receptors (GPCRs) called type 2 taste receptors or T2Rs as bitter taste receptors [65–67].

Several lines of evidence suggested that T2Rs are the receptors for bitter substances. In situ hybridization demonstrated that these receptors are expressed in a subset of taste bud cells. Interestingly, a taste bud cell may have many of these T2Rs [66] and each one of these receptors could be specifically activated by only a few bitter compounds. Therefore, while a T2R may only be able to recognize a few bitter substances with high specificity and sensitivity, a taste bud cell with multiple T2Rs should detect a broad range of bitter and usually toxic compounds with equally high sensitivity. One caveat here is that these cells may not be able to respond differentially to one bitter substance over another. This may explain why many different bitter compounds evoke similar bitterness perception in humans.

Behavioral studies with taster and nontaster mouse strains indicated that variations in the mouse T2R5 gene determine the animal's response to a bitter compound, cycloheximide. T2R5 is located in the CYX locus at the distal end of mouse chromosome 6. Sequence analysis showed that all taster strains (e.g. DBA/2) have the same mT2R5 nucleotide sequences while nontasters (e.g. C57BL/6 and 129/Sv) have missense mutations and consequently amino acid substitutions. A heterologously expressed mT2R5 receptor can be activated by cycloheximide. But a much higher concentration of cycloheximide is required to elicit a response with comparable amplitude via a receptor from a nontaster mouse strain compared to a taster's receptor [67].

Heterologous expression and ligand screening have so far determined bitter compounds for a few receptors. In addition to the cycloheximide receptor mouse T2R5, the rat counterpart, rT2R9, can be stimulated by cycloheximide as well, while the human paralogue hT2R10 can detect the bitter compound strychnine [68]. Human T2R4 and mouse mT2R8 respond to denatonium and to a lesser extent PROP [67]. Human T2R16 can be stimulated by a group of β-glucopyranosides including salicin, which is an extract from willow bark, which has been used as an antipyretic and an analgesic in the treatment of rheumatism for at least 3,500 years [69]. Human T2R43 and T2R44 have similar ligand profiles: both can be activated by aristolochic acid, nitrosaccharin and at higher concentrations, the artificial sweeteners Na-saccharin and acesulfame-K. But T2R44 can also be activated by denatonium and is more sensitive to the artificial sweeteners than T2R43. This observation explains why these sweeteners taste bitter at high concentrations [70, 71]. Surprisingly, human T2R14 can be activated by a number of apparently structurally unrelated chemicals, including picrotoxin and picrotin found in fishberry seeds, (–)-α-thujone, a psychotropic component of absinthe from wormwood, and sodium benzoate found in many plants and frequently used as an antimicrobial preservative [72]. Two of the most commonly used bitter compounds in human bitter taste studies, PTC and its structurally similar proxy PROP, can be recognized via the human receptor T2R38 [73]. However, PROP may be able to activate other T2Rs such as the aforementioned T2R4 receptor [73].

Identification of ligands for each bitter receptor is necessary for establishing the entire bitter tasting profiles for humans. However, only a few of T2Rs have been deorphanized. On the other hand, genomic and evolution analyses of T2R repertoires have provided some insights into bitter taste ability of humans as a species. The human T2R gene repertoire comprises 25 functional TAS2R genes and 8–11 nonfunctional sequences that contain premature stop codons, referred to as pseudogenes [68, 74–76]. In contrast, rodents have a larger T2R repertoire (43 sequences) and a lower proportion of pseudogenes (7 out of 43 or 16% vs. 31% in humans) [75, 77]. The majority of both human and rodent TAS2R genes and pseudogenes are clustered at two locations: human T2Rs reside on chromosomes 7 and 12, with only one gene on chromosome 5; most of mouse and rat TAS2R sequences reside in two regions of one chromosome (chromosome 6 in mice; chromosome 4 in rats). These two regions on the rodent chromosome are syntenic to the corresponding human clusters on two separate chromosomes, indicating that the overall arrangement of TAS2R genes was developed in a common ancestor to the primate and rodent lineages (fig. 2). Genes within the same cluster share higher sequence similarity than those between clusters, suggesting that these genes were generated by duplication [78, 79]. Interspecific analysis of T2R sequences showed that human and mouse T2R genes and pseudogenes can be categorized into three groups. In group A, multiple human genes are orthologous to one mouse gene, i.e., multiple-to-one orthology. Nine human genes and 6 pseudogenes can be classified into this group. In group B, a single human gene is orthologous to multiple mouse genes, i.e., one-to-multiple orthology; three human genes fall into this group. In group C, which consists of 10 human genes and 3 pseudogenes, one human gene is orthologous to one mouse gene. This classification implicates that during the mammalian radiation, humans and rodents underwent species-specific gene duplication to adapt to their unique ecological niches, as shown in group A and B T2R genes, respectively, while group C T2Rs seem essential to both humans and rodents to detect some common bitter substances. This classification seems to be supported by receptor expression and ligand screening data. For example,

Fig. 2. Orthology relationships between human and mouse T2R bitter receptors based on the phylogenetic analyses of amino acid sequences (with permission from Go et al. [78]). Human and mouse receptors are shown as solid and shaded, respectively. Pseudogenes are indicated by asterisks (*). In group A, multiple human receptors are orthologous to one mouse receptor (multiple-to-one orthology). In group B, one human receptor is orthologous to one mouse receptor (one-to-one orthology). In group C, one human receptor is orthologous to multiple mouse receptors (one-to-multiple orthology). Note: a few human T2Rs are uncategorized.

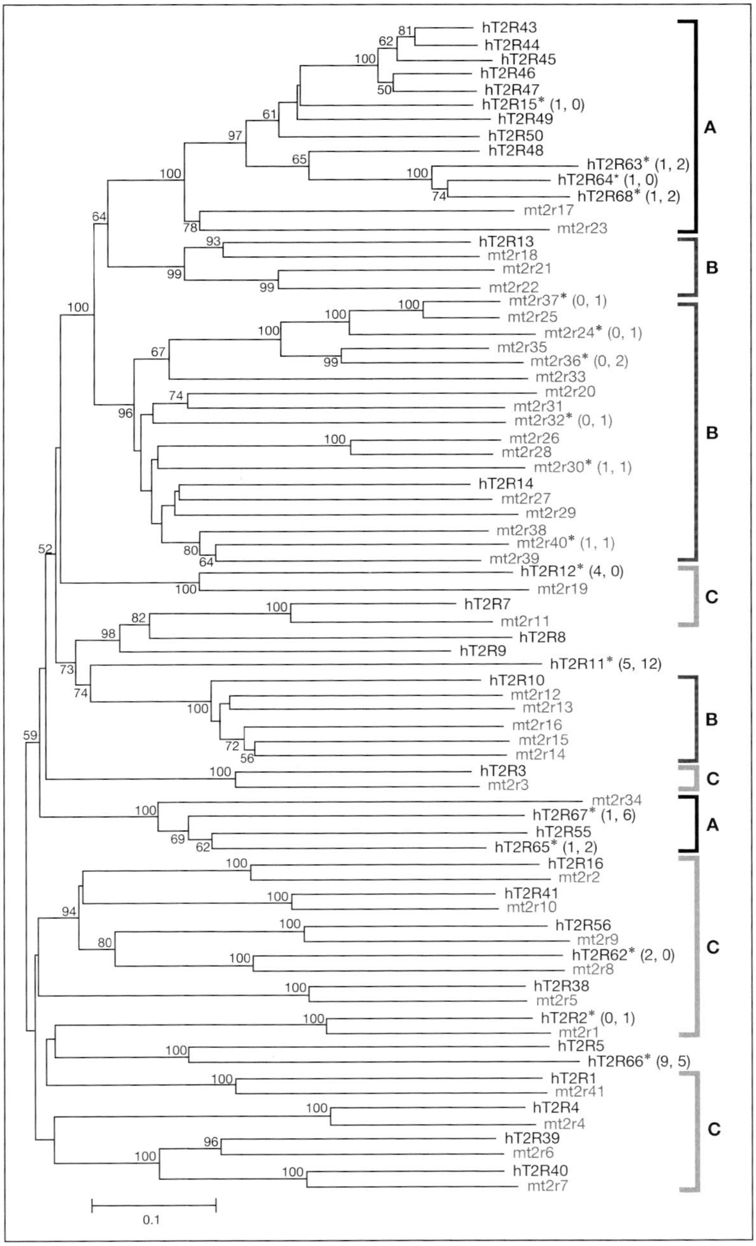
hT2R43
hT2R44
hT2R45
hT2R46
hT2R47
hT2R15* (1, 0)
hT2R49
hT2R50
hT2R48
hT2R63* (1, 2)
hT2R64* (1, 0)
hT2R68* (1, 2)
mt2r17
mt2r23
A
hT2R13
mt2r18
mt2r21
mt2r22
B
mt2r37* (0, 1)
mt2r25
mt2r24* (0, 1)
mt2r35
mt2r36* (0, 2)
mt2r33
mt2r20
mt2r31
mt2r32* (0, 1)
mt2r26
mt2r28
mt2r30* (1, 1)
hT2R14
mt2r27
mt2r29
mt2r38
mt2r40* (1, 1)
mt2r39
B
hT2R12* (4, 0)
mt2r19
hT2R7
mt2r11
C
hT2R8
hT2R9
hT2R11* (5, 12)
hT2R10
mt2r12
mt2r13
mt2r16
mt2r15
mt2r14
B
hT2R3
mt2r3
C
mt2r34
hT2R67* (1, 6)
hT2R55
hT2R65* (1, 2)
A
hT2R16
mt2r2
hT2R41
mt2r10
hT2R56
mt2r9
hT2R62* (2, 0)
mt2r8
hT2R38
mt2r5
hT2R2* (0, 1)
mt2r1
C
hT2R5
hT2R66* (9, 5)
hT2R1
mt2r41
hT2R4
mt2r4
hT2R39
mt2r6
hT2R40
mt2r7
C
0.1

hT2R14 is a group B receptor, which has 17 orthologs in mice. Human T2R14 could be an ancient and promiscuous receptor while the mouse has developed a more sophisticated system to be able to more sensitively and specifically recognize each of these compounds of this group. The same could be true for another group B receptor, human T2R10, which has 5 orthologs in mice. One of the orthologs, mT2R5, may have been more narrowly and specifically tuned for a structurally related but different ligand. Human T2R43 and T2R44 belong to group A, suggesting that they could be newly expanded, human species-specific receptors, which have finely tuned, somewhat overlapping recognitions.

Another striking finding is that like in olfactory and pheromone systems, humans have more T2R pseudogenes. However, no pseudogene is found in group B, in which rodent genes expanded, suggesting that humans maintained a basic but important ability to detect toxic substances that rodents more frequently encounter. However, a large fraction of pseudogenes are present in group A, suggesting that pseudogenization has rapidly been accumulated in the human-specific repertoire of T2R genes, as well as in the group of common genes. This notion is corroborated by the comparative analysis with the chimpanzee T2R repertoire, in which 2 of these 11 human pseudogenes are not mutated and supposedly still functional in the chimp, indicating that since the time that human and chimpanzee lineages split about 6 million years ago, humans lost 2 more bitter receptors [79]. The deterioration of the bitter tasting ability in humans may reflect the environmental and dietary changes during human evolution.

Substantial variation in human taste abilities has also been found among different individuals [80–82]. Single nucleotide polymorphisms have been found in many T2R genes of many individual people. The human population exhibits unusually higher levels of allelic mutations in T2R genes than in other genes. These polymorphisms may bestow on the individual the ability to sense new compounds or ignore certain compounds such as the aforementioned PTC/PROP. Emerging pseudogenes have also been discovered in a particular group of people as well as in the worldwide population, suggesting that loss of function in bitter taste is an ongoing event.

Like many other GPCRs, T2Rs have 7-transmembrane domains as well as some conserved amino acid residues. In contrast to T1Rs (see below), T2Rs have short N- and C-termini. Individual members of the T2R subfamily exhibit a high degree of similarity with 30–70% of the amino acid sequence identical to other family members. The most divergent segments are the extracellular loops, and swapping of these loops between some T2Rs indicated that they participate in the binding of structurally diverse bitter compounds and contribute to the ligand specificity [70].

Comparison of human and rodent genomes reveals that humans have a larger proportion of TAS2R pseudogenes than rodents, which is common for other chemosensory gene families such as olfactory and pheromone receptor genes [83, 84]. The exact roles of these pseudogenes are not known. This pseudogenization may reflect the fact that humans have encountered fewer toxic bitter compounds in their diet during their recent evolution.

Sweetness, Umaminess (Savoriness) and the TAS1R Receptors

Sweet and umami (or savory) tastes are thought to measure the carbohydrate and amino acid contents in food, hence its energy density. We have known for decades that sweet taste is likely transduced by membrane-bound protein receptors, since proteolytic (pronase) treatment of the tongue surface transiently abolishes sweetness perception [85]. Further, gold- or radio-labeled sugars and sweet-tasting peptides and proteins can bind to the membrane portion of lingual epithelium or to the taste pore of taste buds [86, 87], where the presumed sweet receptors reside.

The discovery of sweet and umami receptors was also greatly facilitated by mouse genetic studies and advances in genomic sequencing for model organisms. Animal behavioral assays have reported that inbred mouse strains exhibit a bimodal distribution of sensitivity to sweet compounds including saccharin, i.e., a taster or nontaster trait. The determinant for this trait has been attributed to a Sac locus on the distal end of mouse chromosome 4 that was named after saccharin [59, 88–91]. Data mining of this locus and the syntenous region on human chromosome 1 identified a novel GPCR, T1R3, which has the highest amino acid sequence similarity to two previously isolated orphan receptors from taste tissue, T1R1 and T1R2 [92–97]. Sequence analysis revealed that mutations in this T1R3 gene of nontaster mouse strains are likely to be responsible for their low sensitivity to saccharin and other sweeteners. Introduction of the taster version of the T1R3 gene into the nontaster mice by backcrossing or transgenic manipulation rescues the sweet taste deficit [92, 98], further confirming that T1R3 is responsible for Sac phenotypes.

T1R3, as well as T1R1 and T1R2, is classified based on sequence identity as a member of the class C GPCR subfamily [99]. Other members of this subfamily include calcium-sensing receptor, putative pheromone receptor V2Rs, neurotransmitter receptors mGluRs and $GABA_B$Rs. One of the hallmarks of this subfamily is receptor dimerization. In situ hybridization has shown that T1R1 and T1R2 are indeed co-localized with T1R3 to a subset of taste bud cells while a few taste bud cells express T1R3 alone [95, 98]. Thus, three types of T1R-expressing taste receptor cells exist: cells expressing both T1R1 + T1R3, both T1R2 + T1R3, or T1R3 alone [95, 98]. Coexpression of T1R receptors in heterologous systems has deorphanized all three of them [98, 100, 101]. Human

T1R2/T1R3 heterodimers can be activated by all sweet-tasting compounds tested at physiological concentrations, including sugars: fructose, glucose, lactose sucrose and maltose; amino acids: glycine and *D*-tryptophan; sweet proteins: monellin and thaumatin; synthetic sweeteners: acesulfame-K, aspartame, cyclamate, dulcin, neotame and saccharin. Cells transfected with T1R3 alone can also respond to some of these compounds at much higher concentrations, presumably via T1R3 homodimers.

Interestingly, T1R3 can also form a functional heterodimeric receptor with T1R1. The T1R1 and T1R3 dimer can be activated by many *L*-amino acids, including monosodium glutamate (MSG) and *L*-aspartate. The activation can be potentiated by 5′-ribonucleotides such as inosine monophosphate and guanosine monophosphate (GMP), which is a distinct feature of umami taste. These data strongly suggest that T1R1/T1R3 heteromers are amino acid sensors and function as an umami taste receptor.

Are there other receptors for sweet and umami tastes? Genetic studies indicate that loci other than the Sac locus seem to be associated with the detection of some sweet-tasting amino acids such as glycine [102]. T1R2 or T1R3 null mutant mice exhibited residual preference for sugars [103, 104]. However, the double knockout of both T1R2 and T1R3 eliminated responses to sweeteners [103], suggesting that either T1R2 and T1R3 homomeric receptors or an unknown T1R2/T1R3 interacting moiety was responsible for the residual preference in the single-knockout animals. A truncated form of the metabotropic glutamate receptor taste-mGluR4 has been proposed to be another umami receptor [105, 106]. However, mice lacking this receptor showed enhanced glutamate detection [107], indicating that this receptor may be playing other roles in taste receptor cells. Taste forms of mGluR1 have also been identified in taste receptor cells [108, 109].

How can a single dimeric receptor recognize so many structurally diverse sweet compounds, from simple glucose to glycine to large sweet-tasting proteins? A distinct feature of class C GPCRs is that most of them have a large extracellular amino terminus, which consists of two domains: a clam shell-like ‘Venus flytrap module’ (VFTM) and a cysteine-rich domain, followed by a heptahelical transmembrane domain and an intracellular carboxyl-terminal domain [99]. Molecular studies have elucidated multiple ligand binding sites in each monomer of the T1R2/T1R3 receptor [110–113]. For example, aspartame and neotame interact with the human VFTM domain of T1R2; brazzein and cyclamate bind to the human T1R3 cysteine-rich domain and extracellular loops 2 and 3 of the heptahelical transmembrane domain, respectively; lactisole, a potent inhibitor of sweet taste, docks to the human T1R3 binding pocket formed by transmembrane helices. One of the unexpected but understandable findings is that lactisole can also suppress umami taste by interacting with the T1R3 moiety of the T1R1/T1R3 dimer [114].

All 3 T1R genes have been mapped onto human chromosome 1 [115]. They are organized in the order of T1R1, T1R2 and T1R3. However, T1R1 and T1R2 genes are in the same transcriptional orientation, while T1R3 is in the opposite orientation. All 3 T1R genes have a similar gene structure, consisting of 6 coding exons. This genomic organization has been more or less preserved during evolution. In mice, T1R genes are located on the syntenic region of mouse chromosome 4, although the gene arrangement differs slightly; the T1R1 gene is in the middle, flanked by T1R2 and T1R3. Unlike other GPCRs, which share >90% amino acid identity across mammals, human and rodent umami, sweet, and bitter receptors display only about 70% sequence identity, suggesting that taste receptors have evolutionarily tuned to their species-specific needs and may have contributed to the formation of species-specific behaviors and diets. Mutations and pseudogenization of T1Rs in humans have not been well characterized yet. But in the cat, which has normal T1R1 and T1R3 receptors, microdeletion and stop codons have been found in the T1R2 gene, resulting in the lack of T1R2 expression and the functional T1R2/T1R3 sweet receptor [116]. Pseudogenization of cat T1R2 may have been crucial to the cats' indifference toward sugars and their favoring meat or, more likely, their carnivory may have relaxed pressures to maintain the T1R2 receptor; the direction of the causal arrow is uncertain at this time. Curiously, some fish species possess multiple copies of T1R2, which could generate variants of amino acid or carbohydrate receptors [117]. In some inbred mouse strains, nonsynonymous substitutions in the T1R3 VFTM domain lead to low sensitivity to sweeteners. In addition, rodents are not subject to lactisole inhibition of their sweet taste while humans are. But mutations in other regions of T1R2 and T1R3 made rodents and new world monkeys unable to taste a number of compounds that humans and old world primates deem sweet, including monellin, brazzein, thaumatin, neotame, aspartame, cyclamate and neohesperidin dihydrochalcone [118–123]. These species differences may arise due to relaxed selective pressure on allosteric binding sites relative to the orthosteric sites on VFTMs that evolved to bind canonical saccharides such as glucose, fructose, and sucrose, which virtually all omnivores and herbivores taste [114, 124–126]. Polymorphism studies on these 3 human T1R genes might reveal individual differences in sensitivity to sweet and umami substances, which could help illuminate individual preferences for flavors.

Salty and Sour Taste Receptors

Salt taste is elicited by Na^+ and other cations, which may enable humans and animals to seek out mineral-rich food but avoid oversalty food to maintain ion-water homeostasis. The need to detect ions such as sodium arises from our inability to store these ions in our bodies, unlike calories or calcium. Thus, all terrestrial animals and fresh-water animals must ingest sodium regularly. Like sour taste, saltiness is likely received by ion channels. Permeation of Na^+ and other ions into taste bud cells depolarizes membrane potentials, triggering influx of calcium and then the release of neurotransmitters. It has been postulated for years that a sodium channel, epithelial sodium channel ENaC, is the channel receptor for salty taste [127–130].

ENaC is a heterotetrameric protein, consisting of 2α-, 1β- and 1γ-subunits in vodents. Heterologously expressed ENaCs of all three subunits displayed susceptibility to topical amiloride inhibition, which was in agreement with behavioral and nerve recording data from some rodent strains [127, 131–133]. This was an important discovery as amiloride was believed to be specific to epithelial sodium channels. However, amiloride does not inhibit much of human salt taste, which is rather inhibited by another compound, chlorhexidine, suggesting that the stoichiometry of human ENaCs may be different from rodents' or a totally different channel is responsible for human salt taste [134, 135]. Rodents also possess a component of sodium taste that is not amiloride sensitive [136]. The molecular basis for this salt taste aspect is unclear but the heat- and capsaicin-sensitive TRP receptor VR1 has been implicated [137]. In addition, hormones like aldosterone could change the expression levels of these channels, thus regulating the sensitivity to salty stimuli and to suppression of these inhibitors [138]. At present, the transduction mechanism for human salt taste is unknown.

Sour taste is produced by acids, which play somewhat ambivalent roles affectively. On the one hand, sour taste in some types of food appears to be attractive to humans and animals such as oranges and grapefruits, or sour candy. On the other hand, sourness from spoiled foods and unripe fruits evokes rejection response. Human psychophysical studies and animal nerve recording showed that perceived sourness is proportional to the concentrations of protons, i.e., pH in strong inorganic acids such as HCl, but only a low correlation was observed between sour taste and pH in organic acids such as citric acid, indicating that anions in acidic stimuli also contribute to sour taste intensity [139, 140].

Sour taste is believed to be received by ion channels [55]. However, the identity of these channel receptors has not been firmly established. One reason is that proton is an active agent, can interact with and regulate virtually all ion channels. Another confounding factor is that different animal species appear to use different mechanisms [140].

Several ion channels have been hypothesized to transduce sourness, which include: (1) direct blockage of the apical potassium channel by protons, thus leading to depolarization of membrane potentials [141]; (2) passage of protons through the ENaC [142]; (3) activation of acid-sensing ion channels (ASICs)

[143]; (4) activation of a 5-nitro-2-(3-phenylpropylamino)benzoic acid-sensitive chloride channel, which could also contribute to changes in membrane potentials and regulate the uptake of neurotransmitters into synaptic vesicles [144, 145]; (5) activation of hyperpolarization-activated and cyclic nucleotide-gated ion channels [146]; (6) intracellular acidification [147]. In addition, there are GPCRs that detect protons and could be involved in sour taste.

Much of the effort has been focused on ASICs, and several of them including ASIC2 have been found in human taste buds [143, 148, 149]. In particular, heterologously expressed ASIC2 showed features of acid-induced currents resembling that observed in taste bud cells [149]. However, this channel is not present in mouse taste bud cells and the knockout of this gene did not affect sour taste in mutant mice, suggesting that ASIC2 is not a sour taste channel, at least not in mice [150].

Preliminary studies with human subjects revealed a unimodal distribution of sour sensitivity with a large variation in detection thresholds and some patients exhibiting sour blindness [151]. More data are needed to understand sour taste mechanisms and to find treatments of sour taste disorders.

Signal Transduction Cascades

Bitter stimulus (T2Rs), sweetener (T1R2/T1R3) and umami stimulus (T1R1/T1R3, mGluR1, mGluR4) receptors are GPCRs. Activation of these receptors triggers G-protein-mediated signaling cascades (fig. 3). Several G-protein α-subunits have been identified in taste bud cells, including $G\alpha_{i2}$, $G\alpha_{i3}$, $G\alpha_{14}$, $G\alpha_{15}$, $G\alpha_q$, $G\alpha_s$, α-gustducin and an α-transducin-like subunit, α-gustducin [152–154]. α-Gustducin shares 80% amino acid identity with α-heteromeric, and is selectively expressed in about 30% of taste bud cells. Nearly all T2Rs have been localized to α-gustducin-expressing taste bud cells while additional α-gustducin-expressing cells possess sweet or umami receptors [66, 97]. In the heterologous expression system, α-gustducin or chimeric G-proteins containing part of α-gustducin can relay the signals from activated T2Rs or heteromeric T1Rs to downstream effectors, resulting in an increase in intracellular calcium concentration [67, 110]. In vitro biochemical assays have demonstrated that activated taste receptors can also interact with α-transducin [70, 155]. The knockout of α-gustducin not only markedly reduces animals' sensitivity to bitter stimuli, but also moderately decreases sweet and umami tastes [156]. Double-knockout mice lacking both α-gustducin and α-transducin further reduced umami sensitivity, suggesting that α-transducin is involved only in umami taste [157]. From these data, it can be concluded that bitter signal transduction is largely mediated by α-gustducin, which is also partially involved in

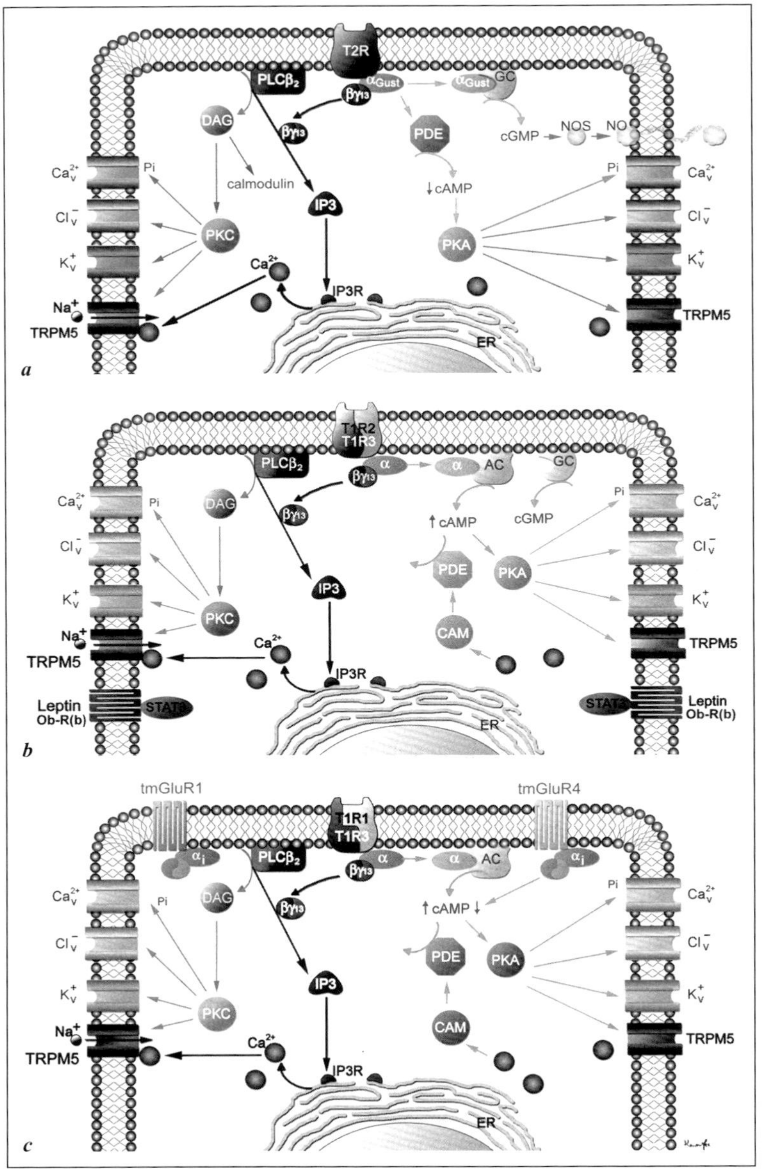

Fig. 3. Signal transduction pathways for bitter, sweet and umami tastes. The main signaling cascades seem to be shared by all three tastes: activation of T2R, T1R2/T1R3 and T1R1/T1R3 receptors by bitter (***a***), sweet (***b***) and umami (***c***) stimuli dissociates the heterotrimeric G-proteins into α and βγ moieties. The βγ moiety in turn activates the effector enzyme PLCβ_2, which generates the second messengers IP3 and DAG. IP3 binds to its receptor

sweet and umami transduction. Several G-protein $\beta\gamma$-subunits have been isolated from taste bud cells. Among them are $G\beta_3$ and $G\gamma_{13}$, which are localized to a subset of taste bud cells, including those expressing α-gustducin [158, 159]. Biochemical interaction indicated that $G\beta_3\gamma_{13}$ is a likely $\beta\gamma$ partner for α-gustducin [159]. Effector enzymes and other downstream components have also been identified, which include phospholipase $C\beta_2$ ($PLC\beta_2$), the inositol 1,4,5-triphosphate (IP3) receptor IP3R3 and a transient receptor potential ion channel, TRPM5 [160–163].

Therefore, activation of T2R bitter receptors stimulates the heterotrimeric G-protein consisting of α-gustducin and $G\beta_3\gamma_{13}$, which dissociates into α-moiety and $\beta\gamma$-moiety, generating bifurcate signaling pathways. α-Gustducin is believed to activate phosphodiesterase and perhaps guanylyl cyclase as well, which regulate intracellular cAMP and cGMP levels, respectively, consequently the activity of protein kinase A and NO synthase [164]. Protein kinase A functionally modulates an array of voltage-gated ion channels, changing the membrane potentials. However, the $G\beta_3\gamma_{13}$ branch appears to be the principal signaling cascade. $G\beta_3\gamma_{13}$ activates $PLC\beta_2$, which hydrolyzes phosphatidylinositol-4, 5-biphosphate and produces two second messengers, diacylglycerol (DAG) and IP3. DAG activates protein kinase C, which phosphorylates a number of intracellular proteins including voltage-gated ion channels. IP3 binds to the IP3 receptor IP3R3, which releases calcium from intracellular stores. The increase in free intracellular calcium opens a nonselective monovalent cation channel, TRPM5 [165–167].

In addition to α-gustducin, other G-protein α-subunits such as α_{i2} and α_s are likely involved in sweet and umami signal transductions. Generation of another second messenger, cAMP, has been reported in response to sweet stimuli [168]. cAMP is known to regulate protein kinase A activity, which in turn phosphorylates other intracellular proteins. However, it is firmly established that $PLC\beta_2$ and TRPM5 are the two indispensable components in all three GPCR-mediated bitter, sweet and umami taste transductions since knocking out

IP3R3, which releases Ca^{2+} from intracellular stores. Increase in the free cytosolic Ca^{2+} opens a nonselective monovalent cation channel TRPM5. Influx of cations leads to depolarization of the taste cell membrane potential. DAG activates protein kinase C (PKC), which phosphorylates and modulates other proteins including voltage-gated ion channels. In bitter receptor cells (***a***), the $G\alpha$ subunit may activate phosphodiesterase (PDE) and/or guanylgl cyclase (GC), regulate cAMP and cGMP levels, thus activity of protein kinase A (PKA) and NO generation. ER = endoplasmic reticulum. In sweet receptor cells (***b***), the $G\alpha$ subunit may regulate the activity of adenylyl cyclase (AC) and cAMP level, which stimulates PKA. Leptin receptors may play a role in sweet taste signal transduction. STAT3 = signal transducer and activator of transcription protein 3. In umami receptor cells (***c***), truncated metatropic glutamate receptors tmGluR1 and tmGluR4 may also contribute to umami sensation. Pi = phosphorylation.

either of the two genes diminishes sensitivity to stimuli of all three taste modalities [163]. Therefore, T1R- and T2R-mediated signaling cascades converge onto a common activating event, that is, opening of TRPM5. It is still not known how activation of TRPM5 on taste bud cells leads to generation of action potentials on afferent nerve fibers. However, it is believed that influx of monovalent cations through TRPM5 channels depolarizes taste bud cells. Interestingly, depolarization of receptor cell membrane potential has also been postulated to be the cellular response to sour and salty stimuli. Therefore, interactions of sapid molecules of all five known taste modalities with GPCRs or channel receptors eventually lead to the generation of receptor potential, which regulates the release of neurotransmitters onto afferent gustatory nerve fibers and/or the release of paracrine agents that act on and regulate the neurotransmitter release from adjacent taste bud cells [50].

Taste Bud Modulators

In a taste bud, cells are in close apposition with one another. This unique organization may play an essential role in cell-to-cell communication. In amphibians, gap junctions are present among taste bud cells, which allow the spread of currents and small molecules to neighboring cells [169]. In mammals, a number of bioactive agents and their receptors have been found on taste bud cells, which include serotonin and its receptors, norepinephrine and its receptors, neuropeptide Y, cholecystokinin, vasoactive intestinal peptide and somatostatin and their receptors, acetylcholine and its receptors, glutamate and its metabotropic and ionotropic receptors, ATP and its P2Y receptors [46, 48, 49, 105, 170–175]. These signaling agents can be released from taste bud cells upon stimulation, and act on the producing cells as autocrines or on adjacent cells as paracrines to modulate these cells or to trigger neurotransmitter output onto the cranial nerve fibers. Occurrence of receptors for the circulating regulatory hormones leptin and aldosterone has also been reported [176, 177] (fig. 4). Activation of these receptors is believed to change gene expression patterns such as ENaC subtypes, or even induce cell proliferation and differentiation.

Physiological Responses of Taste Receptor Cells

Taste bud cells are specialized epithelial cells with some neuronal properties. The excitability of taste bud cells and their voltage-dependent currents have been well documented [55]. Action potentials in gustatory receptor cells were first described in amphibians [178, 179], then in mammals, in response to passing currents or to taste stimuli such as sour, salt and sweet stimuli [142, 180–182]. In

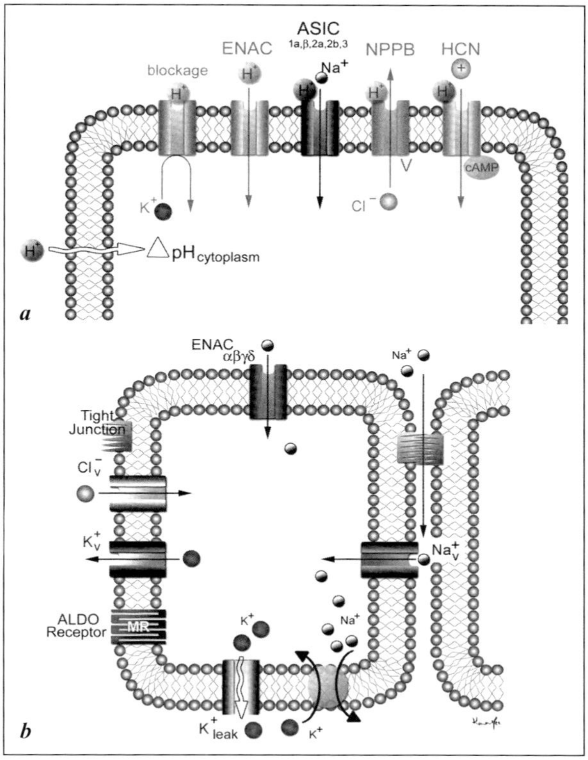

Fig. 4. Signal transduction pathways for ionotropic sour and salty tastes. ***a*** A number of mechanisms may exist for sour transduction, which includes the proton blockage of a K^+ channel, proton permeation of ENACs, activation of ASICs, an NPPB-sensitive Cl^- channel, or a hyperpolarization-activated and cyclic nucleotide-gated ion channel, and intracellular acidification. HCN = Hyperpolarization-activated and cyclic nucleotide-gated ion channel. ***b*** Na^+ and other cations permeate apical multimeric epithelial sodium channels (ENACs), or pass through a tight junction and enter cells via basolateral ion channels, resulting in the depolarization of receptor cell membrane potentials and consequently release of synaptic transmitters. Intracellular ionic equilibrium is restored by Na^+-K^+ pump and ion leak. Aldosterone receptors may co-occur in these cells and activation of this receptor alters the composition of ENAC subunits, thus the susceptibility to amiloride. MR = mineralocorticoid receptors.

rats, as many as 75% of taste bud cells can generate action potentials [183–186]. Two types of action potentials have been observed from taste bud cells [182, 187]: (1) fast action potentials with shorter duration and larger inward and outward currents; (2) slow action potentials with longer duration and smaller inward and outward currents [182, 184, 188]. Analysis of loose patch recording data from hamster taste buds using an artificial neural network suggested that action potentials in taste buds may contribute to taste quality coding [189]. However, the exact role of action potentials in taste signal transmission is still enigmatic since action potentials are usually employed by neurons for long-distance propagation of electrical signals and taste bud cells are short (approx. 100 μm in rodents). Moreover, receptor potentials instead of action potentials may be sufficient to trigger signal transmission. Active investigations are being carried out to solve this puzzle.

Various voltage-gated ion channels have been electrophysiologically described from different taste bud cells, including tetrodotoxin-sensitive Na^+ channels, tetraethylammonium-sensitive delayed rectifier K^+ channels, inward rectifier K^+ channels, outward rectifier Cl^- channels, and low-/high-voltage-activated Ca^{2+} channels [183–186]. The distribution of voltage-gated ion channels across taste bud cells is heterogeneous. In rats, about 57% of taste bud cells possess voltage-gated Na^+ channels but nearly all cells have voltage-gated K^+ channels. Multiple voltage-gated Ca^{2+} and Cl^- conductances have also been reported from subsets of taste bud cells. Developmentally, the complexity buildup of voltage-dependent currents coincides with the maturation of taste bud cells, suggesting that voltage-gated ion channels contribute to the taste bud cells' function [190].

In addition to changes in electrical properties, two more responses of taste bud cells to taste stimuli have also been monitored: changes in intracellular calcium concentration and the release of bioactive agents from taste bud cells [42, 173]. Stimuli of all five taste modalities can induce an increase or decrease in intracellular calcium concentrations in taste bud cells. Initial increase of calcium in some cells may not require the influx of calcium from extracellular sources. However, the presence of extracellular calcium augments the magnitude and prolongs the sustained response. Caution should be taken in interpreting the calcium response data since this response is an intermediate step in the taste signal integration process in a taste bud, which may not lead to the final output of neurotransmitters in some cells, but rather synchronize and reset these cells for following taste stimuli of similar or different modalities.

Biosensors, which have recently been applied to taste research, can detect the release of paracrines and transmitters such as serotonin and ATP from taste bud cells in response to taste stimuli [191]. Further studies have shown that ATP appears to be a neurotransmitter taste buds use to convey the gustatory signal to afferent nerve fibers [50].

Afferent Signal Transmission and Coding

How taste compounds are sensed and ultimately encoded qualitatively is a complex process that involves integration by the system at almost every step of the sensory pathway. The sensory coding of physical information from the chemical stimuli is fundamentally about how the stimuli are registered and filtered during transduction and about how this information continues to be filtered and integrated at various levels up to and including the ultimate cortical levels. The first level at which chemical information is both filtered and integrated is the peripheral receptor molecule. For a given receptor, the initial binding, or conductance event in the case of a channel, is determined by the chemical structure of the ligand and this affects all downstream processes. Yet, when several compounds can activate the same receptor, the information regarding the specific identity of the ligand is lost. In this way, the taste system can only know that one member of a set of chemicals that comprise the receptive field for the receptor has been detected [192]. For example, a human TAS2R receptor filters (or loses) information about the physical stimuli that activate it by the mere fact that more than one stimulus can activate it. An 'observer' of receptor activation cannot necessarily determine which stimulus has activated it simply by examining the fact of activation. This is known as the principle of univariance; that is, any upstream 'observer' of receptor activation, molecular, neural or otherwise, can only know whether the receptor has been activated and to what degree, but not by which member of the set of possible activators [192]. This is assuming the temporal activation of the receptor is not serving as a fingerprint for ligand identity. Conversely, if a receptor is so highly specific that only one ligand activates it, then this activation is equivalent to molecular identification. But this is rarely the case, particularly for taste. It is also important to note that while there is apparent quality coding at the level of the receptor protein based on the receptive field of the receptor, many, if not most ligands, will activate multiple receptors. Perhaps the most famous example is Na-saccharin which tastes both sweet and bitter to most observers and activates TAS1R2-TAS1R3 as well as TAS2Rs, respectively [71, 92].

Beyond the receptor molecule, the receptor cell is also a site for integration, or loss of physical information, if multiple different receptors are expressed on it. Thus, while each receptor may have a distinct receptive field, all receptor activations result in the same outcome – cellular excitation. This is not true if some gustatory stimuli are capable of hyperpolarizing receptor cells while others can depolarize the same cell, creating an opportunity for intracellular opponency; but this has not yet been demonstrated within the mammalian taste system. While it does not appear that TAS1Rs (sweet and savory) and TAS2Rs (bitter) are coexpressed in the same receptor cells, it is clear that members within a class are coexpressed [66, 98, 101, 103, 104, 193]. In situ hybridization and immunohistochemistry have shown that segregated subsets of taste bud cells express T2R bitter taste receptors, T1R1/T1R2 heterodimeric umami receptors, T1R2/T1R3 heterodimeric or T1R3 homodimeric sweet receptors [66, 98].

Multiple TAS2Rs can be found within a single cell type, and while TAS1R3 is coexpressed with TAS1R1 and TAS1R2 [67, 103, 163, 193], TAS1R1 and TAS1R2 do not appear to be coexpressed nor is TAS1R3 always coexpressed with another TAS1R. These data suggest that receptor cells in a taste bud tend to specialize for certain receptor types, specifically for receptors that bind bitter, sweet and umami substances.

In line with this hypothesis, the knockout of T1R1 gene expression only diminished the animal's sensitivity to umami compounds, and did not affect sweet, salty and bitter detection. Likewise, the knockout of T1R2 only reduced sensitivity to sweet stimuli, and did not affect other tastes [103, 104]. A transgenic rescue experiment showed that expression of PLCβ_2 driven by a T2R promoter in PLCβ_2 gene null mice only restored taste sensitivity to bitter compounds, but not to sweet and umami compounds, demonstrating some independence of cells that encode different taste qualities [163]. Additionally, mice expressing transgenically introduced human T1R2 receptors displayed humanized sweet taste preferences, and were able to detect substances that taste sweet to humans, but normally not appetitive to rodents. Furthermore, mice transgenically expressing a human T2R16 receptor for a bitter compound, phenyl-B-*D*-glucopyranoside, in all cells that express sweetener receptors exhibited strong attraction to this bitter compound [193]. These results demonstrate taste quality is encoded by the sets of cells that are 'typed' by the receptors they express on their surface.

Taste receptor cells are typically not the bud cells which synapse onto afferent neural fibers. Therefore, there must be cell-to-cell processing of afferent signals within the taste bud prior to neural signaling. The role that this plays in coding is not clear. Since there are cells within a taste bud that are sensitive to a wide array of different taste stimuli, there must be an intragemmal chemical coding system so that afferent information is not distorted or garbled during bud processing. This could be accomplished if sets of receptor cells that collectively represent a taste stimulus all communicated with the elongated cells that possess neural synapses via a common transmitter, such as serotonin, neuropeptide Y or glutamate. Once the encoding of stimuli has occurred within the bud, the communication by these second- and third-order bud cells with the intragemmal neural fibers does not need to be differentiated chemically. ATP and its receptors P2X2 and P2X3 are now considered the dominant transmitter for communication between taste bud cells and primary afferent neurons [50]. Thus, once the appropriate set of synapsing elongate bud cells are activated, then all sets of cells within a bud need not be further differentiated when activating the afferent neuron; a single mode of synaptic communication suffices.

The neural taste fibers are, however, free to communicate with any cells within the bud and with many buds. Each afferent neural taste fiber is richly arborized in the periphery and contacts several taste buds within a small area

[194]. Also, several different ganglion neural cells may innervate a single taste bud. This creates an opportunity for the coding to be either simplified or increased in its complexity depending on the mixture of bud cell types with which a single afferent neuron synapses. In some species, such as *Pan troglodytes* (chimpanzee), the receptive fields of peripheral afferent fibers appear to be organized along qualitative dimensions similar to the organization of bud receptor cells. How fibers are able to synapse with only those type III cells within a bud, across several buds that are in chemical communication with type II cells of a particular receptor makeup (e.g. TAS2R-expressing cells), is not known [120, 195–198]. On the other hand, some species, such as rats, possess primary afferent taste fibers with a receptive field that includes a qualitatively mixed variety of stimuli [17, 199, 200]. Taste afferent fibers and early taste processing areas are also sensitive to thermal and tactile stimuli and nociception from posterior fields [201, 202]. Whether peripheral fibers are qualitatively refined in their receptive field or whether they respond to a broad array of stimuli, the higher relay areas in the central nervous system may converge and integrate inputs or refine mixed quality signals depending on the region of the brain and species in question [203]. Therefore, it is should be clear that the question of how the sensory system codes for taste quality and thus chemical stimulus class depends largely on where in the system one is looking. How the system processes taste stimuli at a perceptual or behavioral level is discussed below.

Human Psychophysics

Intensity Judgements

In many sensory systems, the ability to detect changes in intensity is described by a fixed function of the ratio of the magnitude of the change to the starting concentration of the stimulus, in other words a percent increase or decrease (a Weber fraction) [204]. For example, a subject may be able to just noticeably detect an increase or a decrease of 10% in concentration regardless of the starting concentration. Taste Weber fractions may be constant over several orders of magnitude in concentration (above-threshold levels), but exceptions to this rule occur at very low and very high concentrations [205, 206].

The rate at which perceived intensity grows with concentration, the exponent of Steven's power function ($I = kC^n$; where I is intensity and C is concentration), helps to determine the input-output intensity function of the compound under study [207]. One might surmise that the exponent for a compound's intensity power function bears some relation to the quality of the sensation. Within a single quality, however, the exponents (n) for several compounds' power functions do not form natural groupings. That is, the exponents of sweeteners and the exponents of bitter compounds are intermingled and do not fall into two neat clusters. Therefore, the rate at which intensity increases is a function of peripheral events in the epithelium (for example individual molecule's binding kinetics) and not psychological events such as the qualitative categorization of sweetness or sourness.

Umami (Savory) as a Distinct Perceptual Quality

The most typical methods of measuring taste quality involve either: (1) the direct scaling of quality intensities by subjects using a variety of scaling techniques and a variable number of scales presented per trial, or (2) the rating of total intensity and the subsequent division of the total intensity into various percent portions of salty, sweet, bitter, and sour. Occasionally, the quality umami is included in these techniques.

Umami is the Japanese term to describe the 'savory' taste elicited by certain foods, including mushrooms (shiitake), seaweeds (sea tangle), fish (bonito), and vegetables (tomato). The prototypical chemical elicitors of umami are MSG mixed with 5′-ribonucleotides like inosine monophosphate or GMP. All the foods described as umami are rich in these compounds. This quality of taste is easily perceived by people of many different cultures. Therefore, there must be cultural reasons why umami is not generally included with the standard four taste qualities: salt, sour, bitter, and sweet. When offered MSG with 5′-ribonucleotides (especially in warm water), Americans or Europeans describe it as brothy, soupy, meaty, and savory. While the term savory has not been included in our parlance of qualitative taste descriptors, the term umami has been accepted in Japan (at least among food scientists and industry). The term is relatively new, however, first coined in 1908 by Ikeda for the taste of a broth made of sea tangle and bonito fish [208].

Adaptation

A fundamental characteristic of taste is the adaptation to background levels of sapid compounds, with the exception of some acids and bitter-tasting compounds. For example, saliva is usually tasteless as it rests in the mouth even though it contains many ions and other potentially sapid chemicals. In psychophysics, adaptation is defined as the decrement in intensity or sensitivity to a compound under constant stimulation by this compound. Adaptation to a small defined portion of the tongue tip is almost complete for a wide variety of taste compounds. After complete or nearly complete adaptation to a low-intensity taste stimulus, a subject will have to receive higher concentrations than the adapting concentration to regain the sensation of the stimulus. Interestingly, concentrations lower than the adapting concentration elicit tastes as well, though they are usually of a different quality than the initial quality of the adapting stimulus.

One very important feature of the taste system (and of other sensory systems as well) is that the exponent of a taste intensity power function increases when the subject is adapted to the stimulus; the higher the degree of adaptation the steeper the function. This is also true when measured with Weber fraction techniques (just noticeable differences). That is, if one adapts to a given concentration, the ability to detect increases or decreases in concentration is greatly enhanced [209]. This means that the point of maximum differential sensitivity varies with the concentration of the adapting stimulus and the degree of adaptation. Interestingly, this suggests that there can be a large change in the detection threshold for a stimulus, as well as an actual enhancement of the detection of suprathreshold concentration changes, following adaptation.

Several reports suggest that adaptation has both a peripheral epithelial basis as well as a central neural component. For example, a small portion of the tongue can be adapted to a stimulus, then neighboring patches of the taste epithelium can be tested. Adaptation impacts the nonstimulated epithelium both across the midline of the tongue and on the same side [210]. These effects cannot be attributed solely to peripheral adaptation of the receptor cells.

Cross-Adaptation

Cross-adaptation occurs when the perceived intensity of a solution is decreased following adaptation to a different compound, relative to adaptation to water. Cross-adaptation is usually not as complete as adaptation and may not be symmetrical. These phenomena are likely due to the complex (multiquality) tastes of mostly all taste stimuli. While select qualities of a stimulus may be cross-adapted, the stimulus as a whole will not. Various salts do not appear to cross-adapt when measuring estimates of total stimulus strength, but do cross-adapt if only the salty quality is rated [211, 212]. This suggests that the saltiness of the different salts share a common pathway. There often is little cross-adaptation between compounds that differ in quality such as between sucrose and quinine, although cross-quality adaptation can occur.

More interestingly, there are stimuli that elicit the same quality of taste sensation but do not cross-adapt one another. Most notable among these are different groupings of intensive sweeteners and the various groupings of bitter-tasting compounds. Thus, three groupings of bitter compounds have been identified as a function of cross-adapting within group but not between groups [213, 214]. For example, quinine is believed to stimulate multiple transduction mechanisms, but it does not cross-adapt PTC, another bitter compound. Thus, PTC is believed to have a transduction mechanism separate from that of quinine, which from the perspective of TAS2R receptors we know to be true [73]. Symmetric cross-adaptation, to the same degree as self-adaptation, might be expected when two different compounds are indistinguishable in a discrimination task, an indication that they may stimulate the same receptors to the same degree.

Taste Mixture Interactions

When taste compounds are mixed together in solution, they often interact with one another so that each tastes different than it would were it presented alone at the same concentration. Virtually all tastes are encountered as mixtures of several taste compounds as we go through life eating and drinking. It is the norm [215].

Enhancement occurs when two (or more) compounds are mixed together and a particular quality of taste is increased in intensity, seen as a leftward shift of the concentration-intensity curve. That is, every point along the concentration axis is perceived as being more intense when in the presence of a fixed concentration of a second compound. Since the concentration-intensity function is generally sigmoidal rather than linear, the magnitude of the effect will be dependent upon which point along the concentration axis the second compound is added. Interactions that occur at low concentrations of compounds tend to show enhancement, where the curve is expansive (concave looking); where the curve is linear (in the middle), there tend to be small linear interactions, and at high concentrations, where the curve is compressive (convex looking), there tend to be suppressive effects [215, 216]. Examples of enhancement are relatively rare for cross-quality compounds but are the general rule for compounds that elicit the same quality.

Synergy is similar to enhancement but is a more potent form of positive interaction. When two (or more) compounds are mixed together and a particular quality of taste is increased both as a leftward shift of the concentration-intensity curve and as a steepening of its slope, then there is synergy. That is, every point along the concentration axis is perceived as being more intense when in the presence of a fixed concentration of a second compound. Synergy is very rare in taste, however. There are two prototypical examples, both involving same-quality mixtures. The first and most clear example comes from the combination of MSG with 5′-ribonucleotides [217]. These two compounds synergize their respective umami tastes. The second is seen with certain intensive sweeteners such as aspartame and acesulfame-K. Together their sweetnesses synergize [218]. It remains to be shown that cross-quality synergy exists.

Suppression is the counterpart to enhancement. When two (or more) compounds are mixed together and a particular quality of taste is decreased as a rightward shift of the concentration-intensity curve, then there is suppression [215]. That is, every point along the concentration axis is perceived as being less intense when in the presence of a fixed concentration of a second compound. Suppression is highly common, especially among compounds of different qualities. For example, when making lemonade, the sourness of the lemons will be suppressed slightly by the sweetness of the sugar and the sweetness of the sugar will be suppressed by the sourness of the lemons. Suppression is usually symmetrical but it can also be asymmetrical [219, 220].

Masking is the counterpart to synergy. Masking is similar to suppression but is a more potent form of negative interaction. When two (or more) compounds are mixed together and a particular quality of taste is decreased, both as a rightward shift of the concentration-intensity curve and as a shallowing of its slope, then there is masking. That is, every point along the concentration axis is perceived as being less intense when in the presence of a fixed concentration of a second compound. Examples of masking come from the taste blocking/inhibiting literature discussed above. NaCl decreases the bitterness of urea not only as a rightward shift of the curve but also as a shallowing of the bitterness function slope [219]. In general, both masking and synergy involve peripheral pharmacological effects on the taste cells. The more general phenomena of enhancement and suppression tend to involve more cognitive interactions, although there could also be a peripheral component to these mixture phenomena.

Release from Suppression

Because interaction effects can be asymmetrical, and compounds almost always interact with one another by one of the four mechanisms mentioned above, there are interesting results when adding a third compound to a binary mixture. For example, if a bitter-tasting compound (urea) is mixed with a sweetener (sucrose), there is likely to be mutual suppression whereby the sweet suppresses the bitter and vice versa. If a sodium salt is added to the binary bitter-sweet mixture, the sodium will have a large masking effect upon the bitter taste but only a very weak suppressive effect upon the sweet. What remains perceptually is predominantly sweet and very slightly bitter, a large relative change. Since the bitter taste suppresses the sweetness of the sucrose, the sweetness will increase in intensity when released from the suppression by the bitter [221, 222]. Similar phenomena have been discussed in the context of employing either sequential adaptation stimuli or taste blockers [223, 224].

Pathological Effects on Taste

For detailed clinical issues regarding taste, the reader is referred to Doty and Bromley [225], Reiter et al. [226], Cowart et al. [227] and Getchell et al. [228], and for aging issues regarding taste, the reader is referred to Bartoshuk [229], Cowart [230] and Murphy and Gilmore [231]. Of the chemosensory disorders people experience, taste problems are in the clear minority and olfactory problems in the majority. Of the patients who presented chemosensory complaints to the UPenn Smell and Taste Center, only 4% were found to have taste deficits (n = 750) [232, 233]. Of these, taste disorders are most frequently quality (or compound) specific, not including all taste sensations, and usually involve taste losses, although taste phantoms also occur [227]. The primary causes of taste dysfunction can be broken into two main categories: (1) drug and toxin effects, and (2) disease effects including: (a) infections/periodontal disease and other local effects, (b) nervous disorders/herpes zoster, (c) nutritional disorders, and (d) endocrine disorders [234–236]. The most common of the two are drug and toxin effects on taste. Drugs may impact taste by direct systemic stimulation of taste receptors by the drug, altering normal function of transduction processes or cellular function, altering salivary function/flow, or perhaps altering central neural processes [225, 227, 237]. At this time, we know little of how drugs impact taste. Phantoms occur for several reasons, including oral yeast infections [238, 239], nerve damage (both mechanical and infectious) [226, 240–243], and head trauma [226, 244]. Finally, aging is also associated with loss in taste sensitivity.

Although taste function decreases with age, the loss of taste is much less pronounced than for olfaction [245–247]. Taste declines specifically for certain qualities or representative compounds with age. In particular, sensitivity to bitter and salty stimuli may decrease [231, 234, 248–251], although large losses in sensitivity to compounds such as citric acid can be shown for localized gustatory areas, such as the tongue tip [229]. Studies of aging have even provided further evidence for differences in transduction mechanisms for different bitter compounds. While young and elderly scale the bitterness of urea similarly, elderly scale the bitterness of quinine sulfate as significantly weaker at high concentrations [248]. This suggests that urea is detected via an age-insensitive mechanism while quinine detection exhibits decreasing function with age.

Young and elderly subjects may differ greatly in detection threshold for standard test stimuli, where elderly require two to nine times greater concentrations to detect the compounds [247]. Usually, much smaller differences exist between the age groups for suprathreshold whole-mouth concentrations (with notable exceptions for quinine [248]). This could result in the false impression that taste losses are of little consequence among the elderly, but when detecting target stimuli in the presence of a background masking taste, elderly are two to three times less sensitive than young subjects [247, 249, 252, 253]. Despite often not being able to detect the presence of key ingredients, like NaCl, in everyday foods, like soups, elderly often seem able to enjoy food and derive pleasure from eating. The chronic overconsumption of NaCl, from oversalting, or of bitter toxins, however, could pose a health risk among the elderly [252, 254].

Conclusions

Taste is surprisingly complex relative to other sensory systems, including its close cousin olfaction. Taste engages a wide variety of transduction sequences, possesses complex processing even within the taste bud, has myriad quality coding systems depending upon which level of anatomy is examined, demonstrates rich interstimulus interactions both within taste and with other sensory modalities, is involved in multiple physiological systems including digestion, and is perhaps the single most important sensory system for life, since loss of taste results in acute decreases in ingestion and feeding behavior and can be life threatening in select patients. Both from research and clinical perspectives, little is known of how taste works at a pan-system level. There has been great progress in recent years at molecular biological levels of understanding taste, but this has not been integrated well with developmental neurophysiological, digestive, higher coding, integrative, or perceptual functions. Our future understandings of taste will depend upon combining molecular, genetic, developmental, and neurophysiological levels of analysis with higher cognitive and perceptual levels of inquiry.

References

1 Katschinski M: Nutritional implications of cephalic phase gastrointestinal responses. Appetite 2000;34:189–196.

2 Teff K: Nutritional implications of the cephalic-phase reflexes: endocrine responses. Appetite 2000;34:206–213.

3 Steiner JE: The gustofacial response: observation on normal and anencephalic newborn infants. Symp Oral Sens Percept 1973;4:254–278.

4 Steiner JE: Discussion paper: innate, discriminative human facial expressions to taste and smell stimulation. Ann NY Acad Sci 1974;237:229–233.

5 Jacquin MF: Gustation and ingestive behavior in the rat. Behav Neurosci 1983;97:98–109.

6 Spector AC: Linking gustatory neurobiology to behavior in vertebrates. Neurosci Biobehav Rev 2000;24:391–416.

7 Maes A, Huygh I, Weltens C, et al: De Gustibus: time scale of loss and recovery of tastes caused by radiotherapy. Radiother Oncol 2002;63:195–201.

8 Kokal WA: The impact of antitumor therapy on nutrition. Cancer 1985;55(1 Suppl):273–278.

9 Langius A, Bjorvell H, Lind MG: Oral- and pharyngeal-cancer patients' perceived symptoms and health. Cancer Nurs 1993;16:214–221.

10 Bray S, Amrein H: A putative Drosophila pheromone receptor expressed in male-specific taste neurons is required for efficient courtship. Neuron 2003;39:1019–1029.

11 Grammer K, Fink B, Neave N: Human pheromones and sexual attraction. Eur J Obstet Gynecol Reprod Biol 2005;118:135–142.

12 Gilbertson TA: Gustatory mechanisms for the detection of fat. Curr Opin Neurobiol 1998;8:447–452.

13 Mattes RD: Fat taste and lipid metabolism in humans. Physiol Behav 2005;86:691–697.

14 Laugerette F, Passilly-Degrace P, Patris B, et al: CD36 involvement in orosensory detection of dietary lipids, spontaneous fat preference, and digestive secretions. J Clin Invest 2005;115: 3177–3184.

15 Delwiche JF, Lera MF, Breslin PAS: Selective removal of a target stimulus localized by taste in humans. Chem Senses 2000;25:181–187.

16 Northcutt RG: Taste buds: development and evolution. Brain Behav Evol 2004;64:198–206.
17 Smith DV, Margolskee RF: Making sense of taste. Sci Am 2001;284:32–39.
18 Finger TE: Evolution of taste and solitary chemoreceptor cell systems. Brain Behav Evol 1997;50:234–243.
19 Bartoshuk LM, Duffy VB, Miller IJ: PTC/PROP tasting: anatomy, psychophysics, and sex effects. Physiol Behav 1994;56:1165–1171.
20 Sbarbati A, Crescimanno C, Osculati F: The anatomy and functional role of the circumvallate papilla/von Ebner gland complex. Med Hypotheses 1999;53:40–44.
21 Spielman AI, D'Abundo S, Field RB, Schmale H: Protein analysis of human von Ebner saliva and a method for its collection from the foliate papillae. J Dent Res 1993;72:1331–1335.
22 Gilbertson TA, Kinnamon SC: Making sense of chemicals. Chem Biol 1996;3:233–237.
23 Bradley RM, Grabauskas G: Neural circuits for taste. Excitation, inhibition, and synaptic plasticity in the rostral gustatory zone of the nucleus of the solitary tract. Ann NY Acad Sci 1998;855: 467–474.
24 Smith DV, Li CS, Davis BJ: Excitatory and inhibitory modulation of taste responses in the hamster brainstem. Ann NY Acad Sci 1998;855:450–456.
25 Sewards TV: Dual separate pathways for sensory and hedonic aspects of taste. Brain Res Bull 2004;62:271–283.
26 Delay RJ, Kinnamon JC, Roper SD: Ultrastructure of mouse vallate taste buds. 2. Cell types and cell lineage. J Comp Neurol 1986;253:242–252.
27 Kinnamon JC, Henzler DM, Royer SM: HVEM ultrastructural analysis of mouse fungiform taste buds, cell types, and associated synapses. Microsc Res Tech 1993;26:142–156.
28 Kinnamon JC, Taylor BJ, Delay RJ, Roper SD: Ultrastructure of mouse vallate taste buds. 1. Taste cells and their associated synapses. J Comp Neurol 1985;235:48–60.
29 Jahnke K, Baur P: Freeze-fracture study of taste bud pores in the foliate papillae of the rabbit. Cell Tissue Res 1979;200:245–256.
30 Simon SA: Influence of tight junctions on the interaction of salts with lingual epithelia: responses of chorda tympani and lingual nerves. Mol Cell Biochem 1992;114:43–48.
31 DeSimone JA, Ye Q, Heck GL: Ion pathways in the taste bud and their significance for transduction. Ciba Found Symp 1993;179:218–234.
32 Miller IJ Jr: Human taste bud density across adult age groups. J Gerontol 1988;43:B26–B30.
33 Miller IJ Jr: Variation in human fungiform taste bud densities among regions and subjects. Anat Rec 1986;216:474–482.
34 Kobayashi K, Kumakura M, Yoshimura K, et al: Comparative morphological studies on the stereo structure of the lingual papillae of selected primates using scanning electron microscopy. Ann Anat 2004;186:525–530.
35 Pumplin DW, Yu C, Smith DV: Light and dark cells of rat vallate taste buds are morphologically distinct cell types. J Comp Neurol 1997;378:389–410.
36 Azzali G: Ultrastructure and immunocytochemistry of gustatory cells in man. Ann Anat 1997;179: 37–44.
37 Azzali G, Gennari PU, Maffei G, Ferri T: Vallate, foliate and fungiform human papillae gustatory cells. An immunocytochemical and ultrastructural study. Minerva Stomatol 1996;45: 363–379.
38 Yoshie S, Kanazawa H, Nishida Y, Fujita T: Occurrence of subtypes of gustatory cells in cat circumvallate taste buds. Arch Histol Cytol 1997;60:421–426.
39 Farbman AI, Hellekant G, Nelson A: Structure of taste buds in foliate papillae of the rhesus monkey, *Macaca mulatta*. Am J Anat 1985;172:41–56.
40 Yee CL, Yang R, Bottger B, Finger TE, Kinnamon JC: 'Type III' cells of rat taste buds: immunohistochemical and ultrastructural studies of neuron-specific enolase, protein gene product 9.5, and serotonin. J Comp Neurol 2001;440:97–108.
41 Yang R, Tabata S, Crowley HH, Margolskee RF, Kinnamon JC: Ultrastructural localization of gustducin immunoreactivity in microvilli of type II taste cells in the rat. J Comp Neurol 2000;425:139–151.
42 Huang YJ, Maruyama Y, Lu KS, et al: Mouse taste buds use serotonin as a neurotransmitter. J Neurosci 2005;25:843–847.

43 Kim DJ, Roper SD: Localization of serotonin in taste buds: a comparative study in four vertebrates. J Comp Neurol 1995;353:364–370.
44 Kinnamon JC, Sherman TA, Roper SD: Ultrastructure of mouse vallate taste buds. 3. Patterns of synaptic connectivity. J Comp Neurol 1988;270:1–10, 56–57.
45 Paran N, Mattern CF, Henkin RI: Ultrastructure of the taste bud of the human fungiform papilla. Cell Tissue Res 1975;161:1–10.
46 Herness S, Zhao FL, Kaya N, Lu SG, Shen T, Sun XD: Adrenergic signalling between rat taste receptor cells. J Physiol 2002;543:601–614.
47 Herness S, Zhao FL, Kaya N, Shen T, Lu SG, Cao Y: Communication routes within the taste bud by neurotransmitters and neuropeptides. Chem Senses 2005;30(Suppl 1):i37–i38.
48 Kaya N, Shen T, Lu SG, Zhao FL, Herness S: A paracrine signaling role for serotonin in rat taste buds: expression and localization of serotonin receptor subtypes. Am J Physiol Regul Integr Comp Physiol 2004;286:R649–R658.
49 Zhao FL, Shen T, Kaya N, Lu SG, Cao Y, Herness S: Expression, physiological action, and coexpression patterns of neuropeptide Y in rat taste-bud cells. Proc Natl Acad Sci USA 2005;102: 11100–11105.
50 Finger TE, Danilova V, Barrows J, et al: ATP signaling is crucial for communication from taste buds to gustatory nerves. Science 2005;310:1495–1499.
51 Farbman AI: Renewal of taste bud cells in rat circumvallate papillae. Cell Tissue Kinet 1980;13: 349–357.
52 Beidler LM, Smallman RL: Renewal of cells within taste buds. J Cell Biol 1965;27:263–272.
53 Conger AD, Wells MA: Radiation and aging effect on taste structure and function. Radiat Res 1969;37:31–49.
54 Stone LM, Tan SS, Tam PP, Finger TE: Analysis of cell lineage relationships in taste buds. J Neurosci 2002;22:4522–4529.
55 Lindemann B: Taste reception. Physiol Rev 1996;76:719–766.
56 Anonymous: Taste blindness. Science 1931;73(suppl):14.
57 Bartoshuk LM, Duffy VB, Miller IJ: PTC/PROP tasting: anatomy, psychophysics, and sex effects. Physiol Behav 1994;56:1165–1171.
58 Reed DR, Nanthakumar E, North M, Bell C, Bartoshuk LM, Price RA: Localization of a gene for bitter-taste perception to human chromosome 5p15. Am J Hum Genet 1999;64:1478–1480.
59 Lush IE, Hornigold N, King P, Stoye JP: The genetics of tasting in mice. 7. Glycine revisited, and the chromosomal location of Sac and Soa. Genet Res 1995;66:167–174.
60 Whitney G, Harder DB: Genetics of bitter perception in mice. Physiol Behav 1994;56:1141–1147.
61 Capeless CG, Whitney G, Azen EA: Chromosome mapping of Soa, a gene influencing gustatory sensitivity to sucrose octaacetate in mice. Behav Genet 1992;22:655–663.
62 Lush IE: The genetics of tasting in mice. 4. The acetates of raffinose, galactose and beta-lactose. Genet Res 1986;47:117–123.
63 Lush IE, Holland G: The genetics of tasting in mice. 5. Glycine and cycloheximide. Genet Res 1988;52:207–212.
64 Lush IE: The genetics of tasting in mice. 3. Quinine. Genet Res 1984;44:151–160.
65 Matsunami H, Montmayeur JP, Buck LB: A family of candidate taste receptors in human and mouse. Nature 2000;404:601–604.
66 Adler E, Hoon MA, Mueller KL, Chandrashekar J, Ryba NJP, Zuker CS: A novel family of mammalian taste receptors. Cell 2000;100:693–702.
67 Chandrashekar J, Mueller KL, Hoon MA, et al: T2Rs function as bitter taste receptors. Cell 2000;100:703–711.
68 Bufe B, Hofmann T, Krautwurst D, Raguse JD, Meyerhof W: The human TAS2R16 receptor mediates bitter taste in response to beta-glucopyranosides. Nat Genet 2002;32:397–401.
69 Vane JR: The fight against rheumatism: from willow bark to COX-1 sparing drugs. J Physiol Pharmacol 2000;51:573–586.
70 Pronin AN, Tang H, Connor J, Keung W: Identification of ligands for two human bitter T2R receptors. Chem Senses 2004;29:583–593.
71 Kuhn C, Bufe B, Winnig M, et al: Bitter taste receptors for saccharin and acesulfame K. J Neurosci 2004;24:10260–10265.

72 Behrens M, Brockhoff A, Kuhn C, Bufe B, Winnig M, Meyerhof W: The human taste receptor hTAS2R14 responds to a variety of different bitter compounds. Biochem Biophys Res Commun 2004;319:479–485.
73 Bufe B, Breslin PA, Kuhn C, et al: The molecular basis of individual differences in phenylthiocarbamide and propylthiouracil bitterness perception. Curr Biol 2005;15:322–327.
74 Shi P, Zhang J, Yang H, Zhang YP: Adaptive diversification of bitter taste receptor genes in mammalian evolution. Mol Biol Evol 2003;20:805–814.
75 Conte C, Ebeling M, Marcuz A, Nef P, Andres-Barquin PJ: Evolutionary relationships of the Tas2r receptor gene families in mouse and human. Physiol Genomics 2003;14:73–82.
76 Conte C, Ebeling M, Marcuz A, Nef P, Andres-Barquin PJ: Identification and characterization of human taste receptor genes belonging to the TAS2R family. Cytogenet Genome Res 2002;98: 45–53.
77 Wu SV, Chen MC, Rozengurt E: Genomic organization, expression and function of bitter taste receptors (T2R) in mouse and rat. Physiol Genomics 2005;22:139–149.
78 Go Y, Satta Y, Takenaka O, Takahata N: Lineage-specific loss of function of bitter taste receptor genes in humans and nonhuman primates. Genetics 2005;170:313–326.
79 Wang X, Thomas SD, Zhang J: Relaxation of selective constraint and loss of function in the evolution of human bitter taste receptor genes. Hum Mol Genet 2004;13:2671–2678.
80 Kim U, Wooding S, Ricci D, Jorde LB, Drayna D: Worldwide haplotype diversity and coding sequence variation at human bitter taste receptor loci. Hum Mutat 2005;26:199–204.
81 Soranzo N, Bufe B, Sabeti PC, et al: Positive selection on a high-sensitivity allele of the human bitter-taste receptor TAS2R16. Curr Biol 2005;15:1257–1265.
82 Ueda T, Ugawa S, Ishida Y, Shibata Y, Murakami S, Shimada S: Identification of coding single-nucleotide polymorphisms in human taste receptor genes involving bitter tasting. Biochem Biophys Res Commun 2001;285:147–151.
83 Rodriguez I, Mombaerts P: Novel human vomeronasal receptor-like genes reveal species-specific families. Curr Biol 2002;12:R409–R411.
84 Mombaerts P: The human repertoire of odorant receptor genes and pseudogenes. Annu Rev Genomics Hum Genet 2001;2:493–510.
85 Hiji T: Selective elimination of taste responses to sugars by proteolytic enzymes. Nature 1975;256:427–429.
86 Cagan RH, Morris RW: Biochemical studies of taste sensation: binding to taste tissue of ^{3}H-labeled monellin, a sweet-tasting protein. Proc Natl Acad Sci USA 1979;76:1692–1696.
87 Farbman AI, Ogden-Ogle CK, Hellekant G, Simmons SR, Albrecht RM, van der Wel H: Labeling of sweet taste binding sites using a colloidal gold-labeled sweet protein, thaumatin. Scanning Microsc 1987;1:351–357.
88 Blizard DA, Kotlus B, Frank ME: Quantitative trait loci associated with short-term intake of sucrose, saccharin and quinine solutions in laboratory mice. Chem Senses 1999;24:373–385.
89 Bachmanov AA, Reed DR, Ninomiya Y, et al: Sucrose consumption in mice: major influence of two genetic loci affecting peripheral sensory responses. Mamm Genome 1997;8:545–548.
90 Fuller JL: Single-locus control of saccharin preference in mice. J Hered 1974;65:33–36.
91 Phillips TJ, Crabbe JC, Metten P, Belknap JK: Localization of genes affecting alcohol drinking in mice. Alcohol Clin Exp Res 1994;18:931–941.
92 Bachmanov AA, Li X, Reed DR, et al: Positional cloning of the mouse saccharin preference (Sac) locus. Chem Senses 2001;26:925–933.
93 Kitagawa M, Kusakabe Y, Miura H, Ninomiya Y, Hino A: Molecular genetic identification of a candidate receptor gene for sweet taste. Biochem Biophys Res Commun 2001;283:236–242.
94 Max M, Shanker YG, Huang L, et al: *Tas1r3,* encoding a new candidate taste receptor, is allelic to the sweet responsiveness locus *Sac.* Nat Genet 2001;28:58–63.
95 Montmayeur JP, Liberles SD, Matsunami H, Buck LB: A candidate taste receptor gene near a sweet taste locus. Nat Neurosci 2001;4:492–498.
96 Sainz E, Korley JN, Battey JF, Sullivan SL: Identification of a novel member of the T1R family of putative taste receptors. J Neurochem 2001;77:896–903.
97 Hoon MA, Adler E, Lindemeier J, Battey JF, Ryba NJ, Zuker CS: Putative mammalian taste receptors: a class of taste-specific GPCRs with distinct topographic selectivity. Cell 1999;96: 541–551.

98 Nelson G, Hoon MA, Chandrashekar J, Zhang Y, Ryba NJP, Zuker CS: Mammalian sweet taste receptors. Cell 2001;106:381–390.
99 Pin JP, Galvez T, Prezeau L: Evolution, structure, and activation mechanism of family 3/C G-protein-coupled receptors. Pharmacol Ther 2003;98:325–354.
100 Li X, Staszewski L, Xu H, Durick K, Zoller M, Adler E: Human receptors for sweet and umami taste. Proc Natl Acad Sci USA 2002;99:4692–4696.
101 Nelson G, Chandrashekar J, Hoon MA, et al: An amino-acid taste receptor. Nature 2002;416: 199–202.
102 Inoue M, Reed DR, Li X, Tordoff MG, Beauchamp GK, Bachmanov AA: Allelic variation of the Tas1r3 taste receptor gene selectively affects behavioral and neural taste responses to sweeteners in the F2 hybrids between C57BL/6ByJ and 129P3/J mice. J Neurosci 2004;24:2296–2303.
103 Zhao GQ, Zhang Y, Hoon MA, et al: The receptors for mammalian sweet and umami taste. Cell 2003;115:255–266.
104 Damak S, Rong M, Yasumatsu K, et al: Detection of sweet and umami taste in the absence of taste receptor T1r3. Science 2003;301:850–853.
105 Chaudhari N, Yang H, Lamp C, et al: The taste of monosodium glutamate: membrane receptors in taste buds. J Neurosci 1996;16:3817–3826.
106 Chaudhari N, Landin AM, Roper SD: A metabotropic glutamate receptor variant functions as a taste receptor. Nat Neurosci 2000;3:113–119.
107 Chaudhari N, Roper SD: Molecular and physiological evidence for glutamate (umami) taste transduction via a G protein-coupled receptor. Ann NY Acad Sci 1998;855:398–406.
108 San Gabriel A, Uneyama H, Yoshie S, Torii K: Cloning and characterization of a novel mGluR1 variant from vallate papillae that functions as a receptor for *L*-glutamate stimuli. Chem Senses 2005;30(Suppl 1):i25–i26.
109 Toyono T, Seta Y, Kataoka S, Kawano S, Shigemoto R, Toyoshima K: Expression of metabotropic glutamate receptor group I in rat gustatory papillae. Cell Tissue Res 2003;313:29–35.
110 Jiang P, Ji Q, Liu Z, et al: The cysteine-rich region of T1R3 determines responses to intensely sweet proteins. J Biol Chem 2004;279:45068–45075.
111 Jiang P, Cui M, Zhao B, et al: Lactisole interacts with the transmembrane domains of human T1R3 to inhibit sweet taste. J Biol Chem 2005;280:15238–15246.
112 Winnig M, Bufe B, Meyerhof W: Valine 738 and lysine 735 in the fifth transmembrane domain of rTas1r3 mediate insensitivity towards lactisole of the rat sweet taste receptor. BMC Neurosci 2005;6:22.
113 Xu H, Staszewski L, Tang H, Adler E, Zoller M, Li X: Different functional roles of T1R subunits in the heteromeric taste receptors. Proc Natl Acad Sci USA 2004;101:14258–14263.
114 Galindo-Cuspinera V, Breslin PA: The liaison of sweet and savory. Chem Senses, 2006;31:221–225.
115 Liao J, Schultz PG: Three sweet receptor genes are clustered in human chromosome 1. Mamm Genome 2003;14:291–301.
116 Li X, Li W, Wang H, et al: Pseudogenization of a sweet-receptor gene accounts for cats' indifference toward sugar. PLoS Genet 2005;1:27–35.
117 Ishimaru Y, Okada S, Naito H, et al: Two families of candidate taste receptors in fishes. Mech Dev 2005;122:1310–1321.
118 Danilova V, Hellekant G, Tinti JM, Nofre C: Gustatory responses of the hamster *Mesocricetus auratus* to various compounds considered sweet by humans. J Neurophysiol 1998;80: 2102–2112.
119 Danilova V, Danilov Y, Roberts T, Tinti JM, Nofre C, Hellekant G: Sense of taste in a new world monkey, the common marmoset: recordings from the chorda tympani and glossopharyngeal nerves. J Neurophysiol 2002;88:579–594.
120 Hellekant G, Ninomiya Y, Danilova V: Taste in chimpanzees. 2. Single chorda tympani fibers. Physiol Behav 1997;61:829–841.
121 Sclafani A, Perez C: Cypha [propionic acid, 2-(4-methoxyphenol) salt] inhibits sweet taste in humans, but not in rats. Physiol Behav 1997;61:25–29.
122 Johnson C, Birch GG, MacDougall DB: The effect of the sweetness inhibitor 2(-4-methoxyphenoxy)propanoic acid (sodium salt) (Na-PMP) on the taste of bitter-sweet stimuli. Chem Senses 1994;19:349–358.
123 Naim M, Rogatka H, Yamamoto T, Zehavi U: Taste responses to neohesperidin dihydrochalcone in rats and baboon monkeys. Physiol Behav 1982;28:979–986.

124 Christopoulos A, Kenakin T: G protein-coupled receptor allosterism and complexing. Pharmacol Rev 2002;54:323–374.
125 Kenakin T: G-protein coupled receptors as allosteric machines. Receptors Channels 2004;10: 51–60.
126 Urizar E, Montanelli L, Loy T, et al: Glycoprotein hormone receptors: link between receptor homodimerization and negative cooperativity. EMBO J 2005;24:1954–1964.
127 Schiffman SS, Lockhead E, Maes FW: Amiloride reduces the taste intensity of Na^+ and Li^+ salts and sweeteners. Proc Natl Acad Sci USA 1983;80:6136–6140.
128 Lin W, Finger TE, Rossier BC, Kinnamon SC: Epithelial Na^+ channel subunits in rat taste cells: localization and regulation by aldosterone. J Comp Neurol 1999;405:406–420.
129 Kretz O, Barbry P, Bock R, Lindemann B: Differential expression of RNA and protein of the three pore-forming subunits of the amiloride-sensitive epithelial sodium channel in taste buds of the rat. J Histochem Cytochem 1999;47:51–64.
130 Li XJ, Blackshaw S, Snyder SH: Expression and localization of amiloride-sensitive sodium channel indicate a role for non-taste cells in taste perception. Proc Natl Acad Sci USA 1994;91:1814–1818.
131 Kellenberger S, Schild L: Epithelial sodium channel/degenerin family of ion channels: a variety of functions for a shared structure. Physiol Rev 2002;82:735–767.
132 Firsov D, Gautschi I, Merillat AM, Rossier BC, Schild L: The heterotetrameric architecture of the epithelial sodium channel (ENaC). EMBO J 1998;17:344–352.
133 Heck GL, Mierson S, DeSimone JA: Salt taste transduction occurs through an amiloride-sensitive sodium transport pathway. Science 1984;223:403–405.
134 Ossebaard CA, Smith DV: Effect of amiloride on the taste of NaCl, Na-gluconate and KCl in humans: implications for Na^+ receptor mechanisms. Chem Senses 1995;20:37–46.
135 Breslin PAS, Tharp CD: Reduction of saltiness and bitterness after a chlorhexidine rinse. Chem Senses 2001;26:105–116.
136 DeSimone JA, Lyall V, Heck GL, et al: A novel pharmacological probe links the amiloride-insensitive NaCl, KCl, and NH(4)Cl chorda tympani taste responses. J Neurophysiol 2001;86: 2638–2641.
137 Lyall V, Heck GL, Vinnikova AK, et al: The mammalian amiloride-insensitive non-specific salt taste receptor is a vanilloid receptor-1 variant. J Physiol 2004;558:147–159.
138 Ma HP, Al-Khalili O, Ramosevac S, et al: Steroids and exogenous gamma-ENaC subunit modulate cation channels formed by alpha-ENaC in human B lymphocytes. J Biol Chem 2004;279: 33206–33212.
139 Ganzevles PG, Kroeze JH: Effects of adaptation and cross-adaptation to common ions on sourness intensity. Physiol Behav 1987;40:641–646.
140 DeSimone JA, Lyall V, Heck GL, Feldman GM: Acid detection by taste receptor cells. Respir Physiol 2001;129:231–245.
141 Kinnamon SC, Dionne VE, Beam KG: Apical localization of K^+ channels in taste cells provides the basis for sour taste transduction. Proc Natl Acad Sci USA 1988;85:7023–7027.
142 Gilbertson TA, Avenet P, Kinnamon SC, Roper SD: Proton currents through amiloride-sensitive Na channels in hamster taste cells. Role in acid transduction. J Gen Physiol 1992;100:803–824.
143 Ugawa S, Minami Y, Guo W, et al: Receptor that leaves a sour taste in the mouth. Nature 1998;395: 555–556.
144 Miyamoto T, Fujiyama R, Okada Y, Sato T: Sour transduction involves activation of NPPB-sensitive conductance in mouse taste cells. J Neurophysiol 1998;80:1852–1859.
145 Huang L, Cao J, Wang H, Vo LA, Brand JG: Identification and functional characterization of a voltage-gated chloride channel and its novel splice variant in taste bud cells. J Biol Chem 2005;280: 36150–36157.
146 Stevens DR, Seifert R, Bufe B, et al: Hyperpolarization-activated channels HCN1 and HCN4 mediate responses to sour stimuli. Nature 2001;413:631–635.
147 Lyall V, Alam RI, Phan DQ, et al: Decrease in rat taste receptor cell intracellular pH is the proximate stimulus in sour taste transduction. Am J Physiol Cell Physiol 2001;281:C1005–C1013.
148 Liu L, Simon SA: Acidic stimuli activates two distinct pathways in taste receptor cells from rat fungiform papillae. Brain Res 2001;923:58–70.

149 Lin W, Ogura T, Kinnamon SC: Acid-activated cation currents in rat vallate taste receptor cells. J Neurophysiol 2002;88:133–141.
150 Richter TA, Dvoryanchikov GA, Roper SD, Chaudhari N: Acid-sensing ion channel-2 is not necessary for sour taste in mice. J Neurosci 2004;24:4088–4091.
151 Breslin PA: Human gustation; in Finger TE, Restrepo D, Silver WL (eds): The Neurobiology of Taste and Smell. New York, Wiley-Liss, 2000, pp 423–461.
152 McLaughlin SK, McKinnon PJ, Margolskee RF: Gustducin is a taste-cell-specific G protein closely related to the transducins. Nature 1992;357:563–569.
153 Ruiz-Avila L, McLaughlin SK, Wildman D, et al: Coupling of bitter receptor to phosphodiesterase through transducin in taste receptor cells. Nature 1995;376:80–85.
154 Kusakabe Y, Yasuoka A, Asano-Miyoshi M, et al: Comprehensive study on G protein α-subunits in taste bud cells, with special reference to the occurrence of $G\alpha_{i2}$ as a major $G\alpha$ species. Chem Senses 2000;25:525–531.
155 Ming D, Ruiz-Avila L, Margolskee RF: Characterization and solubilization of bitter-responsive receptors that couple to gustducin. Proc Natl Acad Sci USA 1998;95:8933–8938.
156 Wong GT, Gannon KS, Margolskee RF: Transduction of bitter and sweet taste by gustducin. Nature 1996;381:796–800.
157 He W, Yasumatsu K, Varadarajan V, et al: Umami taste responses are mediated by α-transducin and α-gustducin. J Neurosci 2004;24:7674–7680.
158 Rossler P, Boekhoff I, Tareilus E, Beck S, Breer H, Freitag J: G protein $\beta\gamma$ complexes in circumvallate taste cells involved in bitter transduction. Chem Senses 2000;25:413–421.
159 Huang L, Shanker YG, Dubauskaite J, et al: $G\gamma_{13}$ colocalizes with gustducin in taste receptor cells and mediates IP3 responses to bitter denatonium. Nat Neurosci 1999;2:1055–1062.
160 Rossler P, Kroner C, Freitag J, Noe J, Breer H: Identification of a phospholipase C β subtype in rat taste cells. Eur J Cell Biol 1998;77:253–261.
161 Clapp TR, Stone LM, Margolskee RF, Kinnamon SC: Immunocytochemical evidence for co-expression of type III IP3 receptor with signaling components of bitter taste transduction. BMC Neurosci 2001;2:6.
162 Perez CA, Huang L, Rong M, et al: A transient receptor potential channel expressed in taste receptor cells. Nat Neurosci 2002;5:1169–1176.
163 Zhang Y, Hoon MA, Chandrashekar J, et al: Coding of sweet, bitter, and umami tastes: different receptor cells sharing similar signaling pathways. Cell 2003;112:293–301.
164 Rosenzweig S, Yan W, Dasso M, Spielman AI: Possible novel mechanism for bitter taste mediated through cGMP. J Neurophysiol 1999;81:1661–1665.
165 Prawitt D, Monteilh-Zoller MK, Brixel L, et al: TRPM5 is a transient Ca^{2+}-activated cation channel responding to rapid changes in $[Ca^{2+}]_i$. Proc Natl Acad Sci USA 2003;100: 15166–15171.
166 Liu D, Liman ER: Intracellular Ca^{2+} and the phospholipid PIP2 regulate the taste transduction ion channel TRPM5. Proc Natl Acad Sci USA 2003;100:15160–15165.
167 Hofmann T, Chubanov V, Gudermann T, Montell C: TRPM5 is a voltage-modulated and Ca^{2+}-activated monovalent selective cation channel. Curr Biol 2003;13:1153–1158.
168 Striem BJ, Pace U, Zehavi U, Naim M, Lancet D: Sweet tastants stimulate adenylate cyclase coupled to GTP-binding protein in rat tongue membranes. Biochem J 1989;260:121–126.
169 Bigiani A, Roper SD: Identification of electrophysiologically distinct cell subpopulations in Necturus taste buds. J Gen Physiol 1993;102:143–170.
170 Delay RJ, Kinnamon SC, Roper SD: Serotonin modulates voltage-dependent calcium current in Necturus taste cells. J Neurophysiol 1997;77:2515–2524.
171 Herness S, Zhao FL, Lu SG, Kaya N, Shen T: Expression and physiological actions of cholecystokinin in rat taste receptor cells. J Neurosci 2002;22:10018–10029.
172 Shen T, Kaya N, Zhao FL, Lu SG, Cao Y, Herness S: Co-expression patterns of the neuropeptides vasoactive intestinal peptide and cholecystokinin with the transduction molecules α-gustducin and T1R2 in rat taste receptor cells. Neuroscience 2005;130:229–238.
173 Ogura T: Acetylcholine increases intracellular Ca^{2+} in taste cells via activation of muscarinic receptors. J Neurophysiol 2002;87:2643–2649.

174 Kataoka S, Toyono T, Seta Y, Ogura T, Toyoshima K: Expression of P2Y1 receptors in rat taste buds. Histochem Cell Biol 2004;121:419–426.
175 Baryshnikov SG, Rogachevskaja OA, Kolesnikov SS: Calcium signaling mediated by P2Y receptors in mouse taste cells. J Neurophysiol 2003;90:3283–3294.
176 Shigemura N, Miura H, Kusakabe Y, Hino A, Ninomiya Y: Expression of leptin receptor (Ob-R) isoforms and signal transducers and activators of transcription (STATs) mRNAs in the mouse taste buds. Arch Histol Cytol 2003;66:253–260.
177 Herness MS: Aldosterone increases the amiloride-sensitivity of the rat gustatory neural response to NaCl. Comp Biochem Physiol Comp Physiol 1992;103:269–273.
178 Roper S: Regenerative impulses in taste cells. Science 1983;220:1311–1312.
179 Kashiwayanagi M, Miyake M, Kurihara K: Voltage-dependent Ca^{2+} channel and Na^{+} channel in frog taste cells. Am J Physiol 1983;244:C82–C88.
180 Avenet P, Lindemann B: Noninvasive recording of receptor cell action potentials and sustained currents from single taste buds maintained in the tongue: the response to mucosal NaCl and amiloride. J Membr Biol 1991;124:33–41.
181 Cummings TA, Powell J, Kinnamon SC: Sweet taste transduction in hamster taste cells: evidence for the role of cyclic nucleotides. J Neurophysiol 1993;70:2326–2336.
182 Chen Y, Sun XD, Herness S: Characteristics of action potentials and their underlying outward currents in rat taste receptor cells. J Neurophysiol 1996;75:820–831.
183 Chen Y, Herness MS: Electrophysiological actions of quinine on voltage-dependent currents in dissociated rat taste cells. Pflugers Arch 1997;434:215–226.
184 Behe P, DeSimone JA, Avenet P, Lindemann B: Membrane currents in taste cells of the rat fungiform papilla. Evidence for two types of Ca currents and inhibition of K currents by saccharin. J Gen Physiol 1990;96:1061–1084.
185 Herness MS, Sun XD: Voltage-dependent sodium currents recorded from dissociated rat taste cells. J Membr Biol 1995;146:73–84.
186 Sun XD, Herness MS: Characterization of inwardly rectifying potassium currents from dissociated rat taste receptor cells. Am J Physiol 1996;271:C1221–C1232.
187 Noguchi T, Ikeda Y, Miyajima M, Yoshii K: Voltage-gated channels involved in taste responses and characterizing taste bud cells in mouse soft palates. Brain Res 2003;982:241–259.
188 Medler KF, Margolskee RF, Kinnamon SC: Electrophysiological characterization of voltage-gated currents in defined taste cell types of mice. J Neurosci 2003;23:2608–2617.
189 Varkevisser B, Peterson D, Ogura T, Kinnamon SC: Neural networks distinguish between taste qualities based on receptor cell population responses. Chem Senses 2001;26:499–505.
190 Ghiaroni V, Fieni F, Pietra P, Bigiani A: Electrophysiological heterogeneity in a functional subset of mouse taste cells during postnatal development. Chem Senses 2003;28:827–833.
191 Huang YJ, Maruyama Y, Lu KS, et al: Using biosensors to detect the release of serotonin from taste buds during taste stimulation. Arch Ital Biol 2005;143:87–96.
192 Feigl H: The mental and the physical; in Feigl H, Scriven M, Maxwell G (eds): Concepts, Theories and the Mind-Body Problem. Minneapolis, University of Minnesota Press, 1958, pp 370–497.
193 Mueller KL, Hoon MA, Erlenbach I, Chandrashekar J, Zuker CS, Ryba NJ: The receptors and coding logic for bitter taste. Nature 2005;434:225–229.
194 Whitehead MC, Ganchrow JR, Ganchrow D, Yao B: Organization of geniculate and trigeminal ganglion cells innervating single fungiform taste papillae: a study with tetramethylrhodamine dextran amine labeling. Neuroscience 1999;93:931–941.
195 Hellekant G, Ninomiya Y: On the taste of umami in chimpanzee. Physiol Behav 1991;49:927–934.
196 Hellekant G, Ninomiya Y: Bitter taste in single chorda tympani taste fibers from chimpanzee. Physiol Behav 1994;56:1185–1188.
197 Hellekant G, Ninomiya Y, Danilova V: Taste in chimpanzees. 3. Labeled-line coding in sweet taste. Physiol Behav 1998;65:191–200.
198 Hellekant G, Ninomiya Y, DuBois GE, Danilova V, Roberts TW: Taste in chimpanzee. 1. The summated response to sweeteners and the effect of gymnemic acid. Physiol Behav 1996;60:469–479.
199 Smith DV, John SJ, Boughter JD: Neuronal cell types and taste quality coding. Physiol Behav 2000;69:77–85.
200 Smith DV, St John SJ: Neural coding of gustatory information. Curr Opin Neurobiol 1999;9: 427–435.

201 Robinson PP: The characteristics and regional distribution of afferent fibres in the chorda tympani of the cat. J Physiol 1988;406:345–357.
202 Travers SP, Norgren R: Organization of orosensory responses in the nucleus of the solitary tract of rat. J Neurophysiol 1995;73:2144–2162.
203 Smith-Swintosky VL, Plata-Salaman CR, Scott TR: Gustatory neural coding in the monkey cortex: stimulus quality. J Neurophysiol 1991;66:1156–1165.
204 Stevens SS: Psychophysics: Introduction to Its Perceptual Neural and Social Prospects. New York, Wiley, 1975.
205 Breslin PA, Beauchamp GK, Pugh EN Jr: Monogeusia for fructose, glucose, sucrose, and maltose. Percept Psychophys 1996;58:327–341.
206 Holway AH, Hurvich LM: On the psychophysics of taste. J Exp Psychol 1938;23:191–198.
207 Stevens SS: Sensory scales of taste intensity. Percept Psychophys 1969;6:302–308.
208 Yamaguchi S: Fundamental properties of umami in human taste sensation; in Kawamura K, Kare MR (eds): Umami: A Basic Taste. New York, Dekker, 1987, pp 41–73.
209 McBurney DH: Temporal properties of the human taste system. Sens Processes 1976;1:150–162.
210 Kroeze JH, Bartoshuk LM: Bitterness suppression as revealed by split-tongue taste stimulation in humans. Physiol Behav 1985;35:779–783.
211 Smith DV, McBurney DH: Gustatory cross-adaptation: does a single mechanism code the salty taste? J Exp Psychol 1969;80:101–105.
212 Smith DV, van der Klaauw NJ: The perception of saltiness is eliminated by NaCl adaptation: implications for gustatory transduction and coding. Chem Senses 1995;20:545–557.
213 McBurney DH, Bartoshuk LM: Interactions between stimuli with different taste qualities. Physiol Behav 1973;10:1101–1106.
214 Yokomukai Y, Cowart BJ, Beauchamp GK: Individual differences in sensitivity to bitter-tasting substances. Chem Senses 1993;18:669–681.
215 Breslin PAS: Interactions among salty, sour and bitter compounds. Trends Food Sci Technol 1996;7:390–399.
216 Bartoshuk LM: Taste mixtures: is mixture suppression related to compression? Physiol Behav 1975;14:643–649.
217 Rivkin B, Bartoshuk LM: Taste synergism between monosodium glutamate and disodium 5′-guanylate. Physiol Behav 1980;24:1169–1172.
218 Lawless HT: Theoretical note: tests of synergy in sweetener mixtures. Chem Senses 1998;23:447–451.
219 Breslin PAS, Beauchamp GK: Suppression of bitterness by sodium: variation among bitter taste stimuli. Chem Senses 1995;20:609–623.
220 Kamen JM, Pilgrim FJ, Gutman NJ, Kroll BJ: Interactions of suprathreshold taste stimuli. J Exp Psychol 1961;62:348–356.
221 Breslin PAS, Beauchamp GK: Salt enhances flavour by suppressing bitterness. Nature 1997; 387:563.
222 Keast RS, Breslin PA: Bitterness suppression with zinc sulfate and na-cyclamate: a model of combined peripheral and central neural approaches to flavor modification. Pharm Res 2005;22: 1970–1977.
223 Lawless H: Paradoxical adaptation to taste mixtures. Physiol Behav 1982;29:149–152.
224 Lawless HT: Evidence for neural inhibition in bittersweet taste mixtures. J Comp Physiol Psychol 1979;93:538–547.
225 Doty RL, Bromley SM: Effects of drugs on olfaction and taste. Otolaryngol Clin North Am 2004;37:1229–1254.
226 Reiter ER, DiNardo LJ, Costanzo RM: Effects of head injury on olfaction and taste. Otolaryngol Clin North Am 2004;37:1167–1184.
227 Cowart BJ, Young IM, Feldman RS, Lowry LD: Clinical disorders of smell and taste; in Beauchamp GK, Bartoshuk LM (eds): Tasting and Smelling. New York, Academic Press, 1997, pp 175–98.
228 Getchell TV, Doty RL, Bartoshuk LM, Snow JB Jr: Smell and Taste in Health and Disease. New York, Raven Press, 1991.
229 Bartoshuk LM: Taste: robust across the age span? Ann NY Acad Sci 1989;561:65–75.
230 Cowart BJ: Development of taste perception in humans: sensitivity and preference throughout the life span. Psychol Bull 1981;90:43–73.

231 Murphy C, Gilmore MM: Quality-specific effects of aging on the human taste system. Percept Psychophys 1989;45:121–128.
232 Deems DA, Doty RL, Settle RG, et al: Smell and taste disorders, a study of 750 patients from the University of Pennsylvania Smell and Taste Center. Arch Otolaryngol Head Neck Surg 1991;117: 519–528.
233 Goodspeed RB, Gent JF, Catalanotto FA: Chemosensory dysfunction: clinical evaluation results from a taste and smell clinic. Postgrad Med 1987;81:251–260.
234 Schiffman SS, Gatlin CA: Clinical physiology of taste and smell. Annu Rev Nutr 1993;13:405–436.
235 Wrobel BB, Leopold DA: Clinical assessment of patients with smell and taste disorders. Otolaryngol Clin North Am 2004;37:1127–1142.
236 Wrobel BB, Leopold DA: Smell and taste disorders. Facial Plast Surg Clin North Am 2004;12: 459–468, vii.
237 Ackerman BH, Kasbekar N: Disturbances of taste and smell induced by drugs. Pharmacotherapy 1997;17:482–496.
238 Brightman VJ, Guggenheimer J: Changes in the oral microbial flora during treatment of recurrent aphthous ulcers (abstract). J Dent Res 1968;47(suppl):126.
239 Osaki T, Ohshima M, Tomita Y, Matsugi N, Nomura Y: Clinical and physiological investigations in patients with taste abnormality. J Oral Pathol Med 1996;25:38–43.
240 Blackburn CW, Bramley PA: Lingual nerve damage associated with the removal of lower third molars. Br Dent J 1989;167:103–107.
241 Grant R, Miller S, Simpson D, Lamey PJ, Bone I: The effect of chorda tympani section on ipsilateral and contralateral salivary secretion and taste in man. J Neurol Neurosurg Psychiatry 1989;52:1058–1062.
242 Kveton JF, Bartoshuk LM: The effect of unilateral chorda tympani damage on taste. Laryngoscope 1994;104:25–29.
243 Yanagisawa K, Bartoshuk LM, Catalanotto FA, Karrer TA, Kveton JF: Anesthesia of the chorda tympani nerve and taste phantoms. Physiol Behav 1998;63:329–335.
244 Costanzo RM, Zassler ZD: Head trauma; in Getchell TV, Doty RL, Bartoshuk LM, Snow JB Jr (eds): Smell and Taste in Health and Disease. New York, Raven Press, 1991, pp 711–730.
245 Cowart BJ: Relationships between taste and smell across the adult life span. Ann NY Acad Sci 1989;561:39–55.
246 Stevens JC, Bartoshuk LM, Cain WS: Chemical senses and aging: taste versus smell. Chem Senses 1984;9:167–179.
247 Stevens JC, Cain WS: Changes in taste and flavor in aging. Crit Rev Food Sci Nutr 1993;33:27–37.
248 Cowart BJ, Yokomukai Y, Beauchamp GK: Bitter taste in aging: compound-specific decline in sensitivity. Physiol Behav 1994;6:1237–1241.
249 Stevens JC: Detection of tastes in mixture with other tastes: issues of masking and aging. Chem Senses 1996;21:211–221.
250 Weiffenbach JM, Baum BJ, Burghauser R: Taste thresholds: quality specific variation with human aging. J Gerontol 1982;37:372–377.
251 Weiffenbach JM, Cowart BJ, Baum BJ: Taste intensity perception in aging. J Gerontol 1986;41: 460–468.
252 Stevens JC, Cain WS, Demarque A, Ruthruff AM: On the discrimination of missing ingredients: aging and salt flavor. Appetite 1991;16:129–140.
253 Stevens JC, Traverzo A: Detection of a target taste in a complex masker. Chem Senses 1997;22: 529–534.
254 Stevens JC, Cruz LA, Hoffman JM, Patterson MQ: Taste sensitivity and aging: high incidence of decline revealed by repeated threshold measures. Chem Senses 1995;20:451–459.

Paul A.S. Breslin
Monell Chemical Senses Center
3500 Market Street
Philadelphia, PA 19104 (USA)
Tel. +1 215 898 5021, Fax +1 215 898 2084, E-Mail breslin@monell.org

Hummel T, Welge-Lüssen A (eds): Taste and Smell. An Update.
Adv Otorhinolaryngol. Basel, Karger, 2006, vol 63, pp 191–220

Central Gustatory Processing in Humans

Dana M. Small

The John B. Pierce Laboratory and Yale University, New Haven, Conn., USA

Abstract

The purpose of this chapter is to provide a general overview of the central representation of gustatory information in the human brain. The anatomical pathways for the two primary animal models (rodent and nonhuman primate) are provided followed by the presumed human gustatory pathway. The section on the gustatory pathway describes what is known about how taste intensity, quality and affective value are represented in the human brain. The chapter concludes with a review of flavor processing.

Contrary to popular belief, things do not 'taste' like chocolate or tomato sauce. One only needs to plug one's nose while eating to realize that taste provides little information regarding the identity of a food. Instead, food identification is usually accomplished before the stimulus is in the mouth via the olfactory and visual modalities and maintained once in the mouth by retronasal olfaction (which occurs when volatiles from the food reach the olfactory epithelium). Most researchers agree that there are five major categories of taste quality – sweet, sour, salty, bitter, and savory – each tuned to identify a specific nutrient or poison and each associated with particular physiological functions, namely, ensuring energy reserves (sweet, savory), maintaining electrolyte balance (salty), guarding pH (sour, bitter), and avoiding toxins (bitter) [1, 2]. Thus, the primary function of the gustatory sense is not to identify foods but rather to identify substances in food and drink that may promote or disrupt homeostasis. Accordingly, toxicity is a better predictor of central neural response than physical structure [3], and taste elicits autonomic responses in addition to perceptual experiences [4]. The purpose of this chapter is to provide a general overview of the central representation of gustatory information in the human brain. The anatomical pathways for the two primary animal models (rodent and nonhuman primate) are provided followed by the presumed human gustatory pathway. The following section on the gustatory pathway describes what is known about how taste intensity, quality and affective value are represented in the human brain. The chapter concludes with a review of flavor processing.

The Gustatory Pathway

Rodents

In rodents, taste information is carried through the cranial nerves VII [chorda tympani (CT), and greater superficial petrosal branches], IX (lingual branch) and X (superior laryngeal branch) and terminates in the rostral division of the nucleus tractus solitarii (NTS) [5]. Many of the neurons that leave the NTS travel to other brainstem nuclei to mediate reflexive and regulatory responses related to feeding, such as salivation and the rejection reflex, leaving only a small portion (roughly 20%) to ascend within the gustatory neuraxis [6]. From the NTS, there are ipsilateral and bilateral [7] ascending fibers that synapse in the gustatory parabrachial nucleus of the pons [8]. Two separate third-order pathways arise from the pons: a dorsal sensory pathway [9], which ascends ipsilaterally to the gustatory cortex after synapsing in the gustatory nucleus of the thalamus [parvocellular division of the ventroposterior medial nucleus (VPMpc)] [10–12], and a ventral affective pathway [13], which projects to the lateral hypothalamus, substantia innominata, central nucleus of the amygdala, and bed nucleus of the stria terminalis in the ventral forebrain [14]. Thalamic gustatory afferents terminate in two regions of the insular cortex [15], one located within the dysgranular insular cortex and the other in the granular insular cortex. Both dorsal and ventral termination loci send back projections to the brainstem taste nuclei [16–21]. Although projections are heaviest ipsilaterally, at least some of them appear to be bilateral [18]. The rat insular cortex can be roughly divided into an anterior gustatory region and a posterior visceral region [22]. However, taste-responsive neurons are widely distributed, and there is a clear overlap with the visceral representation [23].

Nonhuman Primates

The peripheral organization of the gustatory system is similar in rodents and primates with taste information being carried through the cranial nerves XII, IX, and X to the NTS [24]. Second-order gustatory fibers ascend ipsilaterally from the NTS towards the pons, but it is thought that instead of forming a synapse in the parabrachial nucleus, as is the case in rats, taste fibers join the central tegmental tract and project to the VPMpc [24, 25]. Thus, there is currently no evidence for a pontine taste relay in the nonhuman primate and no clear evidence of a separate ventral affective pathway. It is currently not known if there are also contralateral afferent projections, as is the case in rodents. Scott and Plata-Salaman [26] have suggested that the primary manifestation of these anatomical interspecies differences relates to how homeostatic and hedonic factors influence sensory coding of taste. Since the goal of this chapter is to provide an understanding of sensory coding in the human, the primate will serve as the primary model.

The equivalent of the dorsal stream projects ipsilaterally from the thalamus to terminate in the anterior insula/frontal operculum (AI/FO) [27–29]. Using titrated amino acid autoradiography, Pritchard et al. [27] traced the efferent projections of VPMpc. The primary efferent projection was located in the ipsilateral AI/FO cortex adjacent to the superior limiting sulcus and extending rostrally to the caudolateral orbitofrontal cortex (OFC). A second projection was also found that terminated in areas 3a, 3b, 2, and 1 along the lateral margin of the precentral gyrus. They noted that these two regions correspond to the two regions Ogawa et al. [28] identified with electrical stimulation of the peripheral gustatory nerves. Thus, as in the rat, there are multiple gustatory regions within the insula/operculum. However, since AI/FO has few lingual terminations [28] and is relatively more sensitive to stimulation of the taste nerve compared to the lingual nerve [30], it has become known as the primary gustatory area or 'area G' [31]. Cytoarchitectonically, AI/FO can be distinguished from its surrounding cortex because layers II and IV are thicker and contain fine granule cells [28, 32]. Electrophysiological studies [26, 33–38] and one positron emission tomography (PET) [39] study have shown taste responses in the nonhuman primate AI/FO.

The caudal OFC receives direct projections from the insula and opercular taste regions [40]. Baylis et al. [40] isolated taste-responsive neurons and injected horseradish peroxidase into this site. They found substantial labelling in the frontal opercular taste area, anterior dorsal insula, extending into more ventral regions of the insula, as well as in the amygdala, mediodorsal thalamus, rhinal sulcus and substantia nigra. Since no labelling occurred in the VPMpc, Rolls et al. [83] proposed that the caudal OFC region represented the secondary taste cortex. However, it is clear that there are multiple taste-responsive regions in the insula/operculum apart from the AI/FO that do not receive direct thalamic projections and should therefore also be considered higher-order gustatory areas [31].

Although not formally considered part of the gustatory system, the amygdala has reciprocal connections with virtually every level of the gustatory pathway [14, 41–45] and taste-responsive neurons are present within several amygdaloid nuclei [46]. In the monkey, gustatory information reaches the amygdala via direct pathways from cortical taste neurons within the insula and

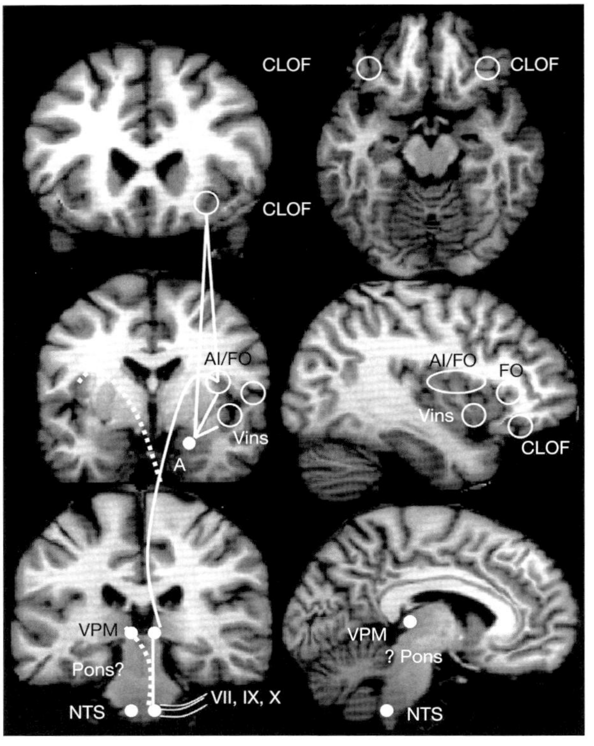

Fig. 1. The human gustatory pathway. Coronal sections on the left show the main pathway. Slices in the right column provide additional views. The question mark in the midbrain demarcates the region where a pontine relay would likely be if it did exist in the human. Most evidence to date suggests that there is not a relay here. CLOF = Caudolateral OFC; dotted lines = possible sites for bilateral projections; FO = other regions where activations are frequently reported in response to stimulation with a taste; solid lines = known projections. Vins, The areas outlined are estimates and are not meant to signify established anatomical boundaries.

operculum [41, 47] and orbitofrontal taste area [45, 48, 49]. The amygdala also sends projections back to the NTS [42] where it may exert an influence on taste processing at this early level of the primate gustatory neuraxis.

Humans

The human gustatory pathway is assumed to be equivalent to the monkey pathway and is depicted in figure 1. There is general agreement that brainstem lesions caudal to the pons lead to decreased sensitivity on the ipsilateral side of the tongue [50–54], supporting ipsilateral ascension of taste fibers. Gustatory

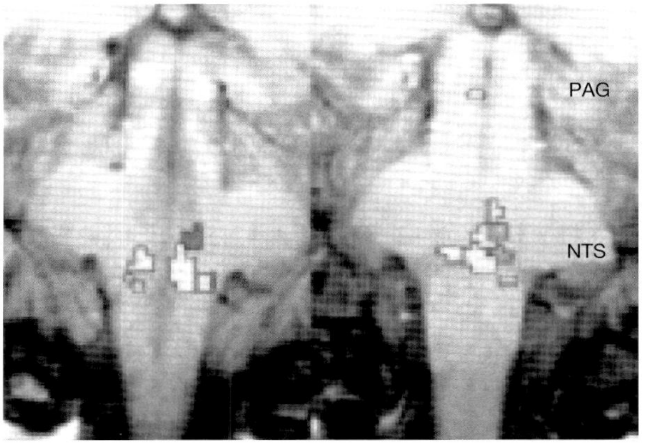

Fig. 2. The results from an fMRI study showing brainstem sites activated during administration of sucrose solution to the tongue. Each outline represents a different human subject. PAG = Periaqueductal gray area. Note the lack of activation in the pons (region immediately caudal to PAG). (Taken and modified from figure 7b in Topolovec et al. [55]; copyright 2004 by Wiley-Liss, Inc. and with permission of the authors.)

disturbance following lesions to the pons are also predominantly ipsilateral, but bilateral and contralateral deficits have been reported [50, 51]. In the case of pontine lesions, it is thought that gustatory changes are caused by disruption of ascending fibers, as opposed to cell body damage (which would imply a pontine taste relay). This is supported by a recent functional magnetic resonance imaging (fMRI) study by Topolovec et al. [55] in which activation in the NTS was observed in all 8 subjects to whole-mouth stimulation with sucrose (2 bilateral, 5 right and 1 left). In contrast, no pontine activation was reported.

Early clinical studies, based on traumatic or cerebral vascular lesions, located the primary gustatory area in area 43 of the parietal operculum [56, 57] or the anterior insula [58]. In their classic paper, Penfield and Faulk [59] reported the results of stimulation of the human insula during epilepsy surgery. Consistent with the reports of the location of taste area in nonhuman animals, most gustatory phenomena occurred following stimulation to anterior portions of the insula, and were interspersed with several other types of responses including olfactory and gustatory hallucinations (all unpleasant), somatosensory, visceral sensory and motor sensations, and feelings of nausea. In a more recent clinical study, Cascino and Karnes [60] reported 3 patients with gustatory hallucinations as part of their seizure disorder. In all 3, high-resolution MRI revealed unilateral lesions in the AI/FO.

Petrides and Pandya [61] performed a comparative cytoarchitectonic study of the human and monkey brain and identified a close correspondence in the

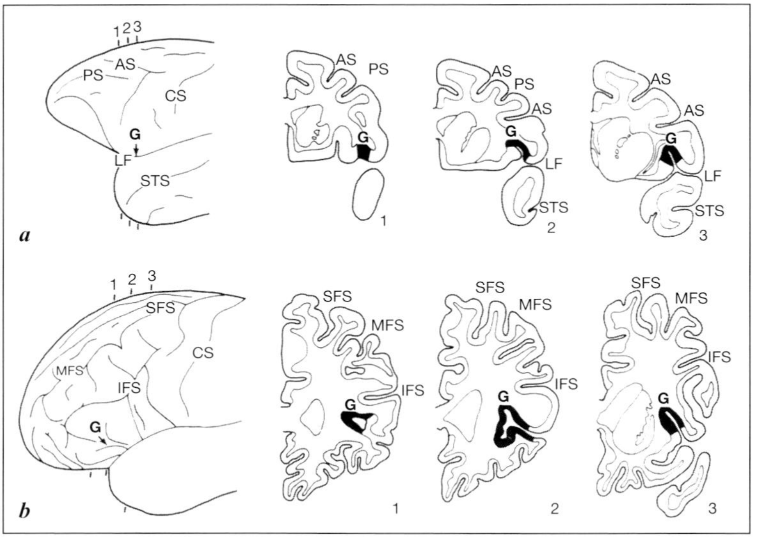

Fig. 3. ***a*** Coronal sections (1, 2, 3) taken at the levels indicated on the outline of the lateral surface of the cerebral hemisphere of the monkey to illustrate the location of the cortical gustatory areas. ***b*** Coronal sections (1, 2, 3) taken at the levels indicated on the outline of the lateral surface of the human brain to illustrate the location of the cortical gustatory areas that are comparable in terms of architecture and topography to the areas identified in the primate and shown in (***a***). AS = Arcuate sulcus; CS = central sulcus; IFS = inferior frontal sulcus; LF = lateral fissure; MFS = middle frontal sulcus; PS = paracingulate sulcus; SFS = superior frontal sulcus; STS = superior temporal sulcus. (Taken from Petrides and Pandya [61]; copyright 1994 by Elsevier press and with permission from the authors.)

architectonic features of the gustatory anterodorsal insula and adjacent frontal opercular cortex (AI/FO in the monkey) (fig. 3). In humans, as in monkeys, this region is located within the cortex of the horizontal ramus of the sylvian fissure, with the rostral limit defined by the end of the ramus. In both species, the region has a uniform cellular distribution in the supragranular layers, which contain small- to medium-sized pyramidal cells [61]. Most functional neuroimaging studies of taste report activation in and around this area [62–78]; however, these same studies also report activations in the mid insula, ventral insula, and parietal and temporal opercula. Evidence from magnetoencephalography suggests that the earliest cortical response to the presentation of a taste occurs in the parietal operculum [77, 78]. Kobayakawa et al. [77, 78] have therefore argued that the parietal operculum represented the true 'primary' taste area. Although, their findings are provocative, this interpretation must be viewed cautiously, as it is inconsistent with primate gustatory organization and the known connectivity of the posterior insula with somatosensory and auditory but not gustatory or olfactory systems [79]. While the location of the 'primary' gustatory area continues to be debated, all data are consistent in showing that there are multiple taste-responsive regions in the insula and surrounding operculum in the human [74], as there are in rodents and monkeys. Whether there is functional specialization within these regions remains to be determined.

To date, there are no published reports of taste perception following orbital lesions. Therefore, neuroimaging studies have been the only source of information about the human orbitofrontal gustatory area. While the OFC is frequently activated in response to gustatory stimulation [67, 68, 70, 72, 73, 75, 76, 80], it is not always activated [69], and often the area of activation is anterior and medial to the region described as the secondary gustatory cortex in monkeys [74]. This raises the possibility of interspecies differences in the precise location of the orbital gustatory area. Alternatively, these anterior activations may represent higher-order processing of taste-related information.

Both contralateral and ipsilateral taste deficits have been reported following lesions to higher levels of the gustatory neuraxis including the thalamus and the insula/operculum [50, 56–58, 81]. However, there is considerable evidence to suggest that there is bilateral representation of taste at the cortical level. First, Aglioti et al. [82] evaluated the ability of a patient with complete callosotomy to either name the quality of a taste or point to the word representing the particular taste quality. They reasoned that if taste was entirely crossed, then the patient would not be able to report the quality of the taste when applied to the right tongue because information would not be able to reach the left 'verbal' hemisphere. Since the patient could report the quality of the taste applied to the right tongue, it was concluded that taste information must reach both the left and right cortex. This conclusion is consistent with the results of an fMRI study in which bilateral activation of the insula, superior temporal lobe, inferior frontal lobe and postcentral gyrus cortex was observed after either the left or right tongue was stimulated with electrogustometry [62]. It is therefore possible that taste fibers proceed ipsilaterally to the NTS, which in turn sends out projections that ascend ipsilaterally until the pons, at which point a small percentage of fibers decussate, thus culminating in bilateral projections to the thalamus and taste cortex.

A separate, but related issue is whether there is cerebral dominance in human gustation. In 1999, Small et al. [74] reviewed all of the PET studies of human gustation that had been performed up until that time. Twenty-two peaks were identified in the insula/opercular area, and of these 16 were in the right hemisphere. In the OFC, only 2 out of the 18 peaks fell in the homologous region of the monkey orbital taste region [40, 83], and both of these were also in the right hemisphere (the rest of the peaks were anterior and medial). It was therefore proposed that the right hemisphere was dominant for taste in human gustation. Since this time, findings have been mixed. In accordance with Small et al. [74], Barry et al. [62] reported that although activation in most regions of the insula was bilateral following stimulation of taste by application of electrogustometry to either the right or left tongue, activation was consistently greater in the right dorsal insular region. Additionally, several studies of taste perception following resection from the anterior temporal lobe have reported greater changes if the resection is from the right hemisphere [80, 84, 85]. However, most studies report bilateral activation of the insula taste regions [67, 70, 72, 75] and still others report evidence for left hemisphere dominance [66, 86]. Various factors have been proposed to influence lateralization including internal state [86], handedness [66, 87], and affective value (discussed below). It is also possible that, as in olfaction, cognitive judgments influence laterality [88, 89]. Cognitive influences upon gustatory processing have yet to be addressed. Thus, the possibility of cerebral dominance and/or lateralization of function in human gustation remains unresolved.

Gustatory Physiology

The neural substrates of taste must code for perceived intensity, pleasantness/unpleasantness (i.e. affective value) and quality. One difficulty in studying the neural representation of these perceptual dimensions is that they are not independent [see chapter 10 by Breslin and Huang, this vol, pp 152–190]. Instead pleasantness/unpleasantness may influence perceived intensity and vice versa. Moreover, the nature of these interactions depends upon quality [1, 90–92]. For example, bitter is perceived as unpleasant at all concentrations, whereas sweet generally becomes more pleasant as concentration increases and then plateaus or decreases. These perceptual interactions are reflected in the underlying physiology by overlapping representation. However, despite the degree of interaction, dissociations have been found and the relative importance of different regions to coding each dimension is beginning to be understood.

Detection and Intensity

Stimulus detection is the most fundamental perceptual process. Gustatory detection thresholds measure the minimum amount of a substance that is required to be present in a solution in order for a subject to detect its presence at above-chance levels. In monkeys, bilateral lesions to the VPMpc result in a profound and persistent elevation in rejection thresholds for quinine hydrochloride [95], indicating that detection of taste relies on structures rostral to the brainstem taste regions. Pribram and Bagshaw [208] found that lesions produced in the AI/FO area of the macaque caused elevated taste thresholds. In humans, Small et al. [80] studied citric acid detection and recognition thresholds in patients with resection from the anterior medial temporal lobe, including the amygdala, for the treatment of epilepsy. These lesions did affect detection thresholds. However, they describe 1 patient whose detection threshold was elevated 5 standard deviations above the group mean. The MRI of this patient revealed that the surgical resection had included a portion of the right anterior ventral insular cortex, whereas for all other patients, anatomical neuroimaging had indicated removal of portions of either temporal lobe without encroachment on the insula. These findings indicate that taste detection in monkeys and humans is likely a result of processing in the insula/opercular gustatory regions.

Processing in the insula/operculum is also critical for suprathreshold taste intensity perception. Although responses to taste stimuli increase with stimulus concentration at all levels of the neuroaxis [46, 72, 83, 96–98], intensity-response functions generated from taste-responsive cells in the AI/FO conform best to the slopes reported in human psychophysical experiments of perceived intensity [36]. Changes in suprathreshold taste intensity perception have also been observed following lesions to the insula in humans. Pritchard et al. [81] asked subjects with insular lesions to rate the intensity of tastes applied to either the right or the left side of the tongue. Patients with right insular damage showed decreased sensitivity to tastes applied to the right side of the tongue and patients with left insular damage showed decreased sensitivity to tastes applied to the left side of the tongue. Simmons et al. [99] also reported decreased taste intensity perception on the side of the tongue ipsilateral to an insular lesion. However, when they compared their data to matched controls they noted that the change was not due to an ipsilateral decrease but rather to a contralateral increase in taste intensity perception. One possible explanation for this result is that a release from inhibition occurred. Similar observations have been reported following damage of the CT in humans [100], suggesting that inhibition may be a general property of the gustatory system. For example, Yanagisawa et al. [101] reported that anesthetizing the CT nerve in humans resulted in increased intensity perception of bitter following application of quinine to the regions of the oral cavity innervated by nerve IX, providing evidence for inhibition between these cranial nerves.

Neural processing in the amygdala may also contribute to intensity coding. Increases in taste intensity perception have been observed following unilateral resection from the anterior medial temporal lobe for the treatment of epilepsy [84, 85]. In all cases, the surgical resection included the amygdala. Interestingly, the effect was largely due to increases in the reported intensity of the unpleasant bitter stimulus. This accords with the finding that rejection thresholds for bitter taste are reduced in rodents following amygdala lesions [102] and suggests that intensity coding in the amygdala may interact differentially with positively compared to negatively valenced gustatory stimuli [85]. Neuroimaging data are also consistent with a role for the insula and the amygdala in taste intensity coding. Responses to the intensity of unpleasant and pleasant taste stimulation have been observed in the mid and dorsal insula, amygdala, cerebellum, pons, and anterior cingulate cortex [72]. In contrast, intensity responses are not observed in the orbital taste region, indicating that this region does not play a direct role in coding taste intensity [72].

Taste Quality Coding

It is widely agreed that there are five major perceptual categories of taste quality: sweet, sour, salty, bitter and savory [103]. An ongoing debate in the field has been whether these taste qualities are coded in specific 'labeled lines' or as patterns emerging across a collection of neurons [26, 98, 104–108]. The labeled-line theory posits that there are specific lines, each carrying and coding information about only one taste quality. The 'across-fiber pattern' theory holds that all neurons contribute equally to the quality coding of all tastes with unique patterns of activity associated with each taste quality. However, a recent chapter written by one of the leading proponents of the labeled-line theory in collaboration with one of the leading proponents of the across-fiber pattern theory concluded that the evidence to date suggests that the various neuron types play a critical role in defining unique across-neuron patterns and that such a system is capable of unambiguously coding taste quality [109].

In monkeys, individual fibers of the peripheral nerves tend to respond 'best' to one taste quality. Thus, hierarchical cluster analysis of the responses of individual fibers in the CT produces four groups of responses corresponding to the four classical taste qualities (sweet, sour, salty and bitter – savory was not studied) [110, 111]. There is also a variety of evidence to suggest differential discriminative capacity of the peripheral taste nerves. Fibers of the CT branch of nerve VII tend to respond best to sweet- and salty-tasting stimuli, moderately to sour tastants and even less to compounds described by humans as bitter [110, 111]. In contrast, taste fibers of nerve IX often respond best to bitter tastants, but also respond well to sweet and savory tastants (monosodium glutamate), though the sweet responses are of a smaller magnitude [111]. This distribution corresponds to psychoperceptual data showing that sensitivity to sweet and salty tastants is most prominent on the anterior portions of the tongue, which is innervated by nerve VII, whereas sensitivity to bitter taste is more acute at the posterior tongue, which is innervated by nerve IX [90]. To date, no single fiber studies of the cranial nerves have been performed in the human.

The specificity of individual responses to chemical compounds has also been described using a breadth of tuning metric developed by Smith and

Travers [113]. This coefficient can range from 0.0, representing complete specificity to one of the stimuli, to 1.0, which indicates an equal response to all tastants. In the primate peripheral taste nerves, a breadth-of-tuning coefficient of 0.54 has been reported, indicative of a fairly high specificity [111]. In contrast, the mean coefficient for NTS neurons was 0.87, which is higher than that reported in rodents and indicates that NTS primate taste neurons respond relatively nonselectively to taste stimulation [96]. However, cells could be classified according to their best response and this was at least partially determined by location (chemotopic organization). Taste-responsive cells are also quite broadly tuned in the thalamus, with an average coefficient of 0.73 and again some indication of chemotopic organization [97].

In monkeys, attempts to divide neurons in the AI/FO into discrete groups indicate that although it is possible to assign neurons to a small number of groups on the basis of their response profiles, the variability of responses within these groups is high [35, 37]. Two basic patterns of activity have been identified: one that characterized sweet and salty stimuli and the other that characterized acids, quinine and water. The breadth-of-tuning coefficient for opercular taste neurons has been reported as 0.67 [37] and 0.56 for anterior insula neurons [35]. These findings indicate increased selectivity with respect to neurons recorded in the NTS and thalamus, which at the very least suggests an intention towards taste quality coding in the AI/FO. Taste-responsive neurons in the OFC are even more finely tuned with coefficients of 0.39 [83].

As mentioned above, taste quality coding is also likely coded in the pattern of activity across neurons. Katz et al. [114] used multiunit recordings in rodents to isolate transient quality-specific cross-correlations in ensembles of taste neurons within the insular taste region. These findings highlight the potential importance of considering across-fiber contributions to taste quality coding. Multiunit recording has not yet been examined in primate electrophysiological studies.

Smith-Swintosky et al. [36] evaluated the relationship between psychophysical studies of taste quality in the human as reported by Kuznicki and Ashbaugh [115] and Schiffman and Erickson [116], and their own electrophysiological results of responses to taste quality in the AI/FO of the alert macaque monkey. The correlation between their data and the data from Kuznicki and Ashbaugh [115] was +0.91, indicating a very close relationship between neural responses evoked in the macaque and the perceptual experience of taste quality in humans. When they performed the same analysis with the results from the study by Schiffman and Erickson [116], however, the correlation (+0.53) was not as high. Interestingly, Smith-Swintosky et al. [36] suggested that this discrepancy arose because they used a higher concentration of salt, which was probably more aversive than the lesser concentration used in the psychophysical study. As a result, the response they recorded to salty stimuli was more akin to the response profiles elicited by aversive stimuli such as quinine. In fact, the correlation between the two data sets rose to +0.85 when they dropped the salty from the analysis. These results suggest that the affective value, determined by concentration, may influence taste quality coding in at least one region of the insula.

It is currently unknown if it is possible to isolate quality-specific responses in humans, and if so, what anatomical structures contribute to taste quality coding. Deficits in identifying or recognizing suprathreshold taste quality have been reported following anterior but not posterior insular lesions [81, 117], indicating that as in the monkey, this area is important in coding taste quality. Elevated taste recognition thresholds have also been observed following unilateral resection from the anterior medial temporal lobe [80, 118], and a case of gustatory agnosia has been described in patients with bilateral anterior medial temporal lobe damage and unilateral insular atrophy [119]. Therefore, although taste detection may be computed in the insula/operculuar region, recognition likely involves integration of the gustatory code with motivational and hedonic networks related to feeding [80].

A separate issue is whether taste-responsive cells are organized in a spatially determined fashion. Several studies have reported evidence for chemotopic organization within the NTS of the rat [120], hamster [121] and monkey [96]. In general, sweet and salty responses are more numerous in the rostral part of the gustatory NTS, prompting Smith and Scott [109] to suggest that the distribution reflects the differential sensitivity of the CT and IXth nerves because these nerves terminate in a rostral to caudal order. Although there is some evidence for chemotopy in the cortical gustatory areas [31, 37, 83, 122], results are inconsistent. Scott et al. [37] recorded taste responses from neurons in the frontal operculum of cynomolgus monkeys and found that sweet-best cells tended to be distributed toward the anterior section, salty-best cells toward the posterior extent and sour-best cells within an intermediate region. In subsequent studies, no evidence for chemotopy was found [36, 123]. Furthermore, when all the data were compiled, a plot of the location of each neuron in the gustatory cortex of the macaque as a function of its most effective basic stimulus revealed no evidence for chemotopic organization [26]. Nevertheless, Scott and Plata-Salaman [26] note in their 1999 review that the probability was higher than chance that a contiguous neuron would have the same sensitivity, indicating that taste sensitivity is not randomly distributed within the cortex. In the OFC, Rolls et al. [83] found some evidence of spatial grouping of taste neurons; however, they too noted that it was not clear whether the basis for this grouping was related to quality or some other attribute of the stimuli, such as affective valence.

The issue of chemotopy has not been investigated in humans. Spatially segregated responses to different gustatory stimuli are frequently reported but interpreted to result from affective rather than qualitative differences in the stimuli [70–72, 76, 124]. However, these studies confound taste pleasantness/unpleasantness with quality by using different taste qualities to elicit the opposite affective response. For example, Small et al. [72] reported that a caudal region of the right OFC responds to sweet irrespective of intensity and not to bitter stimuli, and a region of the left anterior OFC responds to bitter irrespective of intensity and not to sweet stimuli. Since pleasantness/unpleasantness is confounded with quality, the result may be related to quality, valence or a combination of these dimensions.

Affective Value

The affective value of a taste stimulus can be influenced by a variety of factors, including factors intrinsic to the stimulus (quality, intensity, physiological significance) and factors related to the individual (preference, internal state, experience). One intriguing aspect of gustatory hedonics is the possibility that there are innate affective values associated with the different taste qualities [125]. Steiner [126] observed that when a sweet or bitter taste is placed in the mouth of a human infant, stereotyped facial expressions can be observed. He described that a sweet or mildly salty taste elicits a mild rhythmic smacking, slight protrusions of the tongue, and a relaxed expression accompanied sometimes by a slight upturn of the corners of the mouth. In contrast, a bitter, sour, or very salty taste elicits a grimace, a turning away, a gape or gagging movement, pushing out the offending taste, and a pushing away with the hands. This 'innate' response is adaptive for it saves us from the peril of having to learn that bitter signifies poison, and can be thought of as the 'intrinsic affective value' of the quality.

The presence of intrinsic affective values has led some investigators to propose that taste perceptions are primary reinforcers [127]. However, if this were true then affective values should be immutable. They are not. In rodents, diet has been shown to have a profound effect on gustatory organization [128–132], which may in turn be related to preferences [133]. Likewise in humans, diet has been shown to affect the pleasantness perception of taste qualities [134], and the importance of being able to adapt taste preferences to the food available in the environment has been emphasized [135]. The influence of diet upon taste pleasantness perception and preference suggests that although the relationship between the intrinsic affective value and quality is generally very stable, and possibly present at birth, learning can and does occur. This in turn implies that the intrinsic affective value and quality are separable at the neural level.

Genetic differences may also contribute to differences in preferences and hence to the perceived pleasantness of a taste. Deviations in sweet preference have been associated with higher risk for drug abuse [136], and obesity [137–139].

For example, Kampov-Polevoy et al. [140] reported increased liking of concentrated sweet solutions in subjects with a family history of alcoholism. Neurophysiological correlates of taste preference have not been examined in humans but the relationship with obesity and drug abuse suggests that preferences are related to differences in the underlying neural reward circuitry rather than being simply a result of diet.

The affective value of a taste may also change as a result of conditioning, nutrient depletion, or internal state [93, 132, 133, 141–153], as well as by perceived intensity [1, 90–92], with the nature of the interaction depending upon quality. Taken together, these findings underscore the fact that gustatory hedonics are multifaceted, likely involve separate neural substrates and are dependent upon other perceptual dimensions. To date, neuroimaging studies of the affective coding of taste have focused exclusively on the representation of the intrinsic affective value of a taste quality.

In the first neuroimaging study of gustatory hedonics, Zald et al. [76] reported activity in the amygdala, cingulate gyrus, OFC and hippocampus in response to an aversive saline compared to water. All of these regions, plus a peak in the anterior dorsal insula, were also preferentially activated by saline compared to chocolate. Robust activity in the amygdala in response to unpleasant taste was in line with studies indicating a role for this region in fear [154, 155], conditioned taste aversion learning [156–160] and with two earlier chemosensory studies showing amygdala activation to unpleasant flavors [73] and unpleasant odors [161]. Therefore, the authors highlighted the importance of the amygdala in processing unpleasant and potentially threatening tastes. This view has since been challenged by studies reporting equivalent responses in the amygdala to pleasant and unpleasant tastes [70, 72, 75] and another study demonstrating that the gustatory response in the amygdala is driven by taste intensity rather than valence [72]. Consistent with this finding, a retrospective examination of the earlier results showed that the unpleasant tastes and odors were all rated as more intense than the pleasant tastes and odors [162]. Thus, it appears that intensity perception and not affective valence accounted for the early reports of preferential activation to unpleasant taste.

Do these findings mean that the amygdala simply encodes the intensity of taste stimuli without participating in gustatory hedonics? The answer is clearly no. First, as discussed in the section on intensity representation, resections from the anterior medial temporal lobe, which include the amygdala, result in enhanced intensity ratings for unpleasant but not pleasant taste [85]. Second, preferential amygdala and basal forebrain activation has been observed following perception of novel and unpleasant taste-odor pairs versus perception of familiar pleasant taste-odor pairs [73] (fig. 4). Importantly, the differential activation could not result from differences in stimulus concentration because

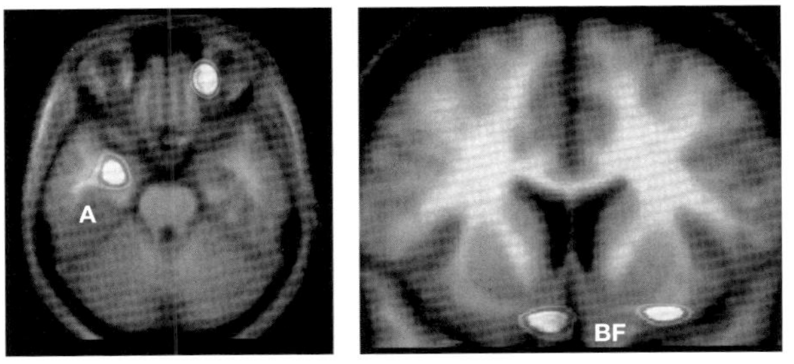

Fig. 4. Activation in the amygdala (A) and basal forebrain (BF) resulting from the summed neural activity evoked during a 60-second PET scan of incongruent taste-odor stimulation (strawberry with salty, soy sauce with sweet, grapefruit with bitter, coffee with sour) compared to a 60-second PET scan of congruent taste-odor stimulation (strawberry with sweet, soy sauce with salty, grapefruit with sour, coffee with bitter and tasteless with odorless). Tastants were presented on tongue shaped filter papers simultaneously with odors that were presented on long-handed cotton wands waved under the nose. (Taken from Small et al. [73]; copyright 1997 by Lippincott, Williams & Wilkins and with permission of the authors.)

identical tastes and odors were used to generate familiar and unfamiliar pairs. Sweet taste paired with strawberry odor and salty taste paired with soy sauce odor created familiar and pleasant flavors whereas salty taste paired with strawberry odor and sweet taste paired with soy sauce odor created novel and somewhat unpleasant flavors. Thus, differential activation could not be related to the concentration of the stimuli. However, it may be related to perceived intensity, which tends to be greater for unpleasant tastes [92].

Taken together the findings indicate that the response in the amygdala reflects an interaction between intensity, novelty and intrinsic affective value. Small et al. [72] therefore proposed that 'the amygdala is important in establishing the saliency of sensory stimuli, which is determined by the interacting dimensions of intensity, valence, and perhaps novelty/familiarity', and that 'one important function of this integrated coding may be in biasing processing in favour of adaptive needs so that subjective experience can be released from dependence upon the physical attributes of the stimulus'. If this were true then lesions to the amygdala may bias the normal interaction between intensity and pleasantness, resulting in enhanced aversion to already unpleasant tastes and a subsequent increase in their perceived intensity. This notion is also consistent with the psychophysical interactions noted above [see also chapter 10 by Breslin and Huang, this vol, pp 152–190] and highlights the potential importance of the amygdala in supporting perceptual interactions that may serve to promote perceptions that insure preservative responses rather than a veridical representation of the stimulus. Parallel findings and conclusions have been reached for the role of the amygdala in representing pure olfactory stimuli [for details, see chapter 3 by Gottfried, this vol, pp 44–69].

The affective value of a taste stimulus can also be modulated by changes in internal state. In monkeys, single-cell recordings of taste-responsive neurons in the amygdala show moderate attenuation of the gustatory response in some neurons after feeding [163]. No study has yet examined the effect of satiety on the neural response evoked by a pure taste stimulus in the human. However, several studies of satiety have been conducted using food and drink [84, 86, 162, 164], and all fail to observe changes in the amygdala as a result of hunger or satiety. The one exception to this comes from studies reporting changes in the amygdala in obese but not lean subjects [165, 166]. These findings suggest that the amygdala of healthy lean subjects is not critical in mediating changes in the reward value of food being consumed. This result contrasts with the known role of the amygdala in tracking changes in the reward value of visual [167, 168] and olfactory stimuli [169, 170] caused by eating [see also chapter 3 by Gottfried, this vol, pp 44–69]; a discrepancy that may be related to differential responsiveness to sensory inputs linked to receipt versus prediction of reward. In support of this notion, the amygdala has been shown to respond more to the anticipation of a sweet taste compared to receipt of that sweet taste [71].

In contrast to the amygdala, valence-specific responses are consistently observed in the OFC [70, 71, 75, 76] and these responses are not sensitive to intensity [72]. They may be localized to the orbital taste area but also extend well beyond this region, and they may be for received as well as for anticipated tastes [71]. Primate OFC taste responses are also greatly attenuated by satiety [171, 172], and in humans neuroimaging studies consistently show differential responsiveness in the OFC to the food when hungry versus when full [86, 124, 164]. Consistent with these results, animals will only self-stimulate here when hungry [173] and lesions to this area lead to gustatory anhedonia [174], indicating that this region is important in representing the intrinsic affective value as well as transient shifts in the affective value associated with satiety.

There is a possibility that an asymmetry exists in gustatory processing in the OFC, with pleasant tastes preferentially activating the right OFC. Figure 5 shows peaks from pleasant or unpleasant taste stimulation color-coded and plotted onto an axial anatomical section. Eight analyses of pleasant taste yielded 13 OFC peaks, and of these 11 were in the right hemisphere and only 2 were in the left hemisphere. Furthermore, the left-hemisphere peaks were from analyses that produced stronger peaks in the right hemisphere. The situation is not as compelling for unpleasant taste. From 8 analyses that yielded 12 peaks, 5 were in the right and 7 in the left hemisphere. It is currently not known why pleasant tastes should preferentially activate the right OFC.

a

b

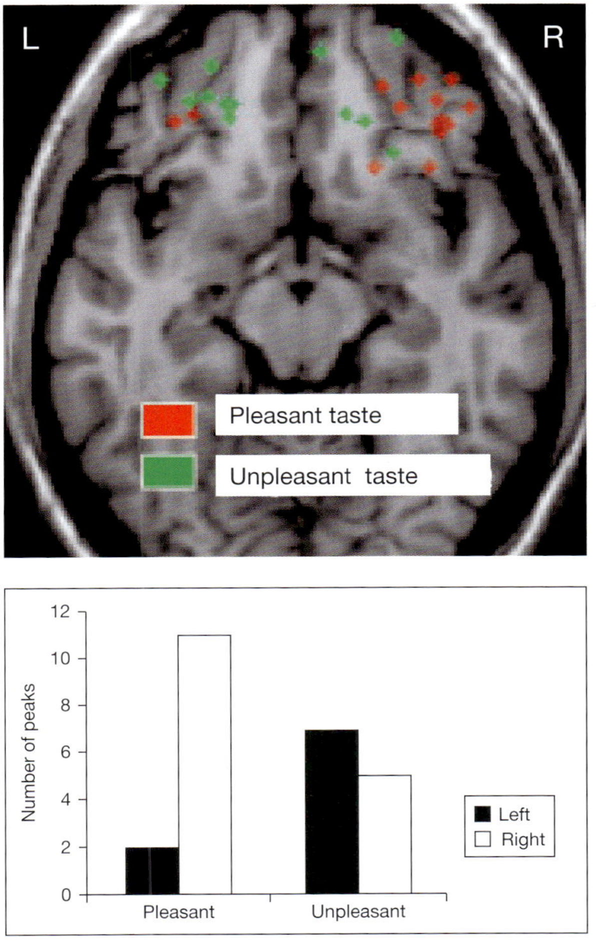

Fig. 5. ***a*** 25 activation foci from 8 analyses of a pleasant taste (either a pleasant taste minus a neutral taste or a pleasant taste minus an unpleasant taste) and 8 analyses of an unpleasant taste (either an unpleasant taste minus a neutral taste or an unpleasant taste minus a pleasant taste) collected from 5 published studies [70–72, 75, 76] and plotted onto an anatomical image. L = Left; R = right. To display peaks in a single plane, the average level of z (superior to inferior = −14 in MNI coordinates) was calculated and the peaks were plotted onto this plane according to their x (medial to lateral) and y (anterior to posterior) coordinate values. ***b*** A bar graph showing the number of peaks (y axis) that fell in the left or right OFC for the pleasant and unpleasant analyses.

Valence-specific responses have also been observed in the anterior portions of the insula [72], and there is some evidence that changes in the responsiveness of this region to repeated exposures of gustatory experience is dependent upon the evolution of the hedonic response [175]. Specifically, if the dislike of a stimulus disappears, there is greater activity following repeated exposure. In contrast, if the liking of a stimulus decreases then there is less activity following repeated exposure. Finally, the insular region (anterior and posterior) is more responsive to food when subjects are hungry compared to when they are full [86, 124, 164, 165, 176]; however, it is not known whether this effect is due to changes within taste-responsive neurons or changes in overlapping neural circuits. Single-cell recording studies in the primate indicate that taste-responsive neurons in the insula/operculum are not modulated by satiety [34, 177].

Perception of pleasant and/or unpleasant taste or food also consistently activates regions of the brain outside the gustatory regions including the striatum, the cingulate gyrus and the midbrain [70–72, 75, 76, 124, 164]. Interestingly, the midbrain and ventral striatum are more likely to be recruited during the anticipation versus the receipt of a pleasant taste [71], whereas the dorsal striatum and anterior cingulate region are more likely to be recruited during the experience of a pleasant taste or food [70, 72, 124, 164].

Flavor

Although it is important to understand how pure taste perceptions are represented in the brain, it is a mistake to consider neurophysiological correlates of taste only within this context. In everyday life, we rarely perceive taste without concomitant experience of oral texture and retronasal olfaction. Unlike early cortical representation of vision, audition and somatosensation, which are represented in the unimodal neocortex, the cortical representation of taste is in the heteromodal paralimbic cortex where there is overlapping representation of orthonasal olfaction [89, 178–181], retronasal olfaction [64, 182], oral movement, oral somatosensation [63, 185] and texture perception [186]. This anatomy reflects the experience of taste, which is almost always accompanied by simultaneous experience of odor and oral somatosensation in the context of feeding. Thus, sensory integration is fundamental to gustation. While one does not give a second thought to hearing without seeing or seeing without hearing, only during a cold does one taste without smelling and this experience is generally described as strange, with food not 'tasting' right.

Neuroimaging studies of olfaction, gustation, and flavor are beginning to isolate a network of regions that are likely responsible for taste/odor integration, and hence flavor perception. Independent presentation of a tastant or an odorant produces overlapping activation in regions of the insula and operculum [63, 74, 89, 178, 182], the OFC [68, 70, 73, 74, 76, 89, 161, 170, 178, 179, 187], and the anterior cingulate cortex [64, 72, 76, 89, 124, 161, 170, 181]. The insula, operculum, OFC and anterior cingulate cortex are also sensitive to somatosensory stimulation of the oral cavity [63, 185, 186]. Similarly, single-cell recording studies in monkeys have identified taste- and smell-responsive cells in the insula/operculum [26] and OFC [188, 189]. The presence of a unimodal representation of taste, odor, and oral touch in the insula, frontal operculum and OFC of the human and nonhuman primate suggests that these regions play a key role in integrating the disparate sensory inputs that give rise to the perception of flavors. Chemosensory responses in the monkey anterior cingulate cortex have yet to be investigated, but the consistency of responses in this region to taste and oral somatosensation in humans is highly suggestive of a role in flavor processing.

Rolls and Baylis [188], Rolls et al. [190, 191], and Verhagen et al. [192] have performed a series of studies in which gustatory, olfactory, visual, and oral somatosensory stimuli were presented to awake behaving monkeys and responses were recorded from single-cell neurons located in the caudal OFC, extending into the ventral insula. They identified unimodal taste, smell, visual, fat, and texture cells that were interspersed with multimodal cells that responded to independent stimulation of two or more modalities. In support of a role for these cells in flavor processing, their best response was often to complex stimuli such as blackcurrant juice. Consistent with the findings in primates and the neuroimaging results discussed above, De Araujo et al. [64] reported activation in the frontal operculum, ventral insula/caudal OFC, amygdala, and anterior cingulate cortex to unimodal stimulation with either a taste or an odor, and to bimodal stimulation with a taste/odor mixture. A study by Small et al. [193], using a similar design, also found activity in the frontal operculum, ventral insula/caudal OFC and anterior cingulate cortex to a taste/odor mixture; but in this study, the response was supra-additive (indicative of neural integration), in that greater activity was observed when the subjects received a taste/odor mixture compared to the summed neural activation evoked by independent stimulation with the taste and the odor components. Thus, in monkeys and humans, it appears that a major function of the core gustatory regions is the integration of taste information with the other sensory constituents of flavor.

Whether taste and smell are integrated into a unitary flavor percept at the perceptual and neural level is contingent upon several factors. These include, but are not limited to, previous experience with taste-odor mixtures, spatial and temporal proximity of the taste and odor and attentional allocation. Behavioral studies of taste-odor integration show that odors can enhance perceived taste intensity, but only if they have previously been experienced with that taste [194–198]. For example, strawberry odor will enhance the perceived intensity of a sweet but not a salty taste solution. Similarly Dalton et al. [199] demonstrated that detection thresholds for an odorant were significantly reduced while subjects held a taste in the mouth, but only if the taste was perceptually congruent. Similar findings of an increased sensitivity to tastes at the threshold level in the presence of, or immediately following the sniffing of, a congruent odor have been reported [200, 201].

In accordance with behavioral studies, bimodal neurons in the primate ventral insula/caudal OFC respond selectively to odors and tastes that have previously been experienced together [188]. For example, a cell may respond to the presentation of a glucose and banana odor but not an onion odor. In humans, evoked potential latency to discrete odor exposure is significantly shorter, and amplitudes greater, when presented with congruent but not incongruent tastes [202]. Additionally, in the study by Small et al. [193], the supra-additive responses were dependent upon the congruency of the taste/odor pair, and significantly greater activation was observed to congruent compared to incongruent mixtures.

The importance of temporal and spatial factors in influencing taste/odor integration lies in their ability to influence whether sensory inputs are perceived as arising from a common event or object, or as two separate events or objects. One critical mechanism promoting unitary perception is the well-known olfactory location illusion, in which retronasal perception of odors is interpreted as originating in the mouth, rather than the nose ('oral capture'). The illusion is so powerful that odors are often mistaken for 'tastes' [203, 204]. For example, the loss of retronasal olfactory inputs causes the 'taste' of foods to change during a head cold. It has been argued that the illusion serves to bring taste and odor into a common spatial registry to facilitate integration [204–206]. Accordingly, massive deactivations in the insula, operculum, caudal OFC and anterior cingulate cortex have been reported following simultaneous delivery of an orthonasally presented odor with a taste [73], even though the stimuli were congruent (e.g. sweet with strawberry odor) (fig. 5). This result is in striking contrast to the supra-additive responses observed in these same regions when odors are given retronasally. Taken together, the data suggest that the more likely a taste and odor are perceived as originating from a common object/source, the greater the ability of one component to influence the other and the greater the likelihood that integrative events will occur such as enhancement and supra-additive neural responses.

Finally, it is probable that attentional allocation affects taste-odor integration. In the first study of taste-odor integration, subjects were asked to sniff a cotton wand and at the same time open their mouths to receive a tongue-shaped filter paper [73]. This rather difficult and unusual task required subjects to divide their attention between the gustatory and the olfactory modalities in order to comply with the experimental procedure. Comparison of brain activity assayed when subjects only needed to focus on receiving the tongue-shaped filter papers or upon sniffing the cotton wands with brain activity assayed when subjects were required to taste and sniff simultaneously revealed deactivations in cortical chemosensory regions. These deactivations may have resulted from the spatial disparity of the odor and taste or from the necessity of dividing attention between these modalities to accomplish the task. Consistent with this interpretation, when subjects are asked to attend to the different qualities within a flavor, enhanced intensity perception of a taste by a congruent odor is eliminated [194, 196, 207]. These data suggest that the way in which attention is allocated to the elements within a flavor stimulus may either compromise or facilitate integrative processes.

Conclusion

In this chapter, data from animal and human neurophysiological and anatomical studies were considered in an effort to understand where and how taste perception is represented in the human brain. The human gustatory pathway is presumed to be similar to the monkey pathway with a first-order synapse in the NTS, a second-order synapse in the thalamus, and thalamic projections terminating in several regions of the insula and overlying operculum. At least some of these projections likely cross the midline, resulting in bilateral representation at the cortical level. Projections from the AI/FO proceed to the caudal OFC, which projects back to the insula/operculum, forward to more anterior regions of the OFC, as well as to the amygdala. Gustatory information also reaches the amygdala directly from the insula. The relative importance of each of these regions to gustatory perception is beginning to be understood. Detection and suprathreshold intensity perception likely rely upon processing in the insula/operculum and affective representation upon processing in the OFC. The amygdala is not involved in taste detection or in representing transient changes in the affective value of taste but it is implicated in representing the intensity, quality and perhaps overall saliency of gustatory stimuli. Although functional dissociations can be found, it is also likely that there is considerable interaction between the neural representation of each of the sensory dimensions. The nature of these interactions is only beginning to be understood. Furthermore, although functional neuroimaging has permitted us to begin to understand central taste organization, there are still fundamental gaps in our knowledge. We know that the gustatory code is modulated by mechanisms promoting homeostasis but we do not know at what level this first occurs or whether the interaction depends upon the nature of the perturbation. We know that the insula is important in representing intensity and the OFC in representing the affective value, but we know virtually nothing about taste quality coding in the human brain or how quality coding interacts with intensity and affective value. We have hints of cerebral dominance and asymmetrical processing but findings are often contradictory and those that are not, are of unknown functional significance. We know nothing of the potential importance of top-down modulation and very little about brain correlates of taste learning, imagery, conditioning or preference. In summary, important first steps have been made towards understanding the representation of taste and flavor in the human brain but there are still many more fundamental questions that remain unanswered.

References

1 Bartoshuk LM: Taste, smell, and pleasure; in Bolles RC (ed): The Hedonics of Taste and Smell. New Jersey, Erlbaum, 1991, pp 15–28.
2 Scott TR, Plata-Salaman CR: Coding of taste quality; in Getchel TV (ed): Smell and Taste in Health and Disease. New York, Raven Press, 1991, pp 345–368.
3 Scott TR, Mark GP: The taste system encodes stimulus toxicity. Brain Res 1987;414:197–203.
4 Norgren R: Taste and the autonomic nervous system. Chem Senses 1985;10:143–161.
5 Torvik A: Afferent connections to the sensory trigeminal nuclei, the nucleus of the solitary tract and adjacent structures: an experimental study in the rat. J Comp Neurol 1955;106:51–141.
6 Ogawa H, Kaisaku J: Physiological characteristics of the solitario-parabrachial relay neurons with tongue afferent inputs in rats. Exp Brain Res 1982;48:362–368.
7 Williams JB, Murphy DM, Reynolds KE, Welch SJ, King MS: Demonstration of a bilateral projection from the rostral nucleus of the solitary tract to the medial parabrachial nucleus in rat. Brain Res 1996;737:231–237.
8 Norgren R, Leonard CM: Taste pathways in rat brainstem. Science 1971;173:1136–1139.
9 Braun JJ, Lasiter PS, Kiefer SW: The gustatory neocortex of the rat. Physiol Psychol 1982;10:13–45.
10 Ganchrow D, Erickson RP: Thalamocortical relations in gustation. Brain Res 1972;36:298–305.
11 Kosar E, Grill HJ, Norgren R: Gustatory cortex in the rat. 2. Thalamocortical projections. Brain Res 1986;379:342–352.
12 Norgren R, Leonard CM: Ascending central gustatory pathways. J Comp Neurol 1973;150:217–237.
13 Pfaffmann C, Norgren R, Grill HJ: Sensory affect and motivation. Ann NY Acad Sci 1977;290: 18–34.
14 Norgren R: Taste pathways to hypothalamus and amygdala. J Comp Neurol 1976;166:17–30.
15 Hayama T, Ogawa H: Afferent connections of granular and dysgranular insular cortices in adult and rat pups. Neurosci Res Suppl 1991;14:S25.
16 Saper CB, Loewy AD, Swanson LW, Cowan WM: Direct hypothalamo-autonomic connections. Brain Res 1976;117:305–312.
17 Saper CB, Loewy AD: Efferent connections of the parabrachial nucleus in the rat. Brain Res 1980;197:291–317.
18 Shipley MT, Sanders MS: Special senses are really special: evidence for a reciprocal, bilateral pathway between insular cortex and nucleus parabrachialis. Brain Res Bull 1982;8:493–501.
19 Takeuchi Y, Hopkins DA: Light and electron microscopic demonstration of hypothalamic projections to the parabrachial nuclei in the cat. Neurosci Lett 1984;46:53–58.
20 Takeuchi Y, McLean JH, Hopkins DA: Reciprocal connections between the amygdala and parabrachial nuclei: ultrastructural demonstration by degeneration and axonal transport of horseradish peroxidase in the cat. Brain Res 1982;239:583–588.

21 van der Kooy D, Koda LY, McGinty JF, Gerfen CR, Bloom FE: The organization of projections from the cortex, amygdala, and hypothalamus to the nucleus of the solitary tract in rat. J Comp Neurol 1984;224:1–24.
22 Cechetto DF, Saper CB: Evidence for a viscerotopic sensory representation in the cortex and thalamus in the rat. J Comp Neurol 1987;262:27–45.
23 Hanamori T, Kunitake T, Kato K, Kannan H: Responses of neurons in the insular cortex to gustatory, visceral, and nociceptive stimuli in rats. J Neurophysiol 1998;79:2535–2545.
24 Beckstead RM, Morse JR, Norgren R: The nucleus of the solitary tract in the monkey: projections to the thalamus and brain stem nuclei. J Comp Neurol 1980;190:259–282.
25 Pritchard TC, Hamilton RB, Norgren R: Projections of the parabrachial nucleus in the old world monkey. Exp Neurol 2000;165:101–117.
26 Scott TR, Plata-Salaman CR: Taste in the monkey cortex. Physiol Behav 1999;67:489–511.
27 Pritchard TC, Hamilton RB, Morse JR, Norgren R: Projections of thalamic gustatory and lingual areas in the monkey, *Macaca fascicularis*. J Comp Neurol 1986;244:213–228.
28 Ogawa H, Ito S, Nomura T: Two distinct projection areas from tongue nerves in the frontal operculum of macaque monkeys as revealed with evoked potential mapping. Neurosci Res 1985;2: 447–459.
29 Mufson EJ, Mesulam MM: Thalamic connections of the insula in the rhesus monkey and comments on the paralimbic connectivity of the medial pulvinar nucleus. J Comp Neurol 1984;227: 109–120.
30 Benjamin RM, Burton H: Projection of taste nerve afferents to anterior opercular-insular cortex in squirrel monkey (*Saimiri sciureus*). Brain Res 1968;7:221–231.
31 Ogawa H: Gustatory cortex of primates: anatomy and physiology. Neurosci Res 1994;20:1–13.
32 Sanides F: The architecture of the cortical taste nerve areas in squirrel monkey (*Saimiri sciureus*) and their relationships to insular, sensorimotor and prefrontal regions. Brain Res 1968;8:97–124.
33 Rolls ET, Scott TR, Sienkiewicz ZJ, Yaxley S: The responsiveness of neurons in the frontal opercular gustatory cortex of the macaque monkey is independent of hunger. J Neurophysiol 1988;397:1–12.
34 Yaxley S, Rolls ET, Sienkiewicz ZJ: The responsiveness of neurons in the insular gustatory cortex of the macaque monkey is independent of hunger. Physiol Behav 1988;42:223–229.
35 Yaxley S, Rolls ET, Sienkiewicz ZJ: Gustatory responses of single neurons in the insula of the macaque monkey. J Neurophysiol 1990;63:689–700.
36 Smith-Swintosky VL, Plata-Salaman CR, Scott TR: Gustatory neural coding in the monkey cortex: stimulus quality. J Neurophysiol 1991;66:1156–1165.
37 Scott TR, Yaxley S, Sienkiewicz ZJ, Rolls ET: Gustatory responses in the frontal opercular cortex of the alert cynomolgus monkey. J Neurophysiol 1986;56:876–890.
38 Scott TR, Plata-Salaman CR, Smith VL, Giza B: Gustatory neural coding in the monkey cortex: stimulus intensity. J Neurophysiol 1991;65:76–86.
39 Kobayashi M, Sasabe T, Takeda M, et al: Functional anatomy of chemical senses in the alert monkey revealed by positron emission tomography. Eur J Neurosci 2002;16:975–980.
40 Baylis LL, Rolls ET, Baylis GC: Afferent connections of the caudolateral orbitofrontal cortex taste area of the primate. Neuroscience 1995;64:801–812.
41 Mufson EJ, Mesulam MM, Pandya DN: Insular interconnections with the amygdala in the rhesus monkey. Neuroscience 1981;6:1231–48.
42 Price JL, Amaral DG: An autoradiographic study of the projections of the central nucleus of the monkey amygdala. J Neurosci 1981;1:1242–1259.
43 Turner BH, Mishkin M, Knapp M: Organization of the amygdalopetal projections from modality-specific cortical association areas in the monkey. J Comp Neurol 1980;191:515–543.
44 Amaral DG, Price JL: Amygdalo-cortical projections in the monkey (*Macaca fascicularis*). J Comp Neurol 1984;230:465–496.
45 Carmichael ST, Price JL: Limbic connections of the orbital and medial prefrontal cortex in macaque monkeys. J Comp Neurol 1995;363:615–641.
46 Scott TR, Karadi Z, Oomura Y, et al: Gustatory neural coding in the amygdala of the alert macaque monkey. J Neurophysiol 1993;69:1810–1820.
47 Aggleton JP, Burton MJ, Passingham RE: Cortical and subcortical afferents to the amygdala of the rhesus monkey (*Macaca mulatta*). Brain Res 1980;190:347–368.

48 Carmichael ST, Clugnet MC, Price JL: Central olfactory connections in the macaque monkey. J Comp Neurol 1994;346:403–434.
49 Carmichael ST, Price JL: Sensory and premotor connections of the orbital and medial prefrontal cortex of macaque monkeys. J Comp Neurol 1995;363:642–664.
50 Onoda K, Ikeda M: Gustatory disturbance due to cerebrovascular disorder. Laryngoscope 1999; 109:123–128.
51 Uesaka Y, Nose H, Ida M, Takagi A: The pathway of gustatory fibers of the human ascends ipsilaterally in the pons. Neurology 1998;50:827–828.
52 Lee BC, Hwang SH, Rison R, Chang GY: Central pathway of taste: clinical and MRI study. Eur Neurol 1998;39:200–203.
53 Nakajima Y, Utsumi H, Takahashi H: Ipsilateral disturbance of taste due to pontine haemorrhage. J Neurol 1983;229:133–136.
54 Goto N, Yamamoto T, Kaneko M, Tomita H: Primary pontine hemorrhage and gustatory disturbance: clinicoanatomic study. Stroke 1983;14:507–511.
55 Topolovec JC, Gati JS, Menon RS, Shoemaker JK, Cechetto DF: Human cardiovascular and gustatory brainstem sites observed by functional magnetic resonance imaging. J Comp Neurol 2004;471: 446–461.
56 Bornstein WS: Cortical representation of taste in man and monkey. 1. Functional and anatomical relations of taste, olfaction, and somatic sensibility. Yale J Biol Med 1940;12:719–736.
57 Bornstein WS: Cortical representation of taste in man and monkey. 2. The localization of the cortical taste area in man and a method of measuring impairment of taste in man. Yale J Biol Med 1940;13:133–156.
58 Motta G: I centri corticali del gusto. Bull Sci Med 1959;131:480–493.
59 Penfield W, Faulk ME: The insula: further observations on its function. Brain 1955;78:445–470.
60 Cascino GD, Karnes WE: Gustatory and second sensory seizures associated with lesions in the insular cortex seen on magnetic resonance imaging. J Epilepsy 1990;3:185–187.
61 Petrides M, Pandya D: Comparative architectonic analysis of the human and macaque frontal cortex; in Boller F, Grafman J (eds): Handbook of Neuropsychology. Amsterdam, Elsevier, 1994, pp 17–58.
62 Barry MA, Gatenby JC, Zeiger JD, Gore JC: Hemispheric dominance of cortical activity evoked by focal electrogustatory stimuli. Chem Senses 2001;26:471–482.
63 Cerf-Ducastel B, van de Moortele PF, MacLeod P, Le Bihan D, Faurion A: Interaction of gustatory and lingual somatosensory perceptions at the cortical level in the human: a functional magnetic resonance imaging study. Chem Senses 2001;26:371–383.
64 de Araujo E, Rolls ET, Kringelbach ML, McGlone F, Phillips N: Taste-olfactory convergence, and the representation of the pleasantness of flavour in the human brain. Eur J Neurosci 2003;18: 2059–2068.
65 Faurion A, Cerf B, Le Bihan D, Pillias AM: fMRI study of taste cortical areas in humans. Ann NY Acad Sci 1998;855:535–545.
66 Faurion A, Cerf B, Van De Moortele PF, Lobel E, Mac Leod P, Le Bihan D: Human taste cortical areas studied with functional magnetic resonance imaging: evidence of functional lateralization related to handedness. Neurosci Lett 1999;277:189–192.
67 Frey S, Petrides M: Re-examination of the human taste region: a positron emission tomography study. Eur J Neurosci 1999;11:2985–2988.
68 Francis S, Rolls ET, Bowtell R, et al: The representation of pleasant touch in the brain and its relationship with taste and olfactory areas. Neuroreport 1999;10:435–459.
69 Kinomura S, Kawashima R, Yamada K, et al: Functional anatomy of taste perception in the human brain studied with positron emission tomography. Brain Res 1994;659:263–266.
70 O'Doherty J, Rolls ET, Francis S, Bowtell R, McGlone F: Representation of pleasant and aversive taste in the human brain. J Neurophysiol 2001;85:1315–1321.
71 O'Doherty JP, Deichmann R, Critchley HD, Dolan RJ: Neural responses during anticipation of a primary taste reward. Neuron 2002;33:815–826.
72 Small DM, Gregory MD, Mak YE, Gitelman DR, Mesulam MM, Parrish TB: Dissociation of neural representation of intensity and affective valuation in human gustation. Neuron 2003;39: 701–711.

73 Small DM, Jones-Gotman M, Zatorre RJ, Petrides M, Evans AC: Flavor processing: more than the sum of its parts. Neuroreport 1997;8:3913–3917.
74 Small DM, Zald DH, Jones-Gotman M, et al: Human cortical gustatory areas: a review of functional neuroimaging data. Neuroreport 1999;10:7–14.
75 Zald DH, Hagen MC, Pardo JV: Neural correlates of tasting concentrated quinine and sugar solutions. J Neurophysiol 2002;87:1068–1075.
76 Zald DH, Lee JT, Fluegel KW, Pardo JV: Aversive gustatory stimulation activates limbic circuits in humans. Brain 1998;121:1143–1154.
77 Kobayakawa T, Endo H, Ayabe-Kanamura S, et al: The primary gustatory area in human cerebral cortex studied by magnetoencephalography. Neurosci Lett 1996;212:155–158.
78 Kobayakawa T, Ogawa H, Kaneda H, Ayabe-Kanamura S, Endo H, Saito S: Spatio-temporal analysis of cortical activity evoked by gustatory stimulation in humans. Chem Senses 1999;24:201–209.
79 Mesulam MM, Mufson EJ: Insula of the old world monkey. 1. Architectonics in the insulo-orbito-temporal component of the paralimbic brain. J Comp Neurol 1982;212:1–22.
80 Small DM, Jones-Gotman M, Zatorre RJ, Petrides M, Evans AC: A role for the right anterior temporal lobe in taste quality recognition. J Neurosci 1997;17:5136–5142.
81 Pritchard TC, Macaluso E, Eslinger PJ: Taste perception in patients with insular cortex lesions. Behav Neurosci 1999;113:663–671.
82 Aglioti S, Tassinari G, Corballis MC, Berlucchi G: Incomplete gustatory lateralization as shown by analysis of taste discrimination after callosotomy. J Cogn Neurosci 2000;12:238–245.
83 Rolls ET, Yaxley S, Sienkiewicz ZJ: Gustatory responses of single neurons in the caudolateral orbitofrontal cortex of the macaque monkey. J Neurophysiol 1990;64:1055–1066.
84 Small DM, Zatorre RJ, Jones-Gotman M: Changes in taste intensity perception following anterior temporal lobe removal in humans. Chem Senses 2001;26:425–432.
85 Small DM, Zatorre RJ, Jones-Gotman M: Increased intensity perception of aversive taste following right anteromedial temporal lobe removal in humans. Brain 2001;124:1566–1575.
86 Del Parigi A, Chen K, Salbe AD, et al: Tasting a liquid meal after a prolonged fast is associated with preferential activation of the left hemisphere. Neuroreport 2002;13:1141–1145.
87 Cerf B, Lebihan D, Van de Moortele PF, Mac Leod P, Faurion A: Functional lateralization of human gustatory cortex related to handedness disclosed by fMRI study. Ann NY Acad Sci 1998;855:575–578.
88 Royet JP, Plailly J: Lateralization of olfactory processes. Chem Senses 2004;29:731–745.
89 Savic I, Gulyas B, Larsson M, Roland P: Olfactory functions are mediated by parallel and hierarchical processing. Neuron 2000;26:735–745.
90 Pfaffmann C, Bartoshuk LM, McBurney DH: Taste psychophysics; in Beidler LM (ed): Handbook of Sensory Physiology, vol IV. Chemical Senses Part 2: Taste. New York, Springer, 1971, pp 77–101.
91 Bartoshuk LM, Beauchamp GK: Chemical senses. Annu Rev Psychol 1994;45:419–449.
92 Kocher EC, Fisher GL: Subjective intensity and taste preference. Percept Mot Skills 1969;28:735–740.
93 Moskowitz HR, et al: Effects of hunger, satiety and glucose load upon taste intensity and taste hedonics. Physiol Behav 1976;16:471–475.
94 Moskowitz H: Sensations, measurement and pleasantness: confessions of a latent introspectionist; in Weiffenbach JB (ed): Taste and Development: The Genesis of Sweet Preference. Bethesda, Department of Health, Education, and Welfare, 1977, pp 282–294.
95 Blum M, Walker AE, Ruch TC: Localization of taste in the thalamus of *Macaca mulatta.* Yale J Biol Med 1943;16:175–191.
96 Scott TR, Yaxley S, Sienkiewicz ZJ, Rolls ET: Gustatory responses in the nucleus tractus solitarius of the alert cynomolgus monkey. J Neurophysiol 1986;55:182–200.
97 Pritchard TC, Hamilton RB, Norgren R: Neural coding of gustatory information in the thalamus of *Macaca mulatta.* J Neurophysiol 1989;61:1–14.
98 Zotterman Y: Action potentials in the glossopharyngeal nerve and in the chorda tympani. Scan Arch Physiol 1935;72:73–77.
99 Simmons KB, Mak YE, Gitelman DR, Small DM: Ipsilateral taste intensity perception deficits in a left insular stroke patient. Chem Senses 2003;28:A123.

100 Kveton JF, Bartoshuk LM: The effect of unilateral chorda tympani damage on taste. Laryngoscope 1994;104:25–29.
101 Yanagisawa K, Bartoshuk LM, Catalanotto FA, Karrer TA, Kveton JF: Anesthesia of the chorda tympani nerve and taste phantoms. Physiol Behav 1998;63:329–335.
102 Touzani K, Taghzouti K, Velley L: Increase of the aversive value of taste stimuli following ibotenic acid lesion of the central amygdaloid nucleus in the rat. Behav Brain Res 1997;88:133–142.
103 Breslin PA: Human gustation; in Finger TE, Singer WL (eds): The Neurobiology of Taste and Smell. San Diego, Wiley-Liss Inc, 2000, pp 423–461.
104 Smith DV, John SJ, Boughter JD: Neuronal cell types and taste quality coding. Physiol Behav 2000;69:77–85.
105 Smith DV, St John SJ: Neural coding of gustatory information. Curr Opin Neurobiol 1999;9: 427–435.
106 Scott TR, Giza BK: Theories of gustatory neural coding; in Doty RL (ed): Handbook of Olfaction and Gustation. New York, Dekker, 1995, pp 573–603.
107 Frank ME: Neuron types, receptors, behavior, and taste quality. Physiol Behav 2000;69:53–62.
108 Pfaffmann C: The afferent code for sensory quality. Am J Psychol 1959;14:226–232.
109 Smith DV, Scott TR: Gustatory neural coding; in Doty RL (ed): Handbook of Olfaction and Gustation. Philadelphia, Dekker, 2003, pp 731–758.
110 Sato M, Ogawa H, Yamashita S: Response properties of macaque monkey chorda tympani fibers. J Gen Physiol 1975;66:781–810.
111 Hellekant G, Danilova V, Ninomiya Y: Primate sense of taste: behavioral and single chorda tympani and glossopharyngeal nerve fiber recordings in the rhesus monkey, *Macaca mulata*. J Neurophysiol 1997;77:978–993.
112 Gordon G, Kitchell R, Strom L, Zotterman Y: The response pattern of taste fibers in the chorda tympani of the monkey. Acta Physiol Scand 1959;46:119–132.
113 Smith DV, Travers JB: A metric for the breadth of tuning of gustatory neurons. Chem Senses 1979;4:215–229.
114 Katz DB, Simon SA, Nicolelis MA: Taste-specific neuronal ensembles in the gustatory cortex of awake rats. J Neurosci 2002;22:1850–1857.
115 Kuznicki JT, Ashbaugh N: Taste quality differences within sweet and salty taste categories. Sens Processes 1979;3:157–182.
116 Schiffman SS, Erickson RP: A psychophysical model for gustatory quality of sodium salts. Physiol Behav 1971;7:617–633.
117 Cereda C, Ghika J, Maeder P, Bogousslavsky J: Strokes restricted to the insular cortex. Neurology 2002;59:1950–1955.
118 Henkin RI, Comiter H, Fedio P, O'Doherty D: Defects in taste and smell recognition following temporal lobectomy. Trans Am Neurol Assoc 1977;102:146–150.
119 Small DM, Bernasconi N, Bernasconi A, Sziklas V, Jones-Gotman M: Gustatory agnosia. Neurology 2005;64:311–317.
120 Halpern BP, Nelson LM: Bulbar gustatory responses to anterior and to posterior tongue stimulation in the rat. Am J Physiol 1965;202:105–110.
121 Dickman JD, Smith DV: Topographic distribution of taste responsiveness in the hamster medulla. Chem Senses 1989;14:231–247.
122 Sudakov K, Maclean PD, Reeves A, Marini R: Unit study of exteroceptive inputs to claustrocortex in awake, sitting squirrel monkeys. Brain Res 1971;28:19–34.
123 Plata-Salaman CR, Smith-Swintosky VL, Scott TR: Gustatory neural coding in the monkey cortex: mixtures. J Neurophysiol 1996;75:2369–2379.
124 Small DM, Zatorre RJ, Dagher A, Evans AC, Jones-Gotman M: Changes in brain activity related to eating chocolate: from pleasure to aversion. Brain 2001;124:1720–1733.
125 Steiner JE, Glaser D, Hawilo ME, Berridge KC: Comparative expression of hedonic impact: affective reactions to taste by human infants and other primates. Neurosci Biobehav Rev 2001;25:53–74.
126 Steiner JE: Facial expressions of the neonate infant indicating the hedonics of food-related stimuli; in Weiffenbach JM (ed): Taste and Development: The Genesis of Sweet Preference. Washington, US Department of Health and Human Services, 1979.
127 Rolls ET: The Brain and Emotion. Oxford, Oxford University Press, 1999.

128 Pittman DW, Contreras RJ: Dietary NaCl influences the organization of chorda tympani neurons projecting to the nucleus of the solitary tract in rats. Chem Senses 2002;27:333–341.
129 Shuler MG, Krimm RF, Hill DL: Neuron/target plasticity in the peripheral gustatory system. J. Comp Neurol 2004;472:183–192.
130 Krimm RF, Hill DL: Early dietary sodium restriction disrupts the peripheral anatomical development of the gustatory system. J Neurobiol 1999;39:218–226.
131 Hendricks SJ, Brunjes PC, Hill DL: Taste bud cell dynamics during normal and sodium-restricted development. J Comp Neurol 2004;472:173–182.
132 Hill DL, Przekop PR Jr: Influences of dietary sodium on functional taste receptor development: a sensitive period. Science 1988;241:1826–1828.
133 Hill DL, Mistretta CM, Bradley RM: Effects of dietary NaCl deprivation during early development on behavioral and neurophysiological taste responses. Behav Neurosci 1986;100: 390–398.
134 Moskowitz HW, Kumaraiah V, Sharma KN: Cross-cultural differences in simple taste preferences. Science 1975;190:1217–1218.
135 Glendinning JI: Is the bitter rejection response always adaptive. Physiol Behav 1994;56: 1217–1227.
136 Grigson PS: Like drugs for chocolate: separate rewards modulated by common mechanisms? Physiol Behav 2002;76:389–395.
137 Drewnowski A, Kurth CL, Rahaim JE: Taste preferences in human obesity: environmental and familial factors. Am J Clin Nutr 1991;54:635–641.
138 Rodin J, Moskowitz HR, Bray GA: Relationship between obesity, weight loss, and taste responsiveness. Physiol Behav 1976;17:591–597.
139 Drewnowski A, Holden-Wiltse J: Taste responses and food preferences in obese women: effects of weight cycling. Int J Obes Relat Metab Disord 1992;16:639–648.
140 Kampov-Polevoy AB, Tsoi MV, Zvartau EE, Nezmanov NG, Kalitov E: Sweet liking and a family history of alcoholism in hospitalized alcoholic and non-alcoholic patients. Alcohol 2001;36: 165–170.
141 Chang FC, Scott TR: Conditioned taste aversions modify neural responses in the rat nucleus tractus solitarius. J Neurosci 1984;4:1850–1862.
142 Bermudez-Rattoni F, McGaugh JL: Insular cortex and amygdala lesions differentially affect acquisition on inhibitory avoidance and conditioned taste aversion. Brain Res 1991;549:165–170.
143 Conover KL, Woodside B, Shizgal P: Effects of sodium depletion on competition and summation between rewarding effects of salt and lateral hypothalamic stimulation in the rat. Behav Neurosci 1994;108:549–558.
144 Giza BK, Ackroff K, McCaughey SA, Sclafani A, Scott TR: Preference conditioning alters taste responses in the nucleus of the solitary tract of the rat. Am J Physiol 1997;273:R1230–R1240.
145 McCaughey SA, Tordoff MG: Calcium-deprived rats sham-drink $CaCl_2$ and NaCl. Appetite 2000;34:305–311.
146 McCaughey SA, Scott TR: Rapid induction of sodium appetite modifies taste-evoked activity in the rat nucleus of the solitary tract. Am J Physiol 2000;279:R1121–R1131.
147 McCaughey SA, Tordoff MG: Calcium deprivation alters gustatory-evoked activity in the rat nucleus of the solitary tract. Am J Physiol 2001;281:R971–R978.
148 Reilly S: The parabrachial nucleus and conditioned taste aversion. Brain Res Bull 1999;48:239–254.
149 Sclafani A, Azzara AV, Touzani K, Grigson PS, Norgren R: Parabrachial nucleus lesions block taste and attenuate flavor preference and aversion conditioning in rats. Behav Neurosci 2001;115: 920–933.
150 Rolls ET, Rolls BJ, Rowe EA: Sensory-specific and motivation-specific satiety for the sight and taste of food and water in man. Physiol Behav 1983;30:185–192.
151 Rolls ET, Rolls JH: Olfactory sensory-specific satiety in humans. Physiol Behav 1997;61:461–473.
152 Berridge KC: Modulation of taste affect by hunger, caloric satiety, and sensory-specific satiety in the rat. Appetite 1991;16:103–120.
153 Cabanac M: Physiological role of pleasure. Science 1971;173:1103–1107.
154 Phillips ML, Young AW, Senior C, et al: A specific neural substrate for perceiving facial expressions of disgust. Nature 1997;389:495–498.

155 Ledoux JE: The emotion and the amygdala; in Aggleton JP (ed): The Amygdala: Neurobiological Aspects of Emotion, Memory, and Mental Dysfunction. New York, Wiley-Liss, 1992, pp 339–357.
156 Aggleton JP, Petrides M, Iversen SD: Differential effects of amygdaloid lesions on conditioned taste aversion learning by rats. Physiol Behav 1981;27:397–400.
157 Buresova O: Neocortico-amygdalar interaction in the conditioned taste aversion in rats. Act Nerv Super (Praha) 1978;20:224–230.
158 Dunn LT, Everitt BJ: Double dissociations of the effects of amygdala and insular cortex lesions on conditioned taste aversion, passive avoidance, and neophobia in the rat using the excitotoxin ibotenic acid. Behav Neurosci 1988;102:3–23.
159 Shimai S, Hoshishima K: Effects of bilateral amygdala lesions on neophobia and conditioned taste aversion in mice. Percept Mot Skills 1982;54:127–130.
160 Yamamoto T, Fujimoto Y: Brain mechanisms of taste aversion learning in the rat. Brain Res Bull 1991;27:403–406.
161 Zald DH, Pardo JV: Emotion, olfaction, and the human amygdala: amygdala activation during aversive olfactory stimulation. Proc Natl Acad Sci USA 1997;94:4119–4124.
162 Zald DH: The human amygdala and the emotional evaluation of sensory stimuli. Brain Res Rev 2003;41:88–123.
163 Yan J, Scott TR: The effect of satiety on responses of gustatory neurons in the amygdala of alert cynomolgus macaques. Brain Res 1996;740:193–200.
164 Kringelbach ML, O'Doherty J, Rolls ET, Andrews C: Activation of the human orbitofrontal cortex to a liquid food is correlated with its subjective pleasantness. Cereb Cortex 2003;13: 1064–1071.
165 Gautier JF, Del Parigi A, Chen K, et al: Effect of satiation on brain activity in obese and lean women. Obes Res 2001;9:676–684.
166 Del Parigi A, Chen K, Hill DO, Wing RR, Reiman E, Tataranni PA: Persistence of abnormal neural responses to a meal in postobese individuals. Int J Obes 2004;28:370–377.
167 Morris JS, Dolan RJ: Involvement of human amygdala and orbitofrontal cortex in hunger-enhanced memory for food stimuli. J Neurosci 2001;21:5304–5310.
168 LaBar KS, Gitelman DR, Parrish TB, Kim YH, Nobre AC, Mesulam MM: Hunger selectively modulates corticolimbic activation to food stimuli in humans. Behav Neurosci 2001;115:493–500.
169 Gottfried JA, O'Doherty J, Dolan RJ: Encoding predictive reward value in human amygdala and orbitfrontal cortex. Science 2003;301:1104–1107.
170 O'Doherty J, Rolls ET, Francis S, et al: Sensory-specific satiety-related olfactory activation of the human orbitofrontal cortex. Neuroreport 2000;11:399–403.
171 Rolls ET, Sienkiewicz ZJ, Yaxley S: Hunger modulates the responses to gustatory stimuli of single neurons in the caudolateral orbitofrontal cortex of the macaque monkey. Eur J Neurosci 1989;1: 53–70.
172 Critchley HD, Rolls ET: Hunger and satiety modify the responses of olfactory and visual neurons in the primate orbitofrontal cortex. J Neurophysiol 1996;75:1673–1686.
173 Mora F, Avrith DB, Phillips AG, Rolls ET: Effects of satiety on self-stimulation of the orbitofrontal cortex in the rhesus monkey. Neurosci Lett 1979;13:141–145.
174 Baylis LL, Gaffan D: Amygdalectomy and ventromedial prefrontal ablation produce similar deficits in food choice and in simple object discrimination learning for an unseen reward. Exp Brain Res 1991;86:617–622.
175 Faurion A, Cerf B, Pillias AM, Boireau N: Increased taste sensitivity by familiarization to novel stimuli: psychophysics, fMRI and electrophysiological techniques suggest modulations at peripheral and central levels; in Rouby C, Schaal B, Dubois D, Gervais R, Holley A (eds): Olfaction, Taste, and Cognition. Cambridge, Cambridge University Press, 2002, pp 350–366.
176 Del Parigi A, Gautier JF, Chen K, et al: Neuroimaging and obesity: mapping the brain responses to hunger and satiation in humans using positron emission tomography. Ann NY Acad Sci 2002;967: 389–397.
177 Rolls ET, Scott TR, Sienkiewicz ZJ, Yaxley S: The responsivenes of neurones in the frontal operculuar gustatory cortex of the macaque monkey is independent of hunger. J Physiol 1988;397: 1–12.

178 Poellinger A, Thomas R, Lio P, et al: Activation and habituation in olfaction – An fMRI study. Neuroimage 2001;13:547–560.
179 Zatorre RJ, Jones-Gotman M, Evans AC, Meyer E: Functional localization and lateralization of human olfactory cortex. Nature 1992;360:339–340.
180 Gottfried JA, Deichmann R, Winston JS, Dolan RJ: Functional heterogeneity in human olfactory cortex: an event-related functional magnetic resonance imaging study. J Neurosci 2002;22: 10819–10828.
181 Royet JP, Plailly J, Delon-Martin C, Kareken DA, Segebarth C: fMRI of emotional responses to odors: influence of hedonic valence and judgment, handedness, and gender. Neuroimage 2003;20: 713–728.
182 Cerf-Ducastel B, Murphy C: fMRI activation in response to odorants orally delivered in aqueous solutions. Chem Senses 2001;26:625–637.
183 Hamdy S, Rothwell JC, Brooks DJ, Bailey D, Aziz Q, Thompson DG: Identification of the cerebral loci processing human swallowing with $H_2{}^{15}O$ PET activation. J Neurophysiol 1999;81: 1917–1926.
184 Zald DH, Pardo JV: The functional neuroanatomy of voluntary swallowing. Ann Neurol 1999;46: 281–286.
185 Pardo JV, Wood TD, Costello PA, Pardo PJ, Lee JT: PET study of the localization and laterality of lingual somatosensory processing in humans. Neurosci Lett 1997;234:23–26.
186 de Araujo E, Rolls ET: Representation in the human brain of food texture and oral fat. J Neurosci 2004;24:3086–3093.
187 Sobel N, Prabhakaran V, Desmond JE, et al: Sniffing and smelling: separate subsystems in the human olfactory cortex. Nature 1998;392:282–286.
188 Rolls ET, Baylis LL: Gustatory, olfactory, and visual convergence within the primate orbitofrontal cortex. J Neurosci 1994;14:5437–5452.
189 Rolls ET, Critchley HD, Treves A: Representation of olfactory information in the primate orbitofrontal cortex. J Neurophysiol 1996;75:1982–1996.
190 Rolls ET, Critchley HD, Mason R, Wakeman EA: Orbitofrontal cortex neurons: role in olfactory and visual association learning. J Neurophysiol 1996;75:1970–1981.
191 Rolls ET, Critchley HD, Browning AS, Hernadi I, Lenard L: Responses to the sensory properties of fat of neurons in the primate orbitofrontal cortex. J Neurosci 1999;19:1532–1540.
192 Verhagen JV, Rolls ET, Kadohisa M: Neurons in the primate orbitofrontal cortex respond to fat texture independently of viscosity. J Neurophysiol 2003;90:1514–1525.
193 Small DM, Voss J, Mak YE, Simmons KB, Parrish TB, Gitelman DR: Experience-dependent neural integration of taste and smell in the human brain. J Neurophysiol 2004;92:1892–1903.
194 Frank RA, van der Klaauw NJ, Schifferstein HN: Both perceptual and conceptual factors influence taste-odor and taste-taste interactions. Percept Psychophys 1993;54:343–354.
195 Frank RA, Byram J: Taste-smell interactions are tastant and odorant dependent. Chem Senses 1988;13:445–455.
196 van de Klaauw NJ, Frank RA: Scaling component intensities of complex stimuli: the influence of response alternatives. Environ Int 1996;22:21–31.
197 Schifferstein HN, Verlegh PW: The role of congruency and pleasantness in odor-induced taste enhancement. Acta Psychol 1996;94:87–105.
198 Sakai N, Kobayakawa T, Gotow N, Saito S, Imada S: Enhancement of sweetness ratings of aspartame by a vanilla odor presented either by orthonasal or retronasal routes. Percept Mot Skills 2001;92:1002–1008.
199 Dalton P, Doolittle N, Nagata H, Breslin PA: The merging of the senses: integration of subthreshold taste and smell. Nat Neurosci 2000;3:431–432.
200 Djordjevic J, Zatorre RJ, Petrides M, Jones-Gotman M: The mind's nose: effects of odor and visual imagery on odor detection. Psychol Sci 2004;15:143–148.
201 Prescott J, Johnstone V, Francis J: Odor/taste interactions: effects of different attentional strategies during exposure. Chem Senses 2004;29:331–340.
202 Welge-Luben A, Drago J, Wolfensberger M, Hummel T: Interaktion von Schmecken und Riechen. HNO 2004;29:123.
203 Murphy C, Cain WS, Bartoshuk LM: Mutual action of taste and olfaction. Sens Processes 1977;1:204–211.

204 Rozin P: 'Taste-smell confusions' and the duality of the olfactory sense. Percept Psychophys 1982;31:397–401.
205 Robinson CJ, Burton H: Organization of somatosensory receptive fields in cortical areas 7b, retroinsula, postauditory and granular insula of *M. fascicularis.* J Comp Neurol 1980;192:69–92.
206 Green BG: Studying taste as a cutaneous sense. Food Qual Pref 2002;14:99–109.
207 Clark CC, Lawless HT: Limiting response alternatives in time-intensity scaling: an examination of the halo-dumping effect. Chem Senses 1994;19:583–594.
208 Pribram KH, Bagshaw M: Further analysis of the temporal lobe syndrome utilizing fronto-temporal ablations. J Comp Neurol 1953;2:347–375.

Dana M. Small, PhD
The John B. Pierce Laboratory
290 Congress Avenue
New Haven, CT 06519 (USA)
Tel. +1 203 401 6204, Fax +1 203 624 4950, E-Mail dsmall@jbpierce.org

Hummel T, Welge-Lüssen A (eds): Taste and Smell. An Update.
Adv Otorhinolaryngol. Basel, Karger, 2006, vol 63, pp 221–241

Modern Psychophysics and the Assessment of Human Oral Sensation

Derek J. Snyder[a,b], *John Prescott*[c], *Linda M. Bartoshuk*[b]

[a]Interdepartmental Neuroscience Program, Yale University, New Haven, Conn., [b]Center for Taste and Smell, University of Florida, Gainesville, Fla., USA; [c]School of Psychology, James Cook University, Cairns, Australia

Abstract

Psychophysical measures attempt to capture and compare subjective experiences objectively. In the chemical senses, these techniques have been instrumental in describing relationships between oral sensation and health risk, but they are often used incorrectly to make group comparisons. This chapter reviews contemporary methods of oral sensory assessment, with particular emphasis on suprathreshold scaling. We believe that these scales presently offer the most realistic picture of oral sensory function, but only when they are used correctly. Using converging methods from psychophysics, anatomy, and genetics, we demonstrate valid uses of modern chemosensory testing in clinical diagnosis and intervention.

Psychophysical measures of experience have played a fundamental role in our understanding of sensory and hedonic processes. In the chemical senses, these measures have revealed the broad impact of oral sensation and dysfunction on health-related behaviors and overall quality of life [1]. Oral sensory disturbances may be relatively benign, but sometimes they are profoundly life altering. As such, chemosensory experience and its consequences represent an important clinical concern.

Assessing this experience, however, is an extremely challenging task. By definition, individual experience is subjective: we can *describe* our experiences and track them over time, but we cannot directly *share* the experiences of another person. Nevertheless, we use comparisons of real-world experience throughout life to communicate what is acceptable (e.g. pleasure, comfort) and what is not (e.g. disease, pain), so it is important to evaluate these experiences carefully. Recent advances in suprathreshold scaling capture sensory and affective differences with improved accuracy, supporting the notion that perceptual experiences can be measured and compared.

In this chapter, we address methodological issues regarding threshold and suprathreshold measures of oral sensation. Using magnitude matching as an example, we argue that suprathreshold intensity scales provide a more complete picture of oral sensory function than do thresholds alone. Our efforts to identify useful suprathreshold tools include an examination of labeled scales, which are used (and misused) to compare experiences between individuals and groups. To confirm our psychophysical results, we demonstrate the use of parallel methods (e.g. multiple standards, genetic and anatomical tools). Finally, using techniques appropriate for comparison, we show how spatial taste testing has advanced our understanding of oral sensory function in health and disease.

Thresholds versus Intensity: How Should Oral Sensation Be Measured?

When we enjoy a meal, we can easily tell if the soup is too salty or the cocktails watered down. These judgments demonstrate that intensity is continuous (i.e. strong or weak variants of stimulus strength) and not binary (i.e. present or absent). Some researchers believe that degrees of suprathreshold intensity are immeasurable or at best ordinal [2], but we contend that suprathreshold measures possess unique diagnostic and predictive capabilities.

Indirect Psychophysics: Threshold Procedures

Thresholds have been used for sensory evaluation ever since Fechner [3] codified them almost 150 years ago. Although thresholds present technical challenges, they are conceptually straightforward: the *absolute threshold* for a stimulus is the lowest concentration at which its presence can be detected as something, whether or not it is qualitatively discernible. The *recognition threshold* is the lowest concentration at which the quality of a stimulus (e.g. sweet, painful) can be identified. Finally, the *difference threshold* is the smallest increase in suprathreshold stimulus concentration that can be detected (i.e. the 'just noticeable difference').

Thresholds enjoy widespread use in research and clinical settings, mainly because they produce values suitable for comparison. Because thresholds are especially sensitive to sensory adaptation, subject fatigue, and criterion shift [4], abbreviated methods have been developed that provide reliable threshold estimates with fewer trials and minimal bias [5–7]. Even so, one of the most discouraging features of thresholds is the time required to measure them: an up-down, forced-choice threshold procedure [7] takes approx. 20 min to administer, yielding only the lower boundary of the taste function; suprathreshold procedures approximate the entire taste function in much less time. Generally, the decision to use thresholds in lieu of suprathreshold measures is reasonable when there is strong concordance between threshold and suprathreshold experiences. However, psychophysical functions for taste stimuli show considerable variation that precludes reasonable predictions of suprathreshold sensation from threshold values alone [8].

As an alternative to chemical measures of taste sensitivity, electrogustometry involves the application of weak anodal electric currents to specific regions of the mouth [9]. Proponents of electrogustometry emphasize its convenience [10]; it is portable, avoids the use of chemical solutions, permits regional stimulation of taste bud fields, and provides values that can be compared across individuals, time points, locations within the mouth, or treatment conditions. Electric taste thresholds show high test-retest reliability and bilateral correspondence [11], and normative data have been described for some groups [12]. Accordingly, electrogustometry has been used to identify sizable taste losses associated with aging, denervation, and disease [11, 13, 14], but the following two disadvantages limit its use in more specific clinical assessments of taste function [15].

- Because saliva is mildly acidic and contains salts, electrogustometry typically evokes sour or salty taste sensations [16]. However, oral sensory alterations are often quality specific, particularly affecting bitter taste [17, 18]. As such, electrogustometry may fail to identify clinically relevant damage.
- Electrogustometric thresholds correlate well with regional [12, 19] but not whole-mouth chemical taste thresholds [11]; suprathreshold functions for electrical and chemical taste also show poor agreement [20]. Thus, as with chemical taste thresholds (see above), electrogustometry cannot reflect real-world taste experience accurately.

Direct Psychophysical Scaling of Suprathreshold Intensity

Thresholds provide only the lower limit of physical energy that can be perceived (e.g. decibels of sound, molar concentration), but suprathreshold or 'direct' scaling methods measure perceived intensity across the full dynamic range of sensation [21]. S.S. Stevens [22] introduced direct scaling methods with ratio properties, the most popular of which is magnitude estimation. In this procedure, subjects provide a number reflecting perceived stimulus intensity; they then give a number twice as large to a stimulus that is twice as intense, a number half as large to a stimulus half as intense, and so on. The size of the numbers is irrelevant; only the ratios among numbers carry meaning. As a result, magnitude estimates describe only how perceived intensity varies with stimulus intensity *within* an individual; they cannot reflect meaningful differences of absolute perceived intensity *between* individuals or groups [23]. Because group comparisons are such a basic element of scientific analysis, this limitation has not been fully appreciated, but its consequences are severe. To illustrate, we now describe studies on individual differences in taste perception; these studies are especially noteworthy in terms of their contributions to comparative suprathreshold scaling.

Genetic Variation in Oral Sensation: The Rise of Magnitude Matching

Taste Blindness: We Live in Different Oral Sensory Worlds

Discovered by the chemist A.L. Fox in 1931 [24], individuals differ significantly in their ability to taste thiourea compounds like phenylthiocarbamide (PTC) and 6-n-propylthiouracil (PROP) [25]; most individuals perceive bitterness (i.e. tasters), but others are 'taste blind' and perceive nothing (i.e. nontasters). Early reports suggested that taste blindness is a recessive trait inherited through a single genetic locus [26], while other studies measured the proportion of tasters by race, sex, and disease [27–29]. In the 1960s, behavioral experiments showed that PTC/PROP threshold sensitivity influences food preferences, alcohol and tobacco use, and body weight [30].

Buoyed by the potential benefits of direct scaling, Bartoshuk sought to compare the suprathreshold bitterness of PTC between nontasters and tasters. However, this comparison presents a problem: magnitude estimates have *relative* meaning when subjects are used as their own controls [31], but how can *absolute* bitterness rating be compared across groups? The answer to this question involves measuring PTC bitterness relative to an unrelated standard. Although magnitude estimates are often multiplied by a constant (i.e. 'normalized') in order to obtain group functions [32], this procedure is qualitatively different.

- To average magnitude estimates while maintaining the ratios among them, ratings must be brought into a common register so that one subject's data are not unduly weighted just because that subject used larger numbers [23]. With this type of normalization, the standard is arbitrary, so group functions convey nothing about absolute perceived intensity.
- When the purpose of normalization is to permit group comparisons, the standard is assumed to be equally intense (on average) to the groups being compared. If this assumption holds, normalization yields valid across-group differences in absolute perceived intensity for stimuli of interest: if '10' denotes the intensity of a standard to nontasters and tasters, PTC ratings of '40' for tasters and '20' for nontasters reflect a twofold intensity difference.

Because thioureas share the N–C=S chemical group [33], Bartoshuk reasoned that the taste intensity of a compound lacking the N–C=S group should be equal, on average, to nontasters and tasters. If so, group averages of PTC bitterness can be compared by rating it relative to, for example, NaCl saltiness. With this procedure, tasters find PTC and PROP more bitter than do nontasters [34, 35], and mounting data indicate that these individual differences reflect entirely different oral sensory worlds: tasters perceive more intense taste and oral tactile sensations overall [36], indicating that taste blindness extends far beyond the N–C=S group [30]. Of particular interest, a subset of tasters known as 'supertasters' consistently give the highest ratings to taste stimuli, oral irritants (e.g. capsaicin), fats, and food-related odors [37, 38].

Magnitude Matching: Non-oral Standards Enable Oral Sensory Comparisons

If supertasters perceive NaCl more intensely than do others, then NaCl is a poor standard for oral sensory comparisons, meaning that observed differences between taster groups are inaccurate (albeit conservatively so). This problem was resolved by experiments on cross-modality matching, in which stimuli from unrelated modalities are compared [39]; by assuming that taste and hearing are unrelated, taste intensity can be rated relative to auditory intensity. This 'magnitude matching' procedure [40] confirmed the suspicion that the saltiness of NaCl varies with taster status [41]. Magnitude matching addresses the problem of group comparisons by changing the task: oral sensations cannot be compared directly across PROP taster groups, so subjects rate stimuli of interest relative to a non-oral sensory standard. As long as variability in the standard remains unrelated to variability in PROP bitterness, oral sensory experiences are comparable across groups.

The ability to observe accurate differences in oral sensation has revealed associations between sensory experience, dietary behavior, and disease risk. For example, PROP bitterness is linked to decreased vegetable preference and intake [42, 43], a known risk factor for colon cancer; it also associates with an increased number of colon polyps [44]. PROP intensity also predicts avoidance of high-fat foods, so supertasters have lower body mass indices and more favorable cardiovascular profiles [45–47].

PTC/PROP Genetics: Supertasting Is Not Explained by a Single Gene

Early family studies indicated that nontasting is a recessive trait with a single locus [26], while the discovery of supertasters led to additive models in which supertasters are homozygous dominant and medium tasters heterozygous [48]. Based on modern genetic analysis, oral sensory variation may in fact involve multiple alleles and/or loci [49]: candidate genes reside on chromosomes 5p15, 7, and 16p [50, 51], and subsequent mapping of chromosome 7q has identified sequence polymorphisms in a putative PTC receptor gene (TAS2R38) that account for observed threshold differences [52]. While medium tasters and supertasters have similar PROP thresholds [36], homozygous dominant individuals for TAS2R38 find PROP slightly more bitter than do heterozygotes, but this relationship is imperfect [53]. In other words, supertasting cannot be explained completely by threshold sensitivity or TAS2R38 expression – additional factors (i.e. oral anatomy, pathology, other genetic markers) must contribute [54]. Specifically, supertasting appears to depend on two conditions: the ability to taste PTC/PROP (which means that taste buds express PTC/PROP receptors) and a high density of fungiform papillae (i.e. structures containing taste buds) on the anterior tongue (which maximizes oral sensory input). As data continue to amass, oral anatomy may prove a better biological index of supertasting than PTC/ PROP receptor expression.

Best Practices and the Perils of Classification

Valid consensus values for PROP classification are lacking, mainly because continuing advances in genetic and psychophysical testing supersede previous estimates. Consequently, existing criteria are idiosyncratic and variable, resulting in a vigorous debate over which classification scheme best reflects differences in oral sensation [45, 55, 56]. At the center of this issue, the validity of any boundary value depends on the instrument used to measure it; when suprathreshold psychophysical tools produce distorted comparisons, the sorting criteria derived from those tools are also distorted. (Thresholds have remained a popular clinical measure for precisely this reason, even though they too distort real-world sensory experience.)

Broadly speaking, the most effective assessment strategies integrate multiple correlates of function. As advances in anatomy and genetics permit more nuanced studies, the best methods for oral sensory evaluation will encompass an array of techniques that complement and enrich sophisticated psychophysical measurement [57]. We have used this multivariate approach to develop the following guidelines for contemporary PROP classification.

- Nontasters and tasters are easily distinguished by genetic analysis of TAS2R38 (i.e. nontasters are recessive, tasters are dominant) [52], which reflects PROP threshold differences (i.e. above 0.2 m*M* for nontasters, below 0.1 m*M* for tasters [58, 59]). In a database of over 1,400 healthy lecture participants living in the USA, these differences roughly correspond to a general Labeled Magnitude Scale (gLMS) boundary value of 'weak' (i.e. approx. 17 out of 100) for filter papers impregnated with saturated PROP (approx. 0.058 *M*) [unpublished data]. Consistent with previous estimates [59, 60], this cutoff yields approx. 25% nontasters in the sample.

- Supertasters are distinguished from medium tasters by psychophysical criteria. Existing population estimates of PROP taster status are based on a single-locus model, so this boundary value will remain arbitrary until all genetic loci related to taste blindness are identified. In the database described above, nontasters represent the lowest 25% of PROP paper ratings, so a working definition of supertasting might include the top 25% of ratings; this logic suggests a gLMS boundary value of approx. 80.
- Individuals with taster genotypes and nontaster PROP ratings probably reflect oral sensory pathology (see below). In these cases, oral anatomy can often be used to identify supertasters [1], who show high fungiform papilla density (i.e. over 100 papillae/cm^2 [59]) and low PROP responses when taste function is compromised.

Despite considerable evidence, some researchers persistently claim that oral sensation has little effect on sensation, food behavior, or health [55, 61, 62]. In nearly every case, these dissenting reports fail to show effects of interest because their methods are incapable of showing effects of interest. Many of these reports involve the inappropriate use of labeled intensity scales.

Labeled Scales: Valid (and Invalid) Comparisons

Measurement scales labeled with intensity descriptors – including Likert, 9-point, and visual analog scales (VAS) [63–65] – are widely used throughout the medical, scientific, and consumer disciplines. Although many category scales have been 'validated', the fact that a scale measures what it was intended to measure does not guarantee its ability to produce valid group comparisons.

Properties of Intensity Labels: Spacing, Relativity, and Elasticity

We commonly use intensity descriptors to compare our experiences with the experiences of those around us (e.g. 'This solution tastes *strong* to me. Does it taste *strong* to you?'). Because we use these words so frequently, they have been incorporated as labels in intensity scales. These labels have special properties that warrant discussion.

- Generally, ratings from category scales have ordinal but not ratio properties [66], because intensity descriptors are not equally spaced [67]. Several investigators have produced scales with labels spaced empirically to provide ratio properties [68]; this spacing has been replicated across multiple sensory and hedonic attributes [69–72], indicating that sensory and hedonic experiences possess similar intensity properties.
- Intensity descriptors are relative by definition: because adjectives modify nouns, they have no absolute meaning until their antecedents are specified. Nevertheless, many group comparisons implicitly assume that scale

descriptors denote the same absolute intensity regardless of the object described [73, 74].

- Intensity descriptor meanings vary among groups of people just as they do among different sensory modalities. In a study assessing the magnitudes denoted by scale descriptors for taste perception [37], the spacing among descriptors appears proportional for nontasters and supertasters, but the supertaster range is expanded (fig. 1a).

In short, intensity labels maintain their relative spacing, but they are elastic in terms of the domain to be measured and individual experiences within that domain. Because labeled scales fail to account for this elasticity, they are inappropriate whenever subject classification (e.g. sex, age, weight, clinical status) produces groups for which scale labels denote different absolute intensities.

Consequences of Invalid Comparisons: Distortion and Reversal

Figure 1b shows errors resulting from the false assumption that intensity descriptors denote the same absolute intensity to everyone. (This figure is idealized, but effects have been verified using taste and food stimuli [1].) The left side shows stimuli that produce equal perceived intensities to nontasters. The diverging lines connecting nontaster and supertaster ratings indicate PROP effects of differing sizes; the intensity difference between groups for the label 'very strong taste' is the same difference shown in figure 1a. When the label 'very strong taste' is treated as if it denotes the same average intensity to nontasters and supertasters, supertaster data are compressed relative to nontaster data, as shown on the right side.

- Stimulus A appears more intense to supertasters than to nontasters, but the magnitude of the effect is blunted.
- The difference between nontasters and supertasters for stimulus C is equal to the difference between the labels, so it disappears.
- For stimulus D, the actual difference between nontasters and tasters is smaller than the difference in meaning for 'very strong taste', so group differences appear to go in the opposite direction. This phenomenon is known as a *reversal artifact* [75].

Despite this problem, some investigators argue that group effects with significant biological impact should be sufficiently robust to be detected with any and all methods [55]. Claims like these are distortions themselves: biological effects exist whether they are measured or not, but measurement tools are useless if they cannot detect those effects realistically. Moreover, the popularity of a scale does not necessarily make it the right tool for the task at hand. Although improved labeled scales show promise, contrary reports arising from invalid scaling methods remain significant obstacles to health-related research efforts.

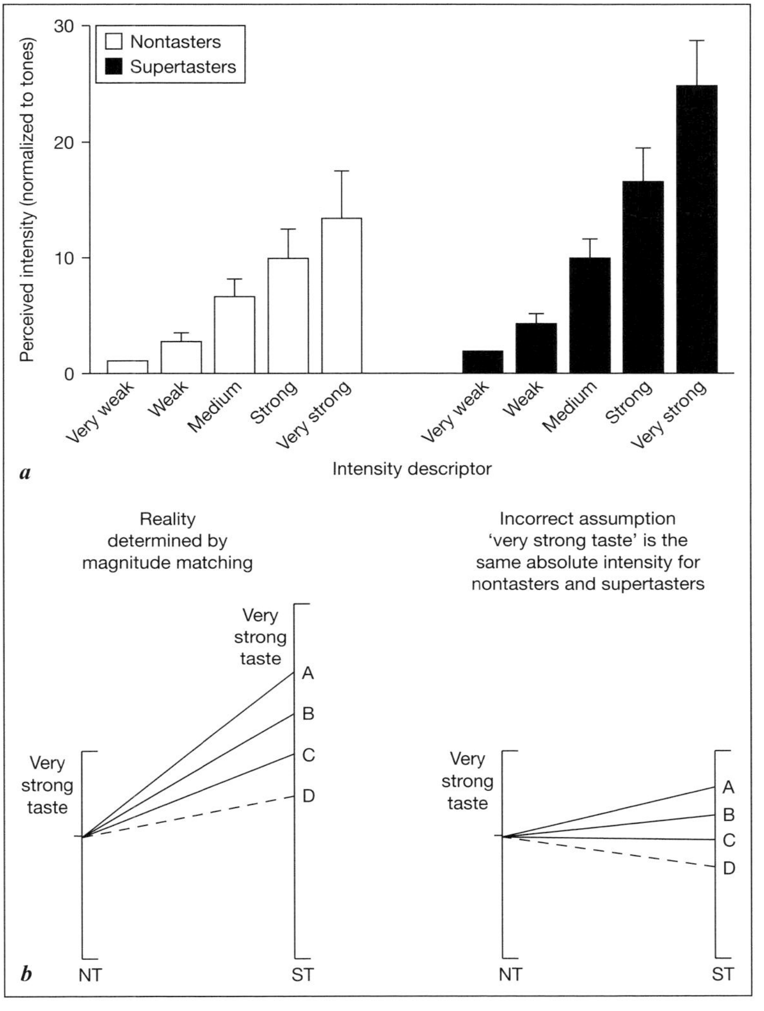

Fig. 1. Nontasters (NT) and supertasters (ST) inhabit different oral sensory worlds. ***a*** The perceived intensity of scale descriptors for NT and ST. ***b*** Consequences of invalid group comparisons. On the left, taste functions reflect actual differences between NT and ST measured with magnitude matching. On the right, the same taste functions are distorted by the incorrect assumption that 'very strong taste' indicates the same absolute perceived intensity to NT and ST: valid effects appear truncated and may reverse direction inaccurately. Modified from Bartoshuk et al. [37].

Building a Better Scale: A Quest for Appropriate Standards

Category scales assume ratio properties when the spacing among categories reflects real-world experience. If this common intensity scale were stretched to its maximum, might it produce a labeled scale allowing valid comparisons of oral sensory intensity? To test this idea, Bartoshuk and colleagues replaced the top anchor of the LMS [71] with the label 'strongest imaginable sensation of any kind'. This scale, now known as the general LMS (gLMS), produces similar group differences in PROP bitterness as magnitude matching [76], indicating that the top anchor functions as a suitable standard for oral sensory assessment. Considering that sensory and affective intensity labels are similarly spaced [68], a bipolar version of the gLMS has proven particularly useful for hedonic measurement [77].

The standards used in the laboratory often require cumbersome and expensive equipment. Because scale labels rely on memories of perceived intensity, remembered sensations have been proposed as standards for magnitude matching. Although the precise relationship between real and remembered intensity is unclear [78], remembered oral sensations appear to reflect effects seen with actual stimuli [79]. Meanwhile, the term 'imaginable' found in many intensity scales is a poor standard, as recent data show individual differences in the intensity of imagined experiences [80]. Thus, in a more radical approach to scaling, perhaps all labels should be abandoned except for those at the ends of the scale. The resulting scale – a line denoting the distance from 'no sensation' to the 'strongest sensation of any kind ever experienced' – is essentially a VAS encompassing all sensory modalities; we have proposed calling it the general/global VAS (gVAS) [81].

Overall, we have used labeled scales with success because we include both real and remembered sensations as standards. By using raw gLMS scores and/or normalizing those scores to other standards, we are able to confirm our conclusions across a variety of assumptions.

Clinical Assessment of Oral Sensory Function

Disorders of oral sensation are both widespread and variable, yet useful resources and appropriate medical treatment are frustratingly sparse [82]. Because taste cues influence nutritional health, metabolism, and affect [83], their loss can be traumatic, yet in other cases taste loss is hardly noticed [84]. Gustatory disturbances are often associated with specific disorders and treatment interventions, but just as often they are of unknown origin and unpredictable onset [85]. Thus, oral sensory evaluations require a thorough examination of physical (e.g. oral anatomy, oral and salivary pathology, neurological damage), sensory (e.g. taste, oral somatosensation, retronasal olfaction), and emotional aspects of chemosensation (e.g. psychopathology, quality of life).

Several afferent nerves carry sensory information from the mouth, each carrying a particular array of information from a particular area. The chorda tympani (CT), a branch of the facial nerve (VII), carries taste information from the anterior tongue; the lingual branch of the trigeminal nerve (V) carries pain, tactile, and temperature information from the same region. The greater superficial petrosal nerve, another branch of nerve VII, carries taste cues from the palate. Multimodal information (i.e. taste, touch, pain, temperature) is carried from the posterior tongue by the glossopharyngeal nerve (IX) and from the throat by the vagus (X) [86]. Taste and oral somatosensory cues combine centrally with retronasal olfaction to produce the composite experience of flavor [87]. This spatial distribution of input has led researchers to consider the clinical relevance of localized oral sensory damage (see below), so modern protocols for oral sensory evaluation typically include judgments of intensity and quality for both regional and whole-mouth stimuli.

Whole-Mouth Oral Sensation

In whole-mouth gustatory testing, chemical stimuli are sampled and moved throughout the mouth, stimulating all oral taste bud fields simultaneously; subjects rinse with water prior to each stimulus. Laboratory tests of oral sensation involve the presentation of chemical solutions at multiple concentrations spanning the functional range of perception [41, 88], but most clinical tests have streamlined this process to a single stimulus for each of the common taste qualities (i.e. sucrose, NaCl, citric acid, quinine hydrochloride) [89]. In addition, multiple concentrations may be used to derive suprathreshold taste functions, and other oral stimuli may be included to evaluate oral tactile sensation (e.g. capsaicin, alcohol) or individual differences (e.g. PROP). Having discussed the disadvantages of thresholds, we favor suprathreshold measurements involving magnitude matching or the gLMS/gVAS with appropriate standards, so stimuli unrelated to oral sensation (e.g. sound, remembered sensations) should be incorporated.

Aqueous solutions are inconvenient for field and clinical use, so alternative methods of stimulus delivery have been explored, including paper strips and tablets [90–92]. With respect to whole-mouth testing, the most enduring example of these methods is the use of PROP papers as a screening tool for taste blindness. In early studies, PTC crystals were placed directly on the tongue [33] or delivered on saturated filter papers [93]. Today, PROP papers are made by soaking laboratory-grade filter papers in a supersaturated PROP solution heated to just below boiling. When dry, each paper contains approx. 1.6 mg PROP [94]; patients with hyperthyroidism are prescribed 100–300 mg daily [95]. Variants of this method have been described [96], but all share the common goal of introducing a small amount of crystalline PROP to the tongue surface. Although PROP papers are convenient and portable, their technical flaws warrant consideration.

- To produce a taste, the filter paper must be completely moistened with saliva, which requires healthy salivary function and a sufficient period of contact with the tongue [55].
- Some studies [35] report excessive false-positive and false-negative responses to PTC/PROP filter papers. These response rates probably result from minor testing variations that affect response bias.
- Filter paper testing shows only moderate concordance with threshold sensitivity [97]. As described above, threshold and suprathreshold measures almost always dissociate when proper scaling is used.

Despite these concerns, filter paper ratings and laboratory PROP assessments show significant agreement and high test-retest reliability [94, 96, 98], perhaps because the effective concentration of PROP is unimportant provided it is high. Comparisons of PROP paper and solution bitterness suggest that the concentration dissolved from paper into saliva approaches maximum solubility [unpublished data]. Because functions of PROP bitterness for nontasters, medium tasters, and supertasters diverge [36], the most efficient way to sort subjects is to use the highest concentration possible, so papers made from saturated PROP [94] may be preferable to those made from lower concentrations.

Oral Sensory Anatomy: Videomicroscopy of the Tongue

Multiple reports indicate that differences in taste bud density account for human oral sensory variation [45, 59]. To explore this idea further, Miller and Reedy [99] developed a method for visualizing the tongue in vivo. Human tongues are coated with large, raised, circular structures (i.e. fungiform papillae) that hold taste buds [100]; blue food coloring applied to the tongue surface fails to stain these papillae, which subsequently appear as pink circles against a blue background. Fungiform papillae can be counted with a magnifying glass and a flashlight, while videomicroscopy allows resolution of pores at the apical tips of taste buds. This method has revealed positive associations between PROP intensity, fungiform papilla density, and taste bud density [59, 99]. Fungiform papillae are dually innervated by CT and nerve V [101], which accounts for the elevated taste and oral tactile sensations experienced by supertasters [45, 102].

Clinically, videomicroscopy is useful in confirming CT damage (see below), which often appears as a discrepancy between high fungiform papilla density and low taste sensation on the anterior tongue [1]. In addition, the association between taste intensity and oral anatomy among healthy subjects can be used to evaluate the ability of various scales to provide valid across-group comparisons: the gLMS and magnitude matching produce robust correlations between taste intensity and fungiform papillae density, but category scales are severely limited in this regard [103].

Spatial Taste Testing

Because different nerves innervate different regions of the oral cavity, oral sensation may be absent in one area but intact in others. Remarkably, individuals with extensive taste damage are often unaware of it unless it is accompanied by tactile loss [84], presumably because taste cues are referred perceptually to sites in the mouth that are touched [104–106]. As a result of this 'tactile referral', regional taste loss rarely produces whole-mouth taste loss, yet it remains clinically significant as a precursor to altered, heightened, and phantom oral sensations. Measures of regional taste function are an important tool for identifying the source of these complaints.

The integrity of specific taste nerves is assessed clinically via spatial testing [8], in which suprathreshold solutions of sweet, sour, salty, and bitter stimuli are applied with cotton swabs onto the anterior tongue tip, foliate papillae (i.e. posterolateral edges of the tongue), circumvallate papillae (i.e. raised circular structures on the posterior tongue), and soft palate. (A spatial taste test involving filter paper strips impregnated with taste stimuli has also been described [92].) Stimuli are presented on the right and left sides at each locus, and subjects make quality and intensity judgments using magnitude matching or the gLMS. Special care must be taken to avoid stimulating both sides of the mouth simultaneously (which impedes localization), triggering a gag reflex during circumvallate stimulation, or allowing palate stimuli to reach the tongue surface (which leads to inflated palate ratings). Following regional testing, subjects swallow a small volume of each solution and rate its intensity, thereby enabling comparisons of regional and whole-mouth sensation.

Regional Taste Sensation: The 'Tongue Map' Is False

Comparisons of psychophysical functions across oral loci indicate that taste is perceived at similar intensity on all tongue areas holding taste buds, but less so on the palate [107]. Thus, oral sensory losses can be detected as significant local variations from otherwise stable perception across the tongue surface.

For many years, the only spatial feature of taste mentioned in textbooks was a map showing the areas on the tongue sensitive to each of the four basic tastes: sweet on the tip, salt and sour on the edges, and bitter on the rear. This inaccurate 'tongue map' arose from a misunderstanding of the work of Hänig [108], who examined taste thresholds at various tongue loci. Hänig showed that thresholds for the four basic tastes vary slightly at different loci, but he did not find that taste modalities are restricted to specific regions of the tongue surface. The misunderstanding occurred years later when Boring [109] plotted the reciprocal of Hänig's threshold values as a measure of regional sensitivity. Subsequent readers failed to realize that the reciprocals actually represented very small threshold differences, and a myth was born.

Clinical Correlates of Localized Taste Loss

Spatial taste testing is most powerful when used in combination with genetic and anatomical data, as it reveals discrepancies between heredity and experience that arise via pathology [57]. The following examples illustrate conditions in which modern oral sensory testing may facilitate diagnosis or intervention.

Disinhibition in the Mouth: A Model for Taste and Oral Pain Phantoms

Dysgeusia refers to a chronic taste that occurs in the absence of obvious stimulation [110]. Many clinical complaints of dysgeusia result from taste stimuli that are not readily apparent to the patient (e.g. medications tasted in saliva, crevicular fluid, or blood [111–113]), but some chronic oral sensations, known as phantoms, appear to arise centrally.

Glossopharyngeal Disinhibition: Taste Phantoms. Neurological disorders can lead to taste phantoms [114], but CT damage appears to be a primary factor in clinical accounts [84]. Moreover, electrophysiological recordings from rodents and dogs show that blocking CT input produces elevated activity in brain regions receiving input from nerve IX [115, 116]. These data indicate that CT inhibits nerve IX normally, so CT loss should disinhibit nerve IX. Human psychophysical data support this model.

- In patient cohorts (e.g. head injury, craniofacial tumors, ear infections) and healthy subjects under anesthesia, unilateral CT loss leads to increased whole-mouth perceived bitterness via increased contralateral taste sensation at nerve IX [18, 117–119]. Oral sensory input rises ipsilaterally into the CNS [120], so these contralateral effects appear to involve central modulation.
- About 40% of healthy subjects experience taste phantoms while under CT anesthesia. These phantom sensations are localized contralaterally to nerve IX, vary in quality and intensity, and fade with the anesthetic. Whole-mouth topical anesthesia abolishes these release-of-inhibition phantoms [119], presumably by suppressing spontaneous neural activity at their source.

- In one report [121], a bitter taste phantom arose bilaterally at nerve IX following tonsillectomy. Spatial testing indicated complete nerve IX loss, yet the phantom became more intense with whole-mouth topical anesthesia. This nerve stimulation phantom was probably caused by surgical damage to nerve IX and disinhibited further by CT anesthesia.

Trigeminal Disinhibition: Oral Pain Phantoms. CT taste input also appears to inhibit cues from nerve V. This interaction may suppress oral pain during intake, and it may facilitate tactile referral of taste information following localized taste damage. Because supertasters have the most taste and trigeminal input, CT damage may lead to adverse sensory consequences due to extreme disinhibition of nerve V: following unilateral CT anesthesia, supertasters show increased ratings for the burn of capsaicin on the contralateral anterior tongue [122].

Oral pain phantoms are another serious consequence of nerve V disinhibition. The identification of burning mouth syndrome (BMS) as such a phantom vividly illustrates the clinical relevance of modern oral sensory testing. BMS, a condition most often found in postmenopausal women, is characterized by severe oral pain in the absence of visible pathology [123]. BMS is often described as psychogenic, but systematic psychophysical testing tells a different story. Most BMS patients show significantly reduced bitterness for quinine on the anterior tongue, consistent with CT damage [124]. Nearly 50% of BMS patients experience taste phantoms at nerve IX [17]; topical anesthesia usually intensifies BMS-related taste and oral pain [125]. Finally, the peak intensity of BMS pain correlates with fungiform papillae density, indicating that BMS is most prevalent among supertasters. Taken together, these data strongly suggest that BMS is an oral pain phantom generated by CT damage. Grushka et al. [126] have shown that agonists to the inhibitory neurotransmitter γ-aminobutyric acid suppress BMS pain, presumably by restoring lost inhibition from absent taste cues.

Clinical Considerations

Laboratory and clinical data support the use of topical anesthesia in the mouth to determine the locus of oral sensory dysfunction. However, interpretations of topical anesthesia must be made carefully, as incomplete anesthesia will impede differential diagnosis. In topical anesthesia, patients hold approx. 5 ml of 0.5% dyclone in the mouth for 60 s, rest for 60 s, rinse with water, and describe any oral sensations experienced for the duration of the sensory block [121].

If a taste or oral pain complaint becomes more intense with oral anesthesia, it does not arise from normal stimulation of oral sensory receptors. Certain therapeutic agents promote venous taste and other dysgeusias, so medication and supplement use should be reviewed. Another possibility is that the nerve innervating the region of sensory disturbance has sustained physical damage. If damage is peripheral to nerve cell bodies, the resulting neuroma may produce a nerve stimulation phantom; topical anesthesia exacerbates nerve stimulation phantoms via central disinhibition. Conclusions involving nerve damage should be confirmed by further neurological examination.

When local anesthesia abolishes a taste or oral pain complaint, an actual stimulus may be present in the mouth. To test for the presence of such a stimulus, the patient should attempt to rinse it from the mouth; if the offending sensation subsides at all, an actual stimulus should be considered. Alternatively, nerve damage unrelated to the complaint may be disinhibiting input related to it; topical anesthesia suppresses these central release-of-inhibition phantoms, presumably by inhibiting spontaneous activity. Spatial testing should reveal localized taste loss at a site distant from the phantom.

Conclusion

Human psychophysics is a powerful aspect of clinical and basic science that offers a window onto neurobehavioral processes often inaccessible by other means. As such, our goal in exploring psychophysical methodology is to craft measurement tools that reflect individual differences accurately and allow adaptive use in clinical, research, and other assessment settings. Conservative approaches to this task emphasize threshold measures, but we have embraced suprathreshold techniques in the hope of measuring biologically relevant sensations, and we have carefully evaluated these techniques in the process. Our overall approach has been to refine existing methods continuously, incorporating real-world reference points in order to represent perception as faithfully as possible. These refinements have posed a constant challenge to remain user friendly; our use of sophisticated scaling tools in laboratory research shows that untrained subjects learn to use them quickly and skillfully, and our clinical research indicates that these tools are accessible to patients. Finally, our systematic use of techniques from psychophysics, anatomy, neurology, and genetics has allowed us to explore complex relationships between oral sensation, affect, behavior, and disease at multiple levels of analysis. In our view, these effects confirm that our methods reflect highly predictive and highly comparable aspects of sensory and hedonic experience.

Acknowledgements

This research is supported by a grant from the US National Institutes of Health (DC 00283) to L.M.B.. D.J.S. is supported by funding from the US National Science Foundation and the Rose Marie Pangborn Sensory Science Fund.

References

1 Bartoshuk LM, Duffy VB, Chapo AK, Fast K, Yiee JH, Hoffman HJ, Ko C-W, Snyder DJ: From psychophysics to the clinic: missteps and advances. Food Qual Pref 2004;15:617–632.
2 Brindley GS: Introduction to sensory experiments; in Brindley GS (ed): Physiology of the Retina and the Visual Pathway. London, Arnold, 1960, pp 144–150.
3 Fechner GT: Elements of Psychophysics (translated by Adler HE; edited by Howes DH, Boring EG). New York, Holt, Rinehart and Winston, 1966 (original 1860).
4 Engen T: Psychophysics. 1. Discrimination and detection; in Kling JW, Riggs LA (eds): Woodworth and Schlosberg's Experimental Psychology. New York, Holt, Rinehart and Winston, 1972, vol 1: Sensation and Perception, pp 11–46.
5 Harris H, Kalmus H: The measurement of taste sensitivity to phenylthiourea (PTC). Ann Eugen 1949;15:24–31.
6 Henkin RI, Gill JR, Bartter FC: Studies on taste thresholds in normal man and in patients with adrenal cortical insufficiency: the role of adrenal cortical steroids and of serum sodium concentration. J Clin Invest 1963;42:727–735.
7 Wetherill GB, Levitt H: Sequential estimation of points on a psychometric function. Br J Math Stat Psychol 1965;18:1–10.
8 Bartoshuk LM: Clinical evaluation of the sense of taste. Ear Nose Throat J 1989;68:331–337.
9 Mackenzie ICK: A simple method of testing taste. Lancet 1955;ii:377–378.
10 Frank ME, Smith DV: Electrogustometry: a simple way to test taste; in Getchell TV, Doty RL, Bartohsuk LM, Snow JB (eds): Smell and Taste in Health and Disease. New York, Raven Press, 1991, pp 503–514.
11 Murphy C, Quiñonez C, Nordin S: Reliability and validity of electrogustometry and its application to young and elderly persons. Chem Senses 1995;20:499–503.
12 Tomita H, Ikeda M, Okuda Y: Basis and practice of clinical taste examinations. Auris Nasus Larynx 1986;13(Suppl I):S1–S15.
13 Grant R, Ferguson MM, Strang R, Turner JW, Bone I: Evoked taste thresholds in a normal population and the application of electrogustometry to trigeminal nerve disease. J Neurol Neurosurg Psychiatry 1987;50:12–21.
14 Le Floch JP, Le Lievre G, Verroust J, Phillippon C, Peynegre R, Perlemuter L: Factors related to the electric taste threshold in type 1 diabetic patients. Diabet Med 1990;7:526–531.
15 Stillman JA, Morton RP, Hay KD, Ahmad Z, Goldsmith D: Electrogustometry: strengths, weaknesses, and clinical evidence of stimulus boundaries. Clin Otolaryngol 2003;28:406–410.
16 Bujas Z: Electrical taste; in Beidler LM (ed): Handbook of Sensory Physiology. Berlin, Springer, 1971, vol 4: Chemical Senses, part 2: Taste, pp 180–199.
17 Grushka M, Sessle BJ, Howley TP: Psychophysical evidence of taste dysfunction in burning mouth syndrome. Chem Senses 1986;11:485–498.
18 Lehman CD, Bartoshuk LM, Catalanotto FC, Kveton JF, Lowlicht RA: The effect of anesthesia of the chorda tympani nerve on taste perception in humans. Physiol Behav 1995;57:943–951.
19 Krarup B: Taste reactions of patients with Bell's palsy. Acta Otolaryngol 1958;49:389–399.
20 Salata JA, Raj JM, Doty RL: Differential sensitivity of tongue areas and palate to chemical stimulation: a suprathreshold cross-modal matching study. Chem Senses 1991;16:483–489.
21 Stevens SS: On the theory of scales of measurement. Science 1946;103:677–680.
22 Stevens SS: The direct estimation of sensory magnitudes: loudness. Am J Psychol 1956;69:1–25.
23 Marks LE: Sensory Processes: The New Psychophysics. New York, Academic Press, 1974.
24 Fox AL: Six in ten 'tasteblind' to bitter chemical. Science 1931;73:14a.
25 Barnicot NA, Harris H, Kalmus H: Taste thresholds of further eighteen compounds and their correlation with PTC thresholds. Ann Eugen 1951;16:119–128.
26 Blakeslee AF: Genetics of sensory thresholds: taste for phenyl-thio-carbamide. Proc Natl Acad Sci USA 1932;18:120–130.
27 Parr LW: Taste blindness and race. J Hered 1934;25:187–190.
28 Kalmus H, Farnsworth D: Impairment and recovery of taste following irradiation of the oropharynx. J Laryngol Otol 1959;73:180–182.

29 Whissell-Buechy D, Wills C: Male and female correlations for taster (PTC) phenotypes and rate of adolescent development. Ann Hum Biol 1989;16:131–146.
30 Fischer R: Gustatory, behavioral, and pharmacological manifestations of chemoreception in man; in Ohloff G, Thomas AF (eds): Gustation and Olfaction. New York, Academic Press, 1971, pp 187–237.
31 McBurney DH, Bartoshuk LM: Interactions between stimuli with different taste qualities. Physiol Behav 1973;10:1101–1106.
32 Bartoshuk LM, Duffy VB: Taste and smell; in Masoro EJ (ed): Handbook of Physiology. 11. Aging. New York, Oxford University Press, 1995, pp 363–375.
33 Fox AL: The relationship between chemical constitution and taste. Proc Natl Acad Sci USA 1932;18:115–120.
34 Hall MJ, Bartoshuk LM, Cain WS, Stevens JC: PTC taste blindness and the taste of caffeine. Nature 1975;253:442–443.
35 Lawless HT: A comparison of different methods used to assess sensitivity to the taste of phenylthiocarbamide (PTC). Chem Senses 1980;5:247–256.
36 Bartoshuk LM: Comparing sensory experiences across individuals: recent psychophysical advances illuminate genetic variation in taste perception. Chem Senses 2000;25:447–460.
37 Bartoshuk LM, Duffy VB, Fast K, Snyder DJ: Genetic differences in human oral perception: advanced methods reveal basic problems in intensity scaling; in Prescott J, Tepper BJ (eds): Genetic Variation in Taste Sensitivity. New York, Dekker, 2004, pp 1–42.
38 Prescott J, Bartoshuk LM, Prutkin JM: 6-n-Propylthiouracil tasting and the perception of nontaste oral sensations; in Prescott J, Tepper BJ (eds): Genetic Variation in Taste Sensitivity. New York, Dekker, 2004, pp 89–104.
39 Stevens JC: Cross-modality validation of subjective scales for loudness, vibration, and electric shock. J Exp Psychol 1959;57:201–209.
40 Marks LE, Stevens JC, Bartoshuk LM, Gent JG, Rifkin B, Stone VK: Magnitude matching: the measurement of taste and smell. Chem Senses 1988;13:63–87.
41 Bartoshuk LM, Duffy VB, Lucchina LA, Prutkin JM, Fast K: PROP (6-n-propylthiouracil) supertasters and the saltiness of NaCl; in Murphy C (ed): Olfaction and Taste XII. New York, New York Academy of Sciences, 1998, pp 793–796.
42 Drewnowski A, Henderson SA, Hann CS, Berg WA, Ruffin MT: Genetic taste markers and preferences for vegetables and fruit of female breast care patients. J Am Diet Assoc 2000;100: 191–197.
43 Dinehart ME, Hayes JE, Bartoshuk LM, Lanier SL, Duffy VB: Bitter taste markers explain variability in vegetable sweetness, bitterness, and intake. Physiol Behav 2006;87:304–313.
44 Basson MD, Bartoshuk LM, Dichello SZ, Panzini L, Weiffenbach JM, Duffy VB: Association between 6-n-propylthiouracil (PROP) bitterness and colonic neoplasms. Dig Dis Sci 2005;50: 483–489.
45 Prutkin JM, Duffy VB, Etter L, Fast K, Gardner E, Lucchina LA, Snyder DJ, Tie K, Weiffenbach JM, Bartoshuk LM: Genetic variation and inferences about perceived taste intensity in mice and men. Physiol Behav 2000;69:161–173.
46 Tepper BJ, Ullrich NV: Influence of genetic taste sensitivity to 6-n-propylthiouracil (PROP), dietary restraint, and disinhibition on body mass index in middle-aged women. Physiol Behav 2002;75:305–312.
47 Duffy VB, Lucchina LA, Bartoshuk LM: Genetic variation in taste: potential biomarker for cardiovascular disease risk? in Prescott J, Tepper BJ (eds): Genetic Variation in Taste Sensitivity. New York, Dekker, 2004, pp 195–228.
48 Reed DR, Bartoshuk LM, Duffy VB, Marino S, Price RA: Propylthiouracil tasting: determination of underlying threshold distributions using maximum likelihood. Chem Senses 1995;20:529–533.
49 Guo S-W, Reed DR: The genetics of phenylthiocarbamide perception. Ann Hum Biol 2001;28: 111–142.
50 Reed DR, Nanthakumar E, North M, Bell C, Bartoshuk LM, Price RA: Localization of a gene for bitter taste perception to human chromosome 5p15. Am J Hum Genet 1999;64:1478–1480.
51 Drayna D, Coon H, Kim U-K, Elsner T, Cromer K, Otterud B, Baird L, Peiffer AP, Leppert M: Genetic analysis of a complex trait in the Utah Genetic Reference Project: a major locus for PTC

taste ability on chromosome 7q and a secondary locus on chromosome 16p. Hum Genet 2003;112:567–572.
52 Kim U-K, Jorgenson E, Coon H, Leppert M, Risch N, Drayna D: Positional cloning of the human quantitative trait locus underlying taste sensitivity to phenylthiocarbamide. Science 2003;299: 1221–1225.
53 Bartoshuk LM, Davidson AC, Kidd JR, Kidd KK, Speed WC, Pakstis AJ, Reed DR, Snyder DJ, Duffy VB: Supertasting is not explained by the PTC/PROP gene. Chem Senses 2005;30:A87.
54 Bartoshuk LM, Duffy VB, Fast K, Green BG, Snyder DJ: Hormones, age, genes, and pathology: how do we assess variation in sensation and preference? in Anderson GH, Blundell JE, Chiva MM (eds): Food Selection: From Genes to Culture. Levallois-Parret, Danone Institute, 2002, pp 173–187.
55 Drewnowski A: Genetics of human taste perception; in Doty RL (ed): Handbook of Olfaction and Gustation, ed 2. New York, Dekker, 2003, pp 847–860.
56 Rankin KM, Godinot N, Christensen CM, Tepper BJ, Kirkmeyer SV: Assessment of different methods for 6-n-propylthiouracil status classification; in Prescott J, Tepper BJ (eds): Genetic Variation in Taste Sensitivity. New York, Dekker, 2004, pp 63–88.
57 Duffy VB, Davidson AC, Kidd JR, Kidd KK, Speed WC, Pakstis AJ, Reed DR, Snyder DJ, Bartoshuk LM: Bitter receptor gene (TAS2R38), 6-n-propylthiouracil (PROP) bitterness, and alcohol intake. Alcohol Clin Exp Res 2004;28:1629–1637.
58 Bartoshuk LM: Bitter taste of saccharin: related to the genetic ability to taste the bitter substance 6-n-propylthiouracil (PROP). Science 1979;205:934–935.
59 Bartoshuk LM, Duffy VB, Miller IJ: PTC/PROP tasting: anatomy, psychophysics, and sex effects. Physiol Behav 1994;56:1165–1171.
60 Harris H, Kalmus H: Chemical sensitivity in genetical differences of taste sensitivity. Ann Eugen 1949;15:32–45.
61 Kranzler HR, Skipsey K, Modesto-Lowe V: PROP taster status and parental history of alcohol dependence. Drug Alcohol Depend 1998;52:109–113.
62 Mattes RD: 6-n-Propylthiouracil taster status: dietary modifier, marker, or misleader? in Prescott J, Tepper BJ (eds): Genetic Variation in Taste Sensitivity. New York, Dekker, 2004, pp 229–250.
63 Likert R: A technique for the measurement of attitudes. Arch Psychol 1932;22:1–55.
64 Jones LV, Peryam DR, Thurstone LL: Development of a scale for measuring soldiers' food preferences. Food Res 1955;20:512–520.
65 Aitken RCB: Measurement of feelings using visual analogue scales. Proc R Soc Med 1969;62: 989–993.
66 Stevens SS, Galanter EH: Ratio scales and category scales for a dozen perceptual continua. J Exp Psychol 1957;54:377–411.
67 Lasagna L: The clinical measurement of pain. Ann NY Acad Sci 1960;86:28–37.
68 Moskowitz HR: Magnitude estimation: notes on what, how, when, and why to use it. J Food Qual 1977;1:195–228.
69 Gracely RH, McGrath P, Dubner R: Validity and sensitivity of ratio scales of sensory and affective verbal pain descriptors: manipulation of affect by diazepam. Pain 1978;5:19–29.
70 Borg GAV: Psychophysical scaling with applications in physical work and the perception of exertion. Scand J Work Environ Health 1990;16(Suppl 1):55–58.
71 Green BG, Shaffer GS, Gilmore MM: A semantically labeled magnitude scale of oral sensation with apparent ratio properties. Chem Senses 1993;18:683–702.
72 Schutz HG, Cardello AV: A labeled affective magnitude (LAM) scale for assessing food liking/disliking. J Sens Stud 2001;16:117–159.
73 Borg GAV: A category scale with ratio properties for intermodal and interindividual comparisons; in Geissler HG, Petzold P (eds): Psychophysical Judgment and the Process of Perception. Berlin, Deutscher Verlag der Wissenschaften, 1982, pp 25–34.
74 Teghtsoonian R: Range effects in psychophysical scaling and a revision of Stevens' law. Am J Psychol 1973;86:3–27.
75 Bartoshuk LM, Snyder DJ: Psychophysical measurement of human taste experience; in Stricker EM, Woods SC (eds): Handbook of Behavioral Neurobiology. New York, Plenum Press, 2004, vol 14: Neurobiology of Food and Fluid Intake, pp 89–107.

76 Bartoshuk LM, Green BG, Hoffman HJ, Ko C-W, Lucchina LA, Snyder DJ, Weiffenbach JM: Valid across-group comparisons with labeled scales: the gLMS versus magnitude matching. Physiol Behav 2004;82:109–114.
77 Duffy VB, Peterson JM, Bartoshuk LM: Associations between taste genetics, oral sensation, and alcohol intake. Physiol Behav 2004;82:435–445.
78 Algom D: Memory psychophysics: an examination of its perceptual and cognitive prospects; in Algom D (ed): Psychophysical Approaches to Cognition. Amsterdam, Elsevier, 1992, pp 441–513.
79 Stevenson RJ, Prescott J: Judgments of chemosensory mixtures in memory. Acta Psychol (Amst) 1997;95:195–214.
80 Fast K, Green BG, Bartoshuk LM: Developing a scale to measure just about anything: comparisons across groups and individuals. Appetite 2002;39:75.
81 Bartoshuk LM, Fast K, Snyder DJ: Differences in our sensory worlds: invalid comparisons with labeled scales. Curr Dir Psychol Sci 2005;14:122–125.
82 Deems DA, Doty RL, Settle RG, Moore-Gillon V, Shaman P, Mester AF, Kimmelman CP, Brightman VJ, Snow JB: Smell and taste disorders: a study of 750 patients from the University of Pennsylvania Smell and Taste Center. Arch Otolaryngol Head Neck Surg 1991;117:519–528.
83 Mattes RD, Cowart BJ: Dietary assessment of patients with chemosensory disorders. J Am Diet Assoc 1994;94:50–56.
84 Bull TR: Taste and the chorda tympani. J Laryngol Otol 1965;79:479–493.
85 Bromley SM, Doty RL: Clinical disorders affecting taste: evaluation and management; in Doty RL (ed): Handbook of Olfaction and Gustation, ed 2. New York, Dekker, 2003.
86 Pritchard TC, Norgren R: Gustatory system; in Paxinos G, Mai JK (eds): The Human Nervous System, ed 2. Amsterdam, Elsevier, 2004, pp 1171–1196.
87 McBurney DH: Taste, smell, and flavor terminology: taking the confusion out of fusion; in Meiselman HL, Rivlin RS (eds): Clinical Measurement of Taste and Smell. New York, Macmillan, 1986, pp 117–125.
88 Tepper BJ, Christensen CM, Cao J: Development of brief methods to classify individuals by PROP taster status. Physiol Behav 2001;73:571–577.
89 Frank ME, Hettinger TP, Barry MA: Contemporary measurement of human gustatory function; in Doty RL (ed): Handbook of Olfaction and Gustation, ed 2. New York, Dekker, 2003.
90 Hummel T, Erras A, Kobal G: A test for screening of taste function. Rhinology 1997;35:146–148.
91 Ahne G, Erras A, Hummel T, Kobal G: Assessment of gustatory function by means of tasting tablets. Laryngoscope 2000;110:1396–1401.
92 Mueller C, Kallert S, Renner B, Stiassny K, Temmel AF, Hummel T, Kobal G: Quantitative assessment of gustatory function in a clinical context using impregnated 'taste strips'. Rhinology 2003;41:2–6.
93 Blakeslee AF, Fox AL: Our different taste worlds. J Hered 1932;23:97–107.
94 Bartoshuk LM, Duffy VB, Reed DR, Williams A: Supertasting, earaches, and head injury: genetics and pathology alter our taste worlds. Neurosci Biobehav Rev 1996;20:79–87.
95 Cooper DS: Antithyroid drugs. N Engl J Med 2005;352:905.
96 Zhao L, Kirkmeyer SV, Tepper BJ: A paper screening test to assess genetic taste sensitivity to 6-n-propylthiouracil. Physiol Behav 2003;78:625–633.
97 Hartmann G: Application of individual taste difference towards phenyl-thio-carbamide in genetic investigations. Ann Eugen 1939;9:123–135.
98 Ly A, Drewnowski A: PROP (6-n-propylthiouracil) tasting and sensory responses to caffeine, sucrose, neohesperidin dihydrochalcone, and chocolate. Chem Senses 2001;26:41–47.
99 Miller IJ, Reedy FE: Variations in human taste bud density and taste intensity perception. Physiol Behav 1990;47:1213–1219.
100 Miller IJ: Variation in human fungiform taste bud densities among regions and subjects. Anat Rec 1986;216:474–482.
101 Gairns FW: Sensory endings other than taste buds in the human tongue. J Physiol 1953;121: 33P–34P.
102 Essick GK, Chopra A, Guest S, McGlone F: Lingual tactile acuity, taste perception, and the density and diameter of fungiform papillae in female subjects. Physiol Behav 2003;80:289–302.

103 Snyder DJ, Fast K, Bartoshuk LM: Valid comparisons of suprathreshold sensations. J Conscious Stud 2004;11:96–112.
104 Todrank J, Bartoshuk LM: A taste illusion: taste sensation localized by touch. Physiol Behav 1991;50:1027–1031.
105 Delwiche JF, Lera MF, Breslin PA: Selective removal of a target stimulus localized by taste in humans. Chem Senses 2000;25:181–187.
106 Green BG: Studying taste as a cutaneous sense. Food Qual Pref 2002;14:99–109.
107 Bartoshuk LM: Clinical psychophysics of taste. Gerodontics 1988;4:249–255.
108 Hänig DP: Zur Psychophysik des Geschmackssinnes; PhD thesis, Leipzig, 1901.
109 Boring EG: Sensation and Perception in the History of Experimental Psychology. New York, Appleton, 1942.
110 Snow JB, Doty RL, Bartoshuk LM, Getchell TV: Categorization of chemosensory disorders; in Getchell TV, Doty RL, Bartoshuk LM, Snow JB (eds): Smell and Taste in Health and Disease. New York, Raven Press, 1991, pp 445–447.
111 Bradley RM: Electrophysiological investigations of intravascular taste using perfused rat tongue. Am J Physiol 1973;224:300–304.
112 Alfano M: The origin of gingival fluid. J Theor Biol 1974;47:127–136.
113 Fetting JH, Wilcox PM, Sheidler VR, Enterline JP, Donehower RC, Grochow LB: Tastes associated with parenteral chemotherapy for breast cancer. Cancer Treat Rep 1985;69:1249–1251.
114 Hausser-Hauw C, Bancaud J: Gustatory hallucinations in epileptic seizures. Brain 1987;110: 339–359.
115 Halpern BP, Nelson LM: Bulbar gustatory responses to anterior and to posterior tongue stimulation in the rat. Am J Physiol 1965;209:105–110.
116 Ninomiya Y, Funakoshi M: Responsiveness of dog thalamic neurons to taste stimulation of various tongue regions. Physiol Behav 1982;29:741–745.
117 Catalanotto FA, Bartoshuk LM, Östrum KM, Gent JF, Fast K: Effects of anesthesia of the facial nerve on taste. Chem Senses 1993;18:461–470.
118 Kveton JF, Bartoshuk LM: The effect of unilateral chorda tympani damage on taste. Laryngoscope 1994;104:25–29.
119 Yanagisawa K, Bartoshuk LM, Catalanotto FA, Karrer TA, Kveton JF: Anesthesia of the chorda tympani nerve and taste phantoms. Physiol Behav 1998;63:329–335.
120 Norgren R: Gustatory system; in Paxinos G (ed): The Human Nervous System. New York, Academic Press, 1990, pp 845–861.
121 Bartoshuk LM, Kveton JF, Yanagisawa K, Catalanotto FA: Taste loss and taste phantoms: a role of inhibition in taste; in Kurihara K, Suzuki N, Ogawa H (eds): Olfaction and Taste XI. Tokyo, Springer, 1994, pp 557–560.
122 Tie K, Fast K, Kveton JF, Cohen ZD, Duffy VB, Green BG, Prutkin JM, Bartoshuk LM: Anesthesia of chorda tympani nerve and effect on oral pain. Chem Senses 1999;24:609.
123 Grushka M, Sessle BJ: Burning mouth syndrome: a historical review. Clin J Pain 1987;2:245–252.
124 Grushka M, Bartoshuk LM: Burning mouth syndrome and oral dysesthesias. Can J Diagn 2000;17:99–109.
125 Ship JA, Grushka M, Lipton JA, Mott AE, Sessle BJ, Dionne RA: Burning mouth syndrome: an update. J Am Dent Assoc 1995;126:842–853.
126 Grushka M, Epstein J, Mott A: An open-label, dose escalation pilot study of the effect of clonazepam in burning mouth syndrome. Oral Surg Oral Med Oral Path Oral Radiol Endod 1998;86:557–561.

Derek J. Snyder
Center for Smell and Taste
University of Florida, PO Box 100127
Gainesville, FL 32610 (USA)
Tel. +1 352 273 5794, Fax +1 352 273 5257, E-Mail derek.snyder@yale.edu

Hummel T, Welge-Lüssen A (eds): Taste and Smell. An Update.
Adv Otorhinolaryngol. Basel, Karger, 2006, vol 63, pp 242–254

Postoperative/Posttraumatic Gustatory Dysfunction

Basile Nicolas Landis, Jean-Silvain Lacroix

Unité de Rhinologie-Olfactologie, Service d'Oto-Rhinologie-Laryngologie et de Chirurgie Cervico-Faciale, Hôpitaux Universitaires de Genève, Genève, Suisse

Abstract

Clinical taste testing in humans is far from being routinely performed in ear, nose and throat (ENT) clinics. Consequently, most reports on posttraumatic and postoperative taste disorders are case reports and mainly consist of qualitative (e.g. dysgeusia, metallic taste) taste changes after either head injury or ENT surgery. Since quantitative taste deficiencies (ageusia, hypogeusia) often go unnoticed by the patients, the real incidence of ageusia and hypogeusia after head trauma and various surgical procedures remains largely unknown. This lack of reliable clinical data is partly due to the lack of easy, reproducible and rapid clinical taste testing devices. The present chapter tries to resume the current knowledge on postoperative and posttraumatic taste disorders. Despite the sparse literature, the chapter focuses on those ENT surgical procedures where at least some prospective and systematic studies on gustatory dysfunction exist. Accordingly, taste disorders after middle ear surgery, tonsillectomy and dental interventions are largely discussed.

Postoperative Gustatory Dysfunction

History

The scientific debate on the pathways and nerves carrying the taste fibers started about 200 years ago [1]. There has been a controversy for at least 50–70 years about which nerves are the main taste fibers [2, 3]. The glossopharyngeal, trigeminal and facial nerve were all held for 'the' taste nerve [1]. A number of authors such as Bernard [4], Alcock [5], and Bellingeri [7] pointed out that the facial nerve and especially the chorda tympani could be involved in taste function. Others like Lewis and Dandy [3], Lussana [6] and Magendie [8] associated gustatory function with the trigeminal nerve. Lussana [2], in reference to his teacher Panizza (see Witt et al. [1]), first clearly identified the trigeminal nerve to provide somatosensory supply for the oral cavity, the chorda tympani to provide gustatory supply for the anterior part and the glossopharyngeal nerve for the posterior part of the tongue, respectively. This was based on observations in subjects with lesions of either the trigeminal nerve, the facial nerve or the chorda tympani [2]. He further corroborated his findings with animal experiments in dogs with various degrees of surgically induced damage to each of the nerves mentioned above [2]. A very nice review on the different pathways proposed by several authors was provided by Lewis and Dandy [3] in 1930.

To conclude this paragraph, it has to be retained that the major knowledge on the anatomical distribution of taste fibers is rather recent and has been possible by clinical observations and taste tests on patients with distinct lesions of the cranial nerves V, VII, IX and X.

Subjective Postoperative Complaints

Patients seeking help because of a taste alteration usually turn out to be suffering from an olfactory problem rather than an isolated gustatory problem [9, 10]. Since 'taste' and 'flavor' are synonyms in the current language, a decrease in flavor perception will lead to a complaint described as 'taste loss'. This is mainly due to the influence of retronasal olfaction on flavor perception and underlines the importance of psychophysical taste and smell testing. Unfortunately, many studies only rely on the patients' complaints and standardized tests have not been available for a long time. Since taste and smell disorders may often occur simultaneously [11, 12], each chemosensory modality should be evaluated separately before drawing any conclusion.

Quantitative Gustatory Disorder

Similar to olfactory disorders [13], gustatory complaints can be divided into two categories. Taste may either be quantitatively or qualitatively compromised. This categorization has proven to be useful in clinical routine. In analogy to olfaction, a total loss of gustatory function would be termed ageusia, while hypogeusia describes a partial loss and normogeusia stands for normal taste function. Isolated losses of any taste modalities are very rare but have been described [14], such as the inability to perceive sweet, while sour, bitter and salty can be tasted [15].

Qualitative Gustatory Disorder

The two main qualitative gustatory disorders are mainly parageusia and phantogeusia. Parageusia is a bad taste elicited by the nutritional intake, which is otherwise absent [16]. Phantogeusia describes the presence of a permanent intraoral bad taste [17, 18].

Causes

Middle Ear Surgery

The facial nerve carries the gustatory innervation for the anterior two thirds of the tongue and the palate [19]. The gustatory fibers run with the chorda tympani which separates itself from the facial nerve within the temporal bone. The chorda tympani also carries parasympathetic fibers for the main salivary glands of the oral cavity [20]. The chorda tympani further travels unprotected through the middle ear cavity. During middle ear surgery, the chorda is freely exposed within the operative field. It is subject to considerable surgical stress by stretch, injury or dryness, or it is directly sectioned in order to facilitate the surgical approach to the ossicles. Accordingly, following ear surgery, lesions of the gustatory system may produce symptoms such as dysgeusia, hypogeusia, ageusia or mouth dryness [3, 20–25]. Most of these postoperative gustatory affections have been shown to be transitory. However, long-lasting dysgeusia cases have been reported [26].

Systematic investigation of taste functions revealed little and often transitory postoperative subjective complaints, but considerable alterations of measurable ipsilateral taste sensitivity. Saito et al. [26] found the postoperative taste alteration to correlate with the peroperative surgical stress. Although most patients do not present long-term complaints, absent or incomplete taste recovery has been observed in the majority of the investigated subjects. Especially the group with a surgically severed chorda tympani exhibited the worst long-term outcome. Attempts to readapt the sectioned ends of the chorda increase the chances of recovery in taste function after surgery [27, 28]. Macroscopic reexamination of the chorda tympani in operated subjects revealed relatively high rates of intact nerves [29]. However, microscopic analyses of such recovered nerves have shown low numbers of intact and myelinated fibers, but mainly fibrosis.

Considering the frequency of otologic surgery and especially the number of bilaterally operated patients (i.e. for otosclerosis), very few complaints of ageusia, hypogeusia and dry mouth are reported [20, 22]. One of the main reasons for the low frequency of complaints appears to be related to the so-called 'release of inhibition' phenomenon [30, 31]. Gustatory afferent inputs from cranial nerves VII, IX, and X converge within the solitary nucleus of the medulla, whereas input from the anterior or posterior portion of the tongue is topographically separated. When the chorda tympani is anesthetized, activity disappears in brain stem regions usually responding to stimulation of the anterior part of the tongue, while the responsiveness to stimulation of the posterior parts of the tongue increases. Thus, subjective reports of whole-mouth gustatory function may not reflect regional taste function. Even lateralized ageusia may go unrecognized by the patient. However, eating is a whole-mouth experience built of gustatory, somatosensory and trigeminal inputs. Thus, it has been shown that lack of gustatory function can be partly masked by touch [32]. Such 'taste illusions' might also account for the poor subjective recognition of taste deficiencies. Beside the injury due to the surgical procedure, there is some evidence that the chorda tympani is already altered in subjects suffering from chronic inflammatory middle ear diseases [33–36]. This is probably due to the aggressive behavior of some middle ear inflammatory processes such as the cholesteatoma which is known to erode adjacent structures in the middle ear [37].

In conclusion, patients undergoing middle ear surgery mostly exhibit altered ipsilateral taste function. However, in the overwhelming majority of cases, this alteration goes unnoticed by the patients and is transitory. Unfortunately, in some patients, dysgeusia persists following middle ear surgery and the therapeutically available options are limited. Although the mechanism of action is not clear, zinc gluconate has recently been shown in a double-blind, placebo-controlled study to improve taste function in idiopathic dysgeusia [38].

Tonsillectomy/Oropharyngeal Surgery

In contrast to middle ear surgery, taste disorders after tonsillectomy have been systematically investigated only by Tomita and Ohtuka [39]. However, during the last 20 years, several reports have been available about this complication [40, 41]. Compared to the very high frequency at which this surgery is performed, taste complaints are extremely rare. After more than 3,500 tonsillectomies, Tomita and Ohtuka [39] observed only 11 cases (0.3%) of taste problems reported by the patients. These data mainly rely on subjective patient reports. Similar to the gap between subjective and measurable taste disorders following middle ear surgery, the 'release of inhibition' phenomenon probably also accounts for this low occurrence of posttonsillectomy gustatory complaints. However, since eating is a whole-mouth experience built out of gustatory, somatosensory and trigeminal inputs, most quantitative localized taste disorders go unnoticed.

A prospective study measuring the taste acuity of the posterior third of the tongue before and after tonsillectomy might unravel more psychophysical taste disturbance than reported by patients. This assumption is mainly based on a large anatomical study conducted by Ohtsuka et al. [42], who examined over 100 tonsillar beds with respect to their vicinity to the lingual branch of the glossopharyngeal nerve (LBGN). Their study revealed that in approximately a quarter of the cases, the LBGN traveled covered and separated from the tonsil by a muscle layer over its whole course to the base of the tongue. In almost 50% of the cases, the muscle lining of the tonsillar bed was discontinuous and only thin muscle bundles covered the tonsillar capsule and the LBGN. Most interestingly, in nearly 25% of cases, the LBGN was firmly adherent to the tonsillar capsule, due to the complete absence of muscle lining between the tonsillar bed and the LBGN. In these cases, and probably also in a similar percentage of patients undergoing tonsillectomy, taste disturbance may occur on removal of the hypertrophic tonsillar capsule. It may be assumed that such taste alterations, in analogy to chorda tympani-related gustatory disorders, are also transitory. In the few cases so far presented, no therapy could be offered [39–41].

Gustatory dysfunctions after all other kinds of oropharyngeal surgery have been described [43–45]. Beside oncologic surgery (see below), it is mainly the sleep apnea surgery which has been reported to alter chemosensory function [46–51]. The most frequent interventions followed by gustatory complaints are laser uvulopalatoplasty and uvulopalatopharyngoplasty. Although taste perception changes were recorded in all studies, psychophysical taste and smell testing was performed in only one study [50]. No changes in gustatory or olfactory function were reported by these authors. Since no taste buds have been described on the uvula, this finding is not surprising [52]. The obviously present changes in taste perception after sleep apnea surgery could be due to a modification in the retronasal airflow pattern [53–56], modifying 'flavor' perception which is widely attributed to 'taste'. Thus, the reported changes rather reflect olfactory changes. Another possibility is the lingual compression (see below) during surgery which could temporarily alter peripheral taste nerve and bud function. Further studies on patients undergoing sleep apnea surgery could clarify these discrepancies between subjects' reports and measured gustatory function.

Oncologic Surgery and Radiation Therapy

In contrast to the previously mentioned surgical procedures, oncologic surgery is far more devastating and taste disorders are frequently reported side effects [57–60]. However, patients barely complain about chemosensory disorders in view of the serious treatment and the severity of the disease. Resection of extensive parts of the oral and cervical structures usually also involves loss of taste fibers. This chapter is not dedicated to enumerate the various head and neck surgical procedures; however, we would like to focus on a special group of patients. Laryngectomized subjects loose their whole flavor perception due to interruption of the naso-laryngo-tracheal continuity. Since the negative pressure necessary for nasal breathing can no longer be produced by laryngectomized patients, ortho- and retronasal olfaction promptly disappears with laryngectomy [61, 62]. This leads to a subjective loss of 'taste' in these patients, which can potentially influence nutritional habits. Unlike patients who underwent other head and neck operations, laryngectomized patients can be helped by relatively simple devices and techniques to temporarily restore ortho- and retronasal olfaction. Orthonasal olfaction can be restored by so-called larynx bypasses consisting of plastic tubes connecting the tracheostoma and the mouth [63–68], while retronasal olfaction is partly possible by a movement termed as 'polite yawning' [69].

In addition to the usually mutilating and heavy head and neck surgery, radiotherapy aggravates the gustatory function [70, 71]. Taste disorders after radiotherapy are mainly due to fibrosis and/or necrosis of the salivary glands and the taste buds [57, 58, 70–73]. After radiotherapy, taste buds normally regenerate after months while mouth dryness and salivary gland necrosis seem to persist [72, 74]. Recent therapeutic options could remedy this situation [74].

Microlaryngoscopy, Tracheal Intubation and Other Procedures with Lingual Compression

Several surgical procedures related to any kind of compression exerted on the tongue during more than several minutes have been described to produce gustatory disorders. Microlaryngoscopy or suspension laryngoscopy have been reported to alter lingual sensitivity and taste [75–77]. Most taste and somatosensory complaints are temporary and cases of persistent gustatory disorders seem to be rare. Tracheal intubation [45, 78–80] and the use of a laryngeal mask [81–84] seem to be at the origin of several cases with transitory or persistent lingual nerve damage. Considering the high frequency of surgical procedures and anesthesias done, these complications are very rarely reported. Similar to laryngoscopy, the most likely mechanisms include anterior displacement of the mandible during insertion of the oropharyngeal airway tubing, compression of the nerve against the mandible, and stretching of the nerve over the hypoglossus by the cuff of the orotracheal tube. Prospective studies investigating the question of lingual compression and dysgeusia or ageusia should further clarify the real occurrence of such complications.

Dental Procedures

Different types of gustatory complaints have been reported after dental procedures. Unfortunately, most reports are based on case studies and few systematic studies have been undertaken to rule out the frequency and reversibility of such taste disturbances after dental procedures [85, 86]. There is a difference between taste disorders occurring after the use of a dental prosthesis or denture and taste disorders due to an oral surgical act. Compared to the number of dental prostheses, cast alloys and dentures used within the general population, taste problems after such oral devices are very rare [86, 87]. Moreover, Garhammer et al. [86] investigated subjects with dysgeusia after the use of a prosthesis and alloys and found, in approximately 10% of these cases, allergies towards the used materials to be responsible for the taste disorders [86]. Apart from allergic reactions, age and gender seem to influence taste disorders due to dentures, with women over 55 years and elderly having more taste problems [86, 88]. In most cases, the cause of dysgeusia remains unclear. However, a possible explanation in this patient group is that the use of dentures further weakens the trophic oral balance [89, 90].

Although gustatory disorders after oral and dental surgery have frequently been reported, the vast majority of the literature is based on case studies [91–93]. Beside these few reports, some authors have conducted larger studies in order to evaluate the impact of orthognathic surgery (Le Fort I osteotomy and sagittal split osteotomy) [94] as well as surgical removal of all four third molars on taste function [85, 95]. These studies, in which taste was also psychophysically measured, showed a transitory decrease in taste function for the tongue or the palate, respectively. Within 6–9 months, taste function returned to preoperative values; however, injury to the chorda tympani seemed to be accompanied by less disturbances than laceration of the greater superficial petrosal branch of the facial nerve which provides palatal taste function [85, 94–96]. The few systematic reports seem to underline that taste disturbances are a transitory and often unrecognized phenomenon. However, most smell and taste outpatient clinics have some experience with patients complaining of persistent dysgeusia or mouth perception changes, such as changed saliva consistency or mouth burning [91, 93]. Due to the lack of larger studies, it remains speculative whether these disturbances are related to the anesthesia used [91, 96], the reactivation of any viral infections following the surgical stress [97], or the presence of anatomically aberrant branches of the chorda tympani or the glossopharyngeal nerve [93, 98]. Unfortunately, no curative therapy for such taste disturbances exists so far. This is probably closely related to the largely unknown origin of these complaints. However, it has been shown that the mental status tends to influence the long-term regression of such complaints [99] and zinc gluconate has recently been shown to improve chronic idiopathic dysgeusia [38].

Preoperative Patient Information

In light of these gustatory and mouth perception complications after intraoral surgical procedures, minimal preoperative patient information needs to be provided. This is not very time consuming and may prevent medicolegal claims. Fortunately, most complications occur rarely and surgical alteration of taste function often goes unnoticed by the patient. Taken together, the sense of taste has a great capacity to compensate for partial loss of function. Thus, most surgeons may never be confronted with this complication in their patients. However, we strongly recommend to inform patients about the small risk reported in the literature of persistent ageusia, hypogeusia, dysgeusia or even changed mouth perception (e.g. mouth burning). Since these complications occur so rarely, they might be considered to occur independently of the surgeon and his skills. Our experience during the last years in the smell and taste outpatient clinic was that the few cases who presented with dysgeusia after a surgery were particularly upset, not about the complaint itself, but rather about the way the surgeon handled postoperative care. Most patients were told that this would be a transitory disorder. After a while, they felt as if their complaints were considered to be psychiatric symptoms by the surgeon who wanted to get rid of the 'unsuccessful' patients. However, by shortly informing the patients preoperatively that this complication exists, is rare, often transitory but sometimes persistent and not really treatable, such postoperative problems between patients and surgeon could be avoided.

This particularly accounts for middle ear surgery in patients with no preexisting inflammatory process in the middle ear cavity. Patients with otosclerosis have been reported to be more prone [23, 35, 100] to develop dysgeusia than patients with cholesteatoma [35], probably because patients with cholesteatoma already have a damaged chorda and altered taste [33, 36, 101]. In patients with middle ear problems, preoperative assessment of taste might also be considered for medicolegal reasons.

Posttraumatic Gustatory Dysfunction

History and Taste versus Smell Disorder

Ogle [102], who was among the first authors to describe posttraumatic anosmia, stated that the patient he examined '...complained not only of loss of smell, but also of loss of taste'. As previously mentioned, this might solely be attributed to anosmia. Like this report, most cases of chemosensory disorders after brain and head injuries have never undergone psychophysical testing. To our knowledge, the first to investigate the extent of posttraumatic ageusia was Sumner in 1967 [103]. Based on an excellent review of the literature on posttraumatic chemosensory complaints and his own experiments, he concludes that in 9 of 10 cases ageusia is related to anosmia and real ageusia seems to be rather rare. A few years later, Schechter and Henkin [104] examined patients following head injury with quite similar results. Beside several case studies, today no larger study has been reconducted in order to better characterize posttraumatic taste disorders.

Causes and Injuries

Posttraumatic taste disorders can be due to accidentally caustic ingestion [105], brain injury [106] and most often head injury [107, 108]. Beside injury of central structures responsible for taste and smell processing such as the frontotemporal or entorhinal cortex [109], complex and important fractures of the skull base or midface with squeezing or disruption of the cranial nerves VII, IX or X account for posttraumatic taste losses. Anecdotally, 'positive' taste changes have been reported after a head injury. In 2 patients, food aversions disappeared obviously after a blow [110], and another 2 patients developed 'gourmand syndrome' after a head trauma [111]. Taken together, this suggests that even subtle brain lesions may lead to changes in gustatory function, although this seems to be a rare event.

The more peripheral taste injuries due to severe facial and skull base fractures might be more frequent than reported. However, patients who suffer such severe traumas usually present posttraumatic syndromes. Thus, taste disorders may simply go unnoticed due to their relatively minor impact on quality of life compared to other more invalidating complaints. Further studies are needed to strengthen the weak database on posttraumatic taste disorders.

References

1 Witt M, Reutter K, Miller IJ Jr: Morphology of the peripheral taste system; in Doty RL (ed): Handbook of Olfaction and Gustation. New York, Dekker, 2003, pp 651–678.
2 Lussana P: Observations pathologiques sur les nerfs du goût. Arch Physiol Norm Pathol 1869;2:20–32.
3 Lewis D, Dandy WE: The course of the nerve fibers transmitting sensation of taste. Arch Surg 1930;21:249–288.
4 Bernard C: Recherches anatomiques et physiologiques sur la chorde du tympanium. Ann Méd Psychol 1848;1:408.
5 Alcock B: Determination of the question, which are the nerves of taste. Dublin J Med Chem Sci 1836;10:256–279.
6 Lussana P: Observations pathologiques sur les nerfs du goût. Arch Physiol Norm Pathol 1869;2:197–209.
7 Bellingeri: De nervus facialis. Turin, 1818.
8 Magendie F: Précis élémentaire de physiologie. Paris, Méquignon-Marvis, 1822.
9 Deems DA, Doty RL, Settle RG, Moore-Gillon V, Shaman P, Mester AF, Kimmelman CP, Brightman VJ, Snow JB Jr: Smell and taste disorders: a study of 750 patients from the University of Pennsylvania Smell and Taste Center. Arch Otolaryngol Head Neck Surg 1991;117:519–528.
10 Fujii M, Fukazawa K, Hashimoto Y, Takayasu S, Umemoto M, Negoro A, Sakagami M: Clinical study of flavor disturbance. Acta Otolaryngol Suppl 2004;553:109–112.
11 Hummel T, Nesztler C, Kallerf S, Kobal G, Bende M, Nordin S: Gustatory sensitivity in patients with anosmia. Chem Senses 2001;26:118.
12 Probst-Cousin S, Rickert CH, Kunde D, Schmid KW, Gullotta F: Paraneoplastische limbische Enzephalitis. Pathologe 1997;18:406–410.
13 Landis BN, Hummel T, Lacroix JS: Basic and clinical aspects of olfaction. Adv Tech Stand Neurosurg 2005;30:69–106.

14 Fox AL: Six in ten 'tasteblind' to bitter chemical. Sci News Lett 1931;19:249.
15 Henkin RI, Shallenberger RS: Aglycogeusia: the inability to recognize sweetness and its possible molecular basis. Nature 1970;227:965–966.
16 Rutherfoord GS, Mathew B: Xanthogranuloma of the choroid plexus of lateral ventricle, presenting with parosmia and parageusia. Br J Neurosurg 1987;1:285–288.
17 Petzold GC, Einhaupl KM, Valdueza JM: Persistent bitter taste as an initial symptom of amyotrophic lateral sclerosis. J Neurol Neurosurg Psychiatry 2003;74:687–688.
18 Nocentini U, Giordano A, Castriota-Scanderbeg A, Caltagirone C: Parageusia: an unusual presentation of multiple sclerosis. Eur Neurol 2004;51:123–124.
19 Henkin RI, Christiansen RL: Taste localization on the tongue, palate, and pharynx of normal man. J Appl Physiol 1967;22:316–320.
20 Grant R, Miller S, Simpson D, Lamey PJ, Bone I: The effect of chorda tympani section on ipsilateral and contralateral salivary secretion and taste in man. J Neurol Neurosurg Psychiatry 1989;52: 1058–1062.
21 Bull TR: Taste and the chorda tympani. J Laryngol Otol 1965;79:479–493.
22 Moon CNJ, Pullen EW: Effects of chorda tympani section during middle ear surgery. Laryngoscope 1963;73:392–405.
23 House HP: Early and late complications of stapes surgery. Arch Otolaryngol 1963;78:606–613.
24 Ho WYH: Disturbance of taste of otitic origin with special reference to operations on the ear. Arch Otolaryngol 1937;26:146–169.
25 Just T, Homoth J, Graumuller S, Pau HW: Schmeckstörung und Erholung der Schmeckfunktion nach Mittelohroperationen. Laryngorhinootologie 2003;82:494–500.
26 Saito T, Manabe Y, Shibamori Y, Yamagishi T, Igawa H, Tokuriki M, Fukuoka Y, Noda I, Ohtsubo T, Saito H: Long-term follow-up results of electrogustometry and subjective taste disorder after middle ear surgery. Laryngoscope 2001;111:2064–2070.
27 Saito T, Shibamori Y, Manabe Y, Yamagishi T, Igawa H, Yamamoto T, Ohtsubo T, Saito H: Intraoperative identification of regenerated chorda tympani nerve and its relationship to recovered taste function. ORL J Otorhinolaryngol Relat Spec 2001;63:359–365.
28 Saito T, Shibamori Y, Manabe Y, Yamagishi T, Igawa H, Ohtsubo T, Saito H: Incidence of regeneration of the chorda tympani nerve after middle ear surgery. Ann Otol Rhinol Laryngol 2002;111: 357–363.
29 Saito T, Shibamori Y, Manabe Y, Yamagishi T, Yamamoto T, Ohtsubo T, Saito H: Morphological and functional study of regenerated chorda tympani nerves in humans. Ann Otol Rhinol Laryngol 2000;109:703–709.
30 Halpern BP, Nelson LM: Bulbar gustatory responses to anterior and posterior tongue stimulation in the rat. Am J Physiol 1965;209:105–110.
31 Lehman CD, Bartoshuk LM, Catalanotto FC, Kveton JF, Lowlicht RA: Effect of anesthesia of the chorda tympani nerve on taste perception in humans. Physiol Behav 1995;57:943–951.
32 Todrank J, Bartoshuk LM: A taste illusion: taste sensation localized by touch. Physiol Behav 1991;50:1027–1031.
33 Arnold SM: The vulnerability of the chorda tympani nerve to middle ear disease. J Laryngol Otol 1974;88:457–466.
34 Yaginuma Y, Kobayashi T, Takaska T: Predictive value of electrogustrometry in the preoperative diagnosis of the severity of middle ear pathology; in Sanna M (ed): Cholesteatoma and Mastoid Surgery: Proceeding of the Fifth International Conference on Cholesteatoma and Mastoid Surgery. Rome, CIC Edizioni Internazionali, 1997, pp 843–846.
35 Sone M, Sakagami M, Tsuji K, Mishiro Y: Younger patients have a higher rate of recovery of taste function after middle ear surgery. Arch Otolaryngol Head Neck Surg 2001;127:967–969.
36 Landis BN, Beutner D, Frasnelli J, Huttenbrink KB, Hummel T: Gustatory function in chronic inflammatory middle ear diseases. Laryngoscope 2005;115:1124–1127.
37 Schechter G: A review of cholesteatoma pathology. Laryngoscope 1969;79:1907–1920.
38 Heckmann SM, Hujoel P, Habiger S, Friess W, Wichmann M, Heckmann JG, Hummel T: Zinc gluconate in the treatment of dysgeusia – A randomized clinical trial. J Dent Res 2005;84:35–38.
39 Tomita H, Ohtuka K: Taste disturbance after tonsillectomy. Acta Otolaryngol Suppl 2002;546: 164–172.

40 Goins MR, Pitovski DZ: Posttonsillectomy taste distortion: a significant complication. Laryngoscope 2004;114:1206–1213.
41 Uzun C, Adali MK, Karasalihoglu AR: Unusual complication of tonsillectomy: taste disturbance and the lingual branch of the glossopharyngeal nerve. J Laryngol Otol 2003;117:314–317.
42 Ohtsuka K, Tomita H, Murakami G: Anatomy of the tonsillar bed: topographical relationship between the palatine tonsil and the lingual branch of the glossopharyngeal nerve. Acta Otolaryngol Suppl 2002;546:99–109.
43 Zuniga JR, Chen N, Phillips CL: Chemosensory and somatosensory regeneration after lingual nerve repair in humans. J Oral Maxillofac Surg 1997;55:2–14.
44 Scrivani SJ, Moses M, Donoff RB, Kaban LB: Taste perception after lingual nerve repair. J Oral Maxillofac Surg 2000;58:3–6.
45 Lang MS, Waite PD: Bilateral lingual nerve injury after laryngoscopy for intubation. J Oral Maxillofac Surg 2001;59:1497–1499.
46 Croft CB, Golding-Wood DG: Uses and complications of uvulopalatopharyngoplasty. J Laryngol Otol 1990;104:871–875.
47 Rombaux P, Hamoir M, Bertrand B, Aubert G, Liistro G, Rodenstein D: Postoperative pain and side effects after uvulopalatopharyngoplasty, laser-assisted uvulopalatoplasty, and radiofrequency tissue volume reduction in primary snoring. Laryngoscope 2003;113:2169–2173.
48 Walker RP, Gopalsami C: Laser-assisted uvulopalatoplasty: postoperative complications. Laryngoscope 1996;106:834–838.
49 Hagert B, Wikblad K, Odkvist L, Wahren LK: Side effects after surgical treatment of snoring. ORL J Otorhinolaryngol Relat Spec 2000;62:76–80.
50 Badia L, Malik N, Lund VJ, Kotecha BT: The effect of laser assisted uvulopalatoplasty on the sense of smell and taste. Rhinology 2001;39:103–106.
51 Goffart Y: Side effects and complications of velopharyngeal surgery. Acta Otorhinolaryngol Belg 2002;56:163–175.
52 Ambro BT, Pribitkin EA, Wysocki L, Brand JG, Keane WM: Evidence that taste buds are not present on the human adult uvula. Ear Nose Throat J 2002;81:456–457.
53 Heilmann S, Strehle G, Rosenheim K, Damm M, Hummel T: Clinical assessment of retronasal olfactory function. Arch Otolaryngol Head Neck Surg 2002;128:414–418.
54 Landis BN, Giger R, Ricchetti A, Leuchter I, Hugentobler M, Hummel T, Lacroix JS: Retronasal olfactory function in nasal polyposis. Laryngoscope 2003;113:1993–1997.
55 Raudenbush B, Meyer B: Effect of nasal dilators on pleasantness, intensity and sampling behaviors of foods in the oral cavity. Rhinology 2001;39:80–83.
56 Büttner A, Beer A, Hannig C, Settles M: Observation of the swallowing process by application of videofluoroscopy and real-time magnetic resonance imaging – Consequences to aroma perception. Chem Senses 2001;26:1211–1219.
57 de Graeff A, de Leeuw JR, Ros WJ, Hordijk GJ, Blijham GH, Winnubst JA: Long-term quality of life of patients with head and neck cancer. Laryngoscope 2000;110:98–106.
58 Kokal WA: The impact of antitumor therapy on nutrition. Cancer 1985;55(1 Suppl):273–278.
59 Prince S, Bailey BM: Squamous carcinoma of the tongue: review. Br J Oral Maxillofac Surg 1999;37:164–174.
60 Zotti P, Lugli D, Vaccher E, Vidotto G, Franchin G, Barzan L: The EORTC quality of life questionnaire-head and neck 35 in Italian laryngectomized patients. European organization for research and treatment of cancer. Qual Life Res 2000;9:1147–1153.
61 Ritter F: Fate of olfaction after laryngectomy. Arch Otolaryngol 1964;79:169–171.
62 Henkin RI, Hoye RC, Ketcham AS, Gould WJ: Hyposmia following laryngectomy. Lancet 1968;ii:479–481.
63 Mozell MM, Schwartz DN, Youngentob SL, Leopold DA, Hornung DE, Sheehe PR: Reversal of hyposmia in laryngectomized patients. Chem Senses 1986;11: 397–410.
64 Schwartz DN, Mozell MM, Youngentob SL, Leopold DL, Sheehe PR: Improvement of olfaction in laryngectomized patients with the larynx bypass. Laryngoscope 1987;97:1280–1286.
65 Edwards N: Swimming by laryngectomees. J Laryngol Otol 1981;95:535–536.
66 Honeysett J: Swimming aids for laryngectomees. Nurs Times 1981;77:1045–1046.
67 Gray RF: Swimming after laryngectomy. Laryngoscope 1982;92:815–817.

68 Landis BN, Giger R, Lacroix JS, Dulguerov P: Swimming, snorkeling, breathing, smelling, and motorcycling after total laryngectomy. Am J Med 2003;114:341–342.
69 Hilgers FJ, van Dam FS, Keyzers S, Koster MN, van As CJ, Muller MJ: Rehabilitation of olfaction after laryngectomy by means of a nasal airflow-inducing maneuver: the 'polite yawning' technique. Arch Otolaryngol Head Neck Surg 2000;126:726–732.
70 Mossman KL, Henkin RI: Radiation-induced changes in taste acuity in cancer patients. Int J Radiat Oncol Biol Phys 1978;4:663–670.
71 Bartoshuk LM: Chemosensory alterations and cancer therapies. NCI Monogr 1990;9:179–184.
72 Alexander C, Bader JB, Schaefer A, Finke C, Kirsch CM: Intermediate and long-term side effects of high-dose radioiodine therapy for thyroid carcinoma. J Nucl Med 1998;39:1551–1554.
73 Rothwell BR: Prevention and treatment of the orofacial complications of radiotherapy. J Am Dent Assoc 1987;114:316–322.
74 Seikaly H, Jha N, Harris JR, Barnaby P, Liu R, Williams D, McGaw T, Rieger J, Wolfaardt J, Hanson J: Long-term outcomes of submandibular gland transfer for prevention of postradiation xerostomia. Arch Otolaryngol Head Neck Surg 2004;130:956–961.
75 Klussmann JP, Knoedgen R, Wittekindt C, Damm M, Eckel HE: Complications of suspension laryngoscopy. Ann Otol Rhinol Laryngol 2002;111:972–976.
76 Gaut A, Williams M: Lingual nerve injury during suspension microlaryngoscopy. Arch Otolaryngol Head Neck Surg 2000;126:669–671.
77 Woschnagg H, Stollberger C, Finsterer J: Loss of taste is loss of weight. Lancet 2002;359:891.
78 Brimacombe J: Bilateral lingual nerve injury following tracheal intubation. Anaesth Intensive Care 1993;21:107–108.
79 Kadry MA, Popat MT: Lingual nerve injury after use of a cuffed oropharyngeal airway. Eur J Anaesthesiol 2001;18:264–266.
80 Loughman E: Lingual nerve injury following tracheal intubation. Anaesth Intensive Care 1983; 11:171.
81 Ahmad NS, Yentis SM: Laryngeal mask airway and lingual nerve injury. Anaesthesia 1996;51: 707–708.
82 Brain AI, Howard D: Lingual nerve injury associated with laryngeal mask use. Anaesthesia 1998;53:713–714.
83 Gaylard D: Lingual nerve injury following the use of the laryngeal mask airway. Anaesth Intensive Care 1999;27:668.
84 Sommer M, Schuldt M, Runge U, Gielen-Wijffels S, Marcus MA: Bilateral hypoglossal nerve injury following the use of the laryngeal mask without the use of nitrous oxide. Acta Anaesthesiol Scand 2004;48:377–378.
85 Akal UK, Kucukyavuz Z, Nalcaci R, Yilmaz T: Evaluation of gustatory function after third molar removal. Int J Oral Maxillofac Surg 2004;33:564–568.
86 Garhammer P, Schmalz G, Hiller KA, Reitinger T, Stolz W: Patients with local adverse effects from dental alloys: frequency, complaints, symptoms, allergy. Clin Oral Investig 2001;5:240–249.
87 McHenry KR: Oral prosthesis and chemosensory taste function. A review of the literature. NY State Dent J 1992;58:36–38.
88 Wayler AH, Perlmuter LC, Cardello AV, Jones JA, Chauncey HH: Effects of age and removable artificial dentition on taste. Spec Care Dentist 1990;10:107–113.
89 Chauncey HH, Muench ME, Kapur KK, Wayler AH: The effect of the loss of teeth on diet and nutrition. Int Dent J 1984;34:98–104.
90 Wayler AH, Muench ME, Kapur KK, Chauncey HH: Masticatory performance and food acceptability in persons with removable partial dentures, full dentures and intact natural dentition. J Gerontol 1984;39:284–289.
91 Hotta M, Endo S, Tomita H: Taste disturbance in two patients after dental anesthesia by inferior alveolar nerve block. Acta Otolaryngol Suppl 2002;546:94–98.
92 Paxton MC, Hadley JN, Hadley MN, Edwards RC, Harrison SJ: Chorda tympani nerve injury following inferior alveolar injection: a review of two cases. J Am Dent Assoc 1994;125:1003–1006.
93 Schwankhaus JD: Bilateral anterior lingual hypogeusia hypesthesia. Neurology 1993;43:2146.
94 Gent JF, Shafer DM, Frank ME: The effect of orthognathic surgery on taste function on the palate and tongue. J Oral Maxillofac Surg 2003;61:766–773.

95 Shafer DM, Frank ME, Gent JF, Fischer ME: Gustatory function after third molar extraction. Oral Surg Oral Med Oral Pathol Oral Radiol Endod 1999;87:419–428.
96 Rehman K, Webster K, Dover MS: Links between anaesthetic modality and nerve damage during lower third molar surgery. Br Dent J 2002;193:43–45.
97 Tazi M, Soichot P, Perrin D: Facial palsy following dental extraction: report of 2 cases. J Oral Maxillofac Surg 2003;61:840–844.
98 Du Toit DF: Nervus lingualis: applied anatomical relevance to dental practice and oral surgery. SADJ 2003;58:207, 210–212.
99 Deems DA, Yen DM, Kreshak A, Doty RL: Spontaneous resolution of dysgeusia. Arch Otolaryngol Head Neck Surg 1996;122:961–963.
100 Sakagami M, Sone M, Tsuji K, Fukazawa K, Mishiro Y: Rate of recovery of taste function after preservation of chorda tympani nerve in middle ear surgery with special reference to type of disease. Ann Otol Rhinol Laryngol 2003;112:52–56.
101 Griffith IP: Observations on the pathology of chorda tympani nerve in temporal bone sections. J Laryngol Otol 1977;91:151–157.
102 Ogle W: Anosmia: cases illustrating the physiology and pathology of the sense of smell. Lancet 1870;i:230–231.
103 Sumner D: Post-traumatic ageusia. Brain 1967;90:187–202.
104 Schechter PJ, Henkin RI: Abnormalities of taste and smell after head trauma. J Neurol Neurosurg Psychiatry 1974;37:802–810.
105 Wason S: The emergency management of caustic ingestions. J Emerg Med 1985;2:175–182.
106 Yousem DM, Geckle RJ, Bilker WB, Kroger H, Doty RL: Posttraumatic smell loss: relationship of psychophysical tests and volumes of the olfactory bulbs and tracts and the temporal lobes. Acad Radiol 1999;6:264–272.
107 Renzi G, Carboni A, Gasparini G, Perugini M, Becelli R: Taste and olfactory disturbances after upper and middle third facial fractures: a preliminary study. Ann Plast Surg 2002;48:355–358.
108 van Damme PA, Freihofer HP: Disturbances of smell and taste after high central midface fractures. J Craniomaxillofac Surg 1992;20:248–250.
109 Costanzo RM, Zasler ND: Head trauma; in Snow JBJ (ed): Smell and Taste in Health and Disease. New York, Raven Press, 1991, pp 711–730.
110 Fujii M, Fujita K, Hiramatsu H, Miyamoto T: Cases of two patients whose food aversions disappeared following severe traumatic brain injury. Brain Inj 1998;12:709–713.
111 Regard M, Landis T: 'Gourmand syndrome': eating passion associated with right anterior lesions. Neurology 1997;48:1185–1190.

Basile N. Landis
Unité de Rhinologie-Olfactologie, Service d'Oto-Rhino-Laryngologie
Hôpitaux Universitaires de Genève
24, rue Micheli-du-Crest
CH–1211 Geneva (Switzerland)
Tel. +41 22 372 82 62, Fax +41 22 372 82 40, E-Mail Basile.Landis@hcuge.ch

Hummel T, Welge-Lüssen A (eds): Taste and Smell. An Update.
Adv Otorhinolaryngol. Basel, Karger, 2006, vol 63, pp 255–264

Neurological Causes of Taste Disorders

J.G. Heckmann, C.J.G. Lang

Department of Neurology, University of Erlangen-Nuremberg, Erlangen, Germany

Abstract

In caring for patients with taste disorders, the clinical assessment should include complete examination of the cranial nerves and, in particular, gustatory testing. Neurophysiological methods such as blink reflex and masseter reflex allow the testing of trigeminofacial and trigeminotrigeminal pathways. Modern imaging methods (MRI and computed tomography) enable the delineation of the neuroanatomical structures which are involved in taste and their relation to the bony skull base. From a neurological point of view, gustatory disorders can result from damage at any location of the neural gustatory pathway from the taste buds via the peripheral (facial, glossopharyngeal and vagal nerve) and central nervous system (brainstem, thalamus) to its representation within the cerebral cortex. Etiopathogenetically, a large number of causes has to be considered, e.g. drugs and physical agents, cerebrovascular disorders including dissection of the carotid artery and pontine/thalamic lesions, space-occupying processes – in particular tumors compressing the cerebellopontine angle and the jugular foramen of the skull base – head trauma and skull base fractures, isolated cranial mononeuropathy (e.g. Bell's palsy) or polyneuropathy, epilepsy, dementia, multiple sclerosis and major depression. In addition to this, aging can also lead to diminished taste perception. Due to the broad differential diagnostic considerations, it is essential to look for additional, even mild, neurological signs and symptoms. Treatment must relate to the underlying cause. Zinc may be tried in idiopathic dysgeusia.

In caring for patients with taste disorders, an interdisciplinary approach implying neurological, dental, and otorhinolaryngological expertise seems to be of particular importance [1]. The neurologist has to perform a general neurological examination including inspection of the tongue and the oral cavity and testing glossal and oral motor and sensory function. In addition, under his supervision, different ancillary examinations are regularly used to specify the etiology of taste disorders.

Electrophysiological tests can be applied to identify abnormalities in the cranial nerve-brainstem pathways which is of importance in cases of trigeminal neuropathy, multiple sclerosis, or pontine lesions. For example, the blink reflex may be used to evaluate the integrity of the interaction between the trigeminal nerve, pontine brainstem, and facial nerve [2]. The integrity of the trigemino-brainstem-trigeminal pathway can be elaborated using the masseter reflex [3].

Imaging techniques are routinely used to demonstrate lesions in the taste pathway. In particular, using special sequences, MRI allows the cranial nerves to be imaged [4]. Furthermore, MRI applications provide significant information to identify type and etiology of a lesion. Analysis of mucosal blood flow in the oral cavity in combination with the assessment of autonomous, cardiocirculatory parameters appears to be useful in the diagnosis of autonomic disorders in burning mouth syndrome or in patients with inborn autonomic disorder both of which are associated with gustatory dysfunction [5, 6]. Microbiological cultures are indicated when fungal or bacterial infections are suspected.

Analysis of saliva should be performed as it constitutes the environment of taste receptors including both, transport of tastants to the receptor and protection of the taste receptor. Typical clinical investigations are sialometry and sialochemistry. In combination with the clinical assessment, these techniques provide information with regard to the individual salivary status [7].

The search for the cause of taste disorders should be guided by the question whether they are due to damage to the peripheral or central nervous system or whether they are caused by neurological disease with undetermined localization.

Peripheral Neurological Causes

Lesions of the peripheral nervous system can be confined to syndromes affecting the facial and/or the glossopharyngeal nerve. Epidemiologically, affections of the facial nerve are far more frequent than those of the glossopharyngeal nerve with the most frequent disease being idiopathic Bell's palsy. While patients typically complain about insufficient closing of the eyelid and numbness along the affected cheek, gustatory dysfunction can also be a leading, and sometimes even the earliest complaint. The etiopathogenesis of Bell's palsy is a matter of debate. Treatment with corticosteroids has been recommended [8]. Other causes of facial nerve lesions have to be considered, e.g. neuritis due to neuroborreliosis or zoster, or space-occupying processes in the cerebellopontine angle such as meningioma or neurinoma [8, 9]. Other causes include iatrogenic lesions, e.g. lengthy laryngoscopic manipulations [10].

Affections of the glossopharyngeal nerve not only lead to gustatory dysfunction but also to problems in swallowing (sore throat ipsilateral to the lesion) and pain in the depth of the pharynx. Characteristically, a back drop phenomenon can be seen, in combination with decreased sensitivity in the territory of the laryngeal nerve, and a diminished gag reflex. A frequent cause for lesions of the glossopharyngeal nerve is the dissection of cervical arteries. This is probably due to a space-occupying hematoma of the vessel's wall which in turn compresses the cranial nerve [11, 12]. In this situation, the glossopharyngeal nerve is often not affected in isolation, but together with other caudal cranial nerves. This leads to syndromes such as the Collet-Sicard syndrome, if cranial nerves IX, X, XI, and XII are affected [13].

Slow progression of the gustatory disorder may be indicative of neoplastic processes affecting the submandibular region or the skull base [14]. A complete neurological examination is mandatory because in a number of polyneuropathies facial, glossopharyngeal, or vagal nerves can be affected either alone or in combination, e.g. in diphtheria, porphyria, lupus, or amyloidosis [15, 16].

Neuralgia of the glossopharyngeal nerve is a rare condition. Leading complaints are pain attacks in the depth of the oral cavity, often triggered by mechanical stimulation of the oral mucosa. In this case, gustatory disorders are rarely observed, probably because nerve function is intact. The reason for glossopharyngeal neuralgia is thought to be an ephaptic phenomenon. Many patients benefit from neurovascular decompressive surgery [17].

Central Neurological Causes

Gustatory dysfunction due to central lesions is, by definition, caused by a disturbance in the taste pathway originating from the level of brainstem including the solitary nucleus up to its cortical representation. An isolated taste disorder due to a central nervous system lesion is rare. In most cases, gustatory symptoms are accompanied by signs and symptoms which, during the acute phase of the disease, are typically more serious than the gustatory symptoms. Thus, gustatory dysfunction is often not reported and the clinician has to look for it.

Recently, with the improvement of noninvasive imaging methods (e.g. functional MRI), new insights into the central gustatory pathway have been gained by analyzing clinical taste disorder phenomena and their topographical neuroanatomical representation [18, 19]. These findings indicate that, after entering the ipsilateral medulla oblongata and synapsing in the nucleus tractus solitarii, the gustatory pathway ascends in the central tegmental tract and not, as previously thought, in the medial lemniscus to the mesencephalon. At this level, some gustatory sensory fibers cross over to the contralateral side. They ascend further to the thalamus where the ventral posteromedial nucleus is the competent synapsing region. After synapsing at this level, gustatory fibers project to the corresponding hemisphere whereby the insular cortex, the frontal operculum, the opercular part of the superior temporal gyrus, and the inferior part of the pre- and postcentral gyri are crucial projection zones [19, 20]. Recently, the orbitofrontal cortex has received attention as a tertiary center for smell and taste. This area has strong amygdaloid connections and coordinates behavioral responses to attraction and aversion to smell, taste and other sensations [21].

Brainstem taste disorders manifest themselves as ipsilateral hemiageusia or hemihypogeusia due to lesions of the bulbar tegmentum at the level of the solitary tract, or due to a pontine lesion. The most frequent causes are demyelinating or cerebrovascular processes [22–30]. In both conditions, abnormalities of the blink and the masseter reflexes are additionally expected. Lesions in the mesencephalon rarely lead to hypogeusia or ageusia. However, this fact is a strong indicator that at least some of the gustatory fibers cross at this level. Etiologically, besides demyelinating processes, vascular traumatic lesions should also be considered [31].

Thalamic taste disorders have been known since 1934 when they were described by Adler [32]. She reported a patient with right-sided hemihypesthesia of the face and right-sided hypogeusia due to an autoptically diagnosed glioblastoma which infiltrated the left nucleus ventralis posteriomedially. This observation led her to state that the gustatory pathway is contralaterally represented in the thalamus. In recent studies on stroke patients, dysgeusia was detected contralateral to a thalamic or corona radiata infarction, thus supporting the theories that the gustatory fibers ascend contralaterally in the cerebral hemisphere and that the pathway ascends from the thalamus to the cerebral cortex via the posterior part of the corona radiata [25, 26]. However, there are reports which also found that an ipsilateral lesion of the thalamus can result in hemihypogeusia, thus supporting the theory of crossing fibers at the lower brainstem level [33]. With thalamic lesions, hedonic aspects have to be considered. Particularly in bilateral lesions, the loss of hedonic qualities may result in impaired appreciation of foods which, in turn, leads to changes in food intake followed by clinically significant, and unwanted, weight loss control [34].

Cortical taste disorder is more difficult to record by history and clinical examination. In pharmacoresistant epilepsy, approximately 4% of the patients report on gustatory auras probably due to focal abnormalities in the opercular parietal region [35]. These auras are mainly bilateral. In patients treated surgically for hippocampal sclerosis, gustatory auras persisted in many cases [36]. In another series of drug-resistant temporal lobe epilepsy, all seizures were found to invade the insular cortex, and in a minority of the cases the seizures originated in the insula itself. Clinically, it was not possible to differentiate ictal symptoms between the two types of seizures. However, a less accurate estimate of taste intensity was observed in patients with excisions from either the left or the right anteromedial temporal lobe. This emphasized the importance of the anterior temporal lobe in gustatory perception; further, in terms of recognition of 'bitter' taste, the right temporal lobe was superior to the left one [37]. Apart from epilepsy, other causes, mainly cerebrovascular and neoplastic, should be considered [38–41]. It is unclear to which extent gustatory dysfunction related to migraine, schizophrenia, major depression, or eating disorders is based on cortical dysfunction [42–44].

Taste disorders in multiple sclerosis can occur as lesions at the level of the brainstem, thalamus, or cortex, but are sparsely reported. In observational studies, alteration of taste (hypoageusia) has been found at a rate varying from 0.25 to 7% [45–47]. It is, however, assumed that taste disorders are more frequent but not detected because they are often only present for a short time or were not considered worthy of diagnostic investigation [45]. On the other hand, there are a number of case reports or case series on pronounced clinical symptomatology of taste disturbance associated with singular circumscribed plaque morphology [24, 33, 48]. It is important to note that ageusia or even parageusia can occur as one of the first symptoms of multiple sclerosis [45, 49].

Taste disorders in neurodegenerative diseases are mostly due to cortical dysfunction [50–52]. Pathophysiologically, the underlying disease regularly affects brain areas which are involved in gustation and goes along with changes in the neurotransmitter systems such as in acetylcholine or epinephrine which play an important role in the taste information processing. In early stages of Alzheimer's disease, an alteration of the associative level of gustatory information processing was found [51]. The authors used various foods for testing which are commonly found in regular diets. They described the taste disorder of the patients with Alzheimer's disease as associative agnosia. In a study using taste strips, patients with different types of dementia scored significantly lower than matched healthy controls. Furthermore, the severity of dementia correlated positively with the results of the taste strips test [52]. The findings on taste function in Parkinson's disease (PD) patients are not unequivocal. In the PD subgroup investigated by Lang et al. [52], PD patients exhibited a significant reduction of both olfactory and gustatory sensitivity. In contrast, Sienkiewicz-Jarosz et al. [53] reported that perceived intensity, pleasantness, and identification of tastants did not differ between PD patients and controls. These authors actually reported an enhanced taste acuity in terms of electrogustometric threshold. In particular, in PD patients, the aspect of saliva should be considered due to the rule 'without moisture, there is no taste' [21]. PD patients produce significantly less saliva than control subjects [54]. Taken together, there are hints that in neurodegenerative diseases both smell and taste can be impaired. These observations point to the close relation between the chemosensory senses itself and the relation between sensory and cognitive abilities [52].

Taste disorders in acute stroke in turn can occur with lesions at the level of the brainstem, thalamus, or cortex. There are some reports on pronounced hemihypogeusia or ageusia due to a circumscribed cerebrovascular lesion in the brainstem or thalamus [25–30, 34]. Recently, some reports have shown that cerebrovascular lesions of the insula are accompanied by taste disorders [40]. Kim and Choi [38] reported that food preferences may change as a consequence of stroke. In an own prospective, observational study on unselected first ischemic stroke patients, 30.4% of all patients revealed impairment of gustatory function, and 5.5% of all patients exhibited a lateralized impairment of taste (right-left difference >30%) using a standardized, validated test kit, the 'taste strips' [55]. Among the hypogeusic patients, there were significantly more males; moreover, they had a lower NIHSS score, and more frequently swallowing and smell disorders [Heckmann et al., 2005, unpublished data]. This observation led us to suggest that taste disorders following stroke are much more common than previously assumed. Therefore, taste disorders should be considered in acute stroke patients, especially with regard to nutritional aspects of rehabilitation.

Ageing

A discrete taste loss in the elderly is frequent but rarely causes significant clinical problems. Taken alone, the decreased gustatory sensitivity does not warrant an extensive search for a taste disorder due to disease [56, 57]. Following quantitative gustatory testing plus appropriate clinical examination, patients can usually be counseled such that this problem is not serious and that the addition of seasonings to their food, tongue cleaning, or cessation of smoking in smokers might be helpful [58]. However, if other signs and symptoms are associated with the observed gustatory loss, in our view, a thorough workup is warranted.

Neurological Causes with Undetermined Localization

Within the context of clinical neurology, there are numerous conditions which present with gustatory dysfunction, but in which an exact topographical localization within the nervous system is not possible. Thus, there are reports on taste disorders in familiar dysautonomia, Machado-Joseph disease or Guillain-Barré syndrome probably due to disturbance of small nerve fibers [6, 59–61]. Taste disorders due to high-altitude sickness are speculated to be related to hypoxic damage of nerve fibers [62]. In some families with hereditary ataxia, genetically determined global thermoanalgesia or absence of fungiform papillae on the tongue have been reported [63]. Also, in craniofacial trauma, taste disorders are observed albeit being much less frequent than olfactory disorder. However, taste disorders are much more likely to improve spontaneously than olfactory dysfunction [64]. Recently, hypogeusia has been described as a prominent early feature of the new variant Creutzfeldt-Jakob disease probably due to deposits of prions in the central gustatory pathway [65]. In human rabies virus, antigen was demonstrated in the plexuses of the salivary glands. Thus, it can be speculated that taste disorder existed in early rabies before fatal encephalomyelitis progressed [66].

Treatment of Taste Disorders

As with olfactory disturbances, there are few therapeutic options. Treatment with zinc is frequently tried regardless of the fact that results of clinical studies are not unequivocal. In addition, both systemic corticoids and vitamin A have been used to treat taste disorders, despite the lack of convincing clinical studies [67, 68]. Thus, the main focus of therapy of gustatory disorders is the search for, and therapy of, possible underlying diseases. Local causes need appropriate dental, dermatological, or otorhinolaryngological care; underlying schizophrenia or depression demand psychiatric treatment. This approach also includes the thorough revision of drugs taken by the patient.

If there is no specific treatment option, in particular in idiopathic dysgeusia, a treatment course using zinc (140 mg/day for 4 months) may be promising [69].

In conclusion, many neurological diseases may cause taste disorders. While therapy is limited, the diagnostic armamentarium is available to identify such disorders and to determine most causes of gustatory dysfunction.

References

1 Heckmann JG, Heckmann SM, Lang CJG, Hummel T: Neurological aspects of taste disorders. Arch Neurol 2003;60:667–671.

2 Jaaskelainen SK, Forssell H, Tenovuo O: Abnormalities of the blink reflex in burning mouth syndrome. Pain 1997;73:455–460.

3 Fitzek S, Fitzek C, Hopf HC: The masseter reflex: postprocessing methods and influence of age and gender. Normative values of the masseter reflex. Eur Neurol 2001;46:202–205.

4 Lell M, Schmid A, Stemper B, Maihöfner C, Heckmann JG, Tomandl B: Simultaneous involvement of third and sixth cranial nerve in a patient with Lyme disease. Neuroradiology 2003;45:85–87.

5 Heckmann SM, Heckmann JG, Hilz MJ, Popp M, Marthol H, Neundörfer B, Hummel T: Oral mucosal blood flow in patients with burning mouth syndrome. Pain 2001;90:281–286.

6 Gadoth N, Mass E, Gordon CR, Steiner JE: Taste and smell in familial dysautonomia. Dev Med Child Neurol 1997;39:393–397.

7 Matsuo R: Role of saliva in the maintenance of taste sensitivity. Crit Rev Oral Biol Med 2000;11:216–229.

8 Roob G, Fazekas F, Hartung HP: Peripheral facial palsy: etiology, diagnosis and treatment. Eur Neurol 1999;41:3–9.
9 Schnarch A, Markitziu A: Dysgeusia, gustatory sweating, and crocodile tears syndrome induced by a cerebellopontine angle meningioma. Oral Surg Oral Med Oral Pathol 1990;70:711–714.
10 Woschnagg H, Stollberger C, Finsterer J: Loss of taste is loss of weight. Lancet 2002;359:891.
11 Heckmann JG, Tomandl B, Huk W, Neundörfer B: Dissection of extracranial arteries supplying the brain (in German). Dtsch Med Wochenschr 2000;125:1333–1336.
12 Taillibert S, Bazin B, Pierrot-Deseilligny C: Dysgeusia resulting from internal carotid dissection. A limited glossopharyngeal nerve palsy. J Neurol Neurosurg Psychiatry 1998;64:691–692.
13 Heckmann JG, Tomandl B, Duhm C, Stefan H, Neundörfer B: Collet-Sicard syndrome due to coiling and dissection of the internal carotid artery. Cerebrovasc Dis 2000;10:487–488.
14 Strong MJ, Noseworthy JH: Hemiageusia, hemianaesthesia and hemiatrophy of the tongue. Can J Neurol Sci 1986;13:109–110.
15 Fisher CM: Raeder's benign paratrigeminal syndrome with dysgeusia. Trans Am Neurol Assoc 1971;96:234–236.
16 Neundörfer B: Polyneuropathies (in German). Med Welt 2002;53:106–110.
17 Patel A, Kassam A, Horowitz M, Chang YF: Microvascular decompression in the management of glossopharyngeal neuralgia: analysis of 217 cases. Neurosurgery 2002;50:705–711.
18 Cerf B, Le Bihan D, Van de Moortele PF, Mac Leod P, Faurion A: Functional lateralization of human gustatory cortex related to handedness disclosed by fMRI study. Ann NY Acad Sci 1998;855:575–578.
19 Faurion A, Cerf B, Le Bihan D, Pillias AM: fMRI study of taste cortical areas in humans. Ann NY Acad Sci 1998;855:535–545.
20 Sanchez-Juan P, Combarros O: Gustatory nervous pathway syndromes (in Spanish). Neurologia 2001;16:262–271.
21 Hawkes C: Comment on Heckmann JG et al. 'Neurological aspects of taste disorders' (Arch Neurol 2003;60:667–671). J Watch Neurol 2003;5:78.
22 Pascual-Leone A, Altafullah I, Dhuna A: Hemiageusia: an unusual presentation of multiple sclerosis. J Neurol Neurosurg Psychiatry 1991;54:657.
23 Uesaka Y, Nose H, Ida M, Takagi A: The pathway of gustatory fibers of the human ascends ipsilaterally in the pons. Neurology 1998;50:827–828.
24 Combarros O, Sánchez-Juan P, Berciano J, De Pablos C: Hemiageusia from an ipsilateral multiple sclerosis plaque at the midpontine tegmentum. J Neurol Neurosurg Psychiatry 2000;68:796.
25 Fujikane M, Nakazawa M, Ogasawara M, Hirata K, Tsudo N: Unilateral gustatory disturbance by pontine infarction (in Japanese). Rinsho Shinkeigaku 1998;38:342–343.
26 Onoda K, Ikeda M: Gustatory disturbance due to cerebrovascular disorder. Laryngoscope 1999;109:123–128.
27 Sunada I, Akano Y, Yamamoto S, Tashiro T: Pontine hemorrhage causing disturbance of taste. Neuroradiology 1995;37:659.
28 Goto N, Yamamoto T, Kaneko M, Tomita H: Primary pontine haemorrhage and gustatory disturbance: clinicoanatomic study. Stroke 1983;14:507–511.
29 Nakajima Y, Utsimi H, Takahasi H: Ipsilateral disturbance of taste due to pontine haemorraghe. J Neurol 1983;229:133–136.
30 Lee BC, Hwang SH, Rison R, Chang GY: Central pathway of taste: clinical and MRI study. Eur Neurol 1998;39:200–203.
31 Johnson TM: Ataxic hemiparesis and hemiageusia from an isolated post-traumatic midbrain lesion. Neurology 1996;47:1348–1349.
32 Adler A: Zur Topik des Verlaufes der Geschmackssinnsfasern und anderer afferenter Bahnen im Thalamus. Z Ges Neurol Psychiatry 1934;149:208–220.
33 Combarros O, Miró J, Berciano J: Ageusia associated with thalamic plaque in multiple sclerosis. Eur Neurol 1994;34:344–346.
34 Rousseaux M, Muller P, Gahide I, Mottin Y, Romon M: Disorders of smell, taste, and food intake in a patient with a dorsomedial thalamic infarct. Stroke 1996;27:2328–2330.
35 Hausser-Hauw C, Bancaud J: Gustatory hallucinations in epileptic seizures. Electrophysiological, clinical and anatomical correlates. Brain 1987;110:339–359.

36 Fried I, Spencer DD, Spencer SS: The anatomy of epileptic auras: focal pathology and surgical outcome. J Neurosurg 1995;83:60–66.
37 Small DM, Zatorre RJ, Jones-Gotman M: Changes in taste intensity perception following anterior temporal lobe removal in humans. Chem Senses 2001;26:425–432.
38 Kim JS, Choi S: Altered food preference after cortical infarction: Korean style. Cerebrovasc Dis 2002;13:187–191.
39 Andre JM, Beis JM, Morin N, Paysant J: Buccal hemineglect. Arch Neurol 2000;57:1734–1741.
40 Pritchard TC, Macaluso DA, Eslinger PJ: Taste perception in patients with insular cortex lesions. Behav Neurosci 1999;113:663–671.
41 El-Dairy A, McCabe BF: Temporal lobe tumor manifested by localized dysgeusia. Ann Otol Rhinol Laryngol 1990;99:586–587.
42 Blau JN, Solomon F: Smell and other sensory disturbances in migraine. J Neurol 1985;232: 275–276.
43 Miller SM, Naylor GJ: Unpleasant taste – A neglected symptom in depression. J Affect Disord 1989;17:291–293.
44 Fahy TA, DeSilva P, Silverstone P, Russell GF: The effects of loss of taste and smell in a case of anorexia nervosa and bulimia nervosa. Br J Psychiatry 1989;155:860–861.
45 Nocentini U, Giordano A, Catriota-Scanderbeg A, Caltagirone C: Parageusia: An unusual presentation of multiple sclerosis. Eur Neurol 2004;51:123–124.
46 Rollin H: Gustatory disturbances in multiple sclerosis (in German). Laryngol Rhinol Otol (Stuttg) 1976;55:678–681.
47 Catalanotto FA, Dore-Duffy P, Donaldson JO, Testa M, Peterson M, Ostrom KM: Quality-specific taste changes in multiple sclerosis. Ann Neurol 1984;16:611–615.
48 Hisahara S, Makiyama Y, Yoshizawa T, Mizusawa H, Shoji S: A case of unilateral gustatory disturbance produced by the contralateral midbrain lesion (in Japanese). Rinsho Shinkeigaku 1994;34: 1055–1057.
49 Benatru I, Terraux P, Cherasse A, Couvreur G, Giroud M, Moreau T: Gustatory disorders during multiple sclerosis relapse (in French). Rev Neurol (Paris) 2003;159:287–292.
50 Schiffman SS, Clark CM, Warwick ZS: Gustatory and olfactory dysfunction in dementia: not specific to Alzheimer's disease. Neurobiol Aging 1990;11:597–600.
51 Broggio E, Pluchon C, Ingrand P, Gil R: Taste impairment in Alzheimer's disease. Rev Neurol (Paris) 2001;157:409–413.
52 Lang C, Leuschner T, Ulrich K, Stößel C, Heckmann J, Hummel T: Taste and smell in dementing diseases; in Korczyn AD (ed): Mental Dysfunction in Parkinson's Disease. Bologna, Medimond, 2004, pp 151–154.
53 Sienkiewicz-Jarosz H, Scinska A, Kuran W, Ryglewicz D, Rogowski A, Wrobel E, Korkosz A, Kukwa A, Kostowski W, Bienkowski P: Taste responses in patients with Parkinson's disease. J Neurol Neurosurg Psychiatry 2005;76:40–46.
54 Proulx M, De Courval FP, Wiseman MA, Panisset M: Salivary production in Parkinson's disease. Mov Disord 2005;20:204–207.
55 Mueller C, Kallert S, Renner B, Stiassny K, Temmel AFP, Hummel T, Kobal G: Quantitative assessment of gustatory function in a clinical context using impregnated 'taste strips'. Rhinology 2003;41:2–6.
56 Bartoshuk LM: Taste. Robust across the age span? Ann NY Acad Sci 1989;561:65–75.
57 Bartoshuk LM, Rifkin B, Marks LE, Bars P: Taste and aging. J Gerontol 1986;41:51–57.
58 Winkler S, Garg AK, Mekayarajjananonth T, Bakaee LG, Khan E: Depressed taste and smell in geriatric patients. J Am Dent Assoc 1999;130:1759–1765.
59 Uchiyama T, Fukutake T, Arai K, Nakagawa K, Hattori T: Machado-Joseph disease associated with an absence of fungiform papillae on the tongue. Neurology 2001;56:558–560.
60 Combarros O, Pascual J, de Pablos C, Ortega F, Berciano J: Taste loss as an initial symptom of Guillain-Barré syndrome. Neurology 1996;47:1604–1605.
61 Odaka M, Yuki N, Nishimoto Y, Hirata K: Guillain-Barré syndrome presenting with loss of taste. Neurology 2002;58:1437–1438.
62 Kassirer MR, Such RV: Persistent high-altitude headache and ageusia without anosmia. Arch Neurol 1989;46:340–341.

63 Fukutake T, Kita K, Sakakibara R, Takagi K, Tokumaru Y, Kojima S, Hattori T, Hirayama K: Late-onset hereditary ataxia with global thermoanalgesia and absence of fungiform papillae on the tongue in a Japanese family. Brain 1996;119:1011–1021.
64 Wienke A, Walter O: Loss of the sense of taste and smell after head and brain injury (in German). Laryngorhinootologie 2001;80:752.
65 Reuber M, Al-Din AS, Baborie A, Chakrabarty A: New variant Creutzfeldt-Jakob disease presenting with loss of taste and smell. J Neurol Neurosurg Psychiatry 2001;71:412–413.
66 Jackson AC, Ye H, Phelan CC, Ridaura-Sanz C, Zheng Q, Li Z, Wan X, Lopez-Corella E: Extraneural organ involvement in human rabies. Lab Invest 1999;79:945–951.
67 Henkin RI, Schecter PJ, Friedewald WT, Demets DL, Raff M: A double-blind study of the effects of zinc sulfate on taste and smell dysfunction. Am J Med Sci 1976;272:285–299.
68 Stoll AL, Oepen G: Zinc salts for the treatment of olfactory and gustatory symptoms in psychiatric patients: a case series. J Clin Psychiatry 1994;55:309–311.
69 Heckmann SM, Hujoel P, Habiger S, Friess W, Wichmann M, Heckmann JG, Hummel T: Zinc gluconate in the treatment of dysgeusia – A randomized clinical trial. J Dent Res 2005;84:35–38.

Josef G. Heckmann, MD
Department of Neurology, University of Erlangen-Nuremberg
Schwabachanlage 6
DE–91054 Erlangen (Germany)
Tel. +49 9131 853 3001, Fax +49 9131 853 4436
E-Mail josef.heckmann@neuro.imed.uni-erlangen.de

Hummel T, Welge-Lüssen A (eds): Taste and Smell. An Update.
Adv Otorhinolaryngol. Basel, Karger, 2006, vol 63, pp 265–277

Toxic Effects on Gustatory Function

Evan R. Reiter[a], *Laurence J. DiNardo*[a], *Richard M. Costanzo*[b]

[a]Department of Otolaryngology – Head and Neck Surgery, [b]Department of Physiology and Otolaryngology – Head and Neck Surgery, School of Medicine, Virginia Commonwealth University, Richmond, Va., USA

Abstract

A large number of substances and disease processes may impact the sense of taste. Toxic substances may cause taste dysfunction from their effects on the gustatory system from the salivary gland, to the taste bud, to the central neural pathways. A number of external toxins, including industrial compounds, tobacco, and alcohol, may adversely affect taste, most commonly through local effects in the oral cavity. Blood-borne toxins, such as medications and those present in autoimmune and other systemic disorders (e.g. renal or liver failure), have access to all parts of the gustatory system, and thus may exhibit varied effects on taste function. An understanding of these potential toxins and their impact on gustation will help physicians better recognize, and potentially limit the impact of such taste alterations on their patients.

Mechanisms of Toxic Alterations of Gustation

A wide variety of toxins may affect gustatory function. Depending on the nature of the toxin and its route of access into the body, either directly through the oral cavity or via the bloodstream, changes in gustatory function may be caused by a variety of mechanisms. The primary mechanisms include alterations in the composition or quantity of saliva, changes in the oral mucus membranes, direct effects on the taste buds, and modulation of peripheral or central neural pathways. In the following discussion, we review examples of each of these effects.

Alteration of Saliva

Saliva is a complex fluid that coats the surface of the oral cavity and provides the necessary ionic milieu for proper function of the taste receptor cells [1]. Saliva is secreted at a baseline level by three paired major salivary glands, the parotid, submandibular, and sublingual glands, as well as by the widely distributed minor salivary glands. However, production may vary under control of the autonomic nervous system. Interestingly, while saliva clearly influences taste reception, taste perception may also influence salivary secretion through reflex activation of the autonomic nervous system [2].

Saliva plays an important role in taste function [3]. Taste stimuli must diffuse through the saliva layer in order to gain access to the taste receptor cells [4]. Alterations in the amount and composition of saliva may result in changes in taste perception. Since saliva contains sufficient concentrations of sodium, potassium, bicarbonate and chloride to be detected by taste receptors, the taste system must adapt to baseline levels of stimulation. When the constituents of saliva are altered, changes in the response to taste stimuli occur [5]. Detection thresholds for sodium and potassium change when there is an increase or decrease in the saliva concentrations of these ions. Sourness and acid tastes may be altered when the buffering capacity of saliva is compromised by changes in bicarbonate levels. Salivary gland secretions may also decrease in specific autoimmune diseases such as Sjögren's syndrome [6]. In this case, autoimmune destruction of salivary glandular tissue leads to marked reduction in salivary flow, with resultant xerostomia. Patients with Sjögren's syndrome often complain of taste dysfunction most likely due to problems related to the solubilization and delivery of taste molecules to the receptor cells.

Mucosal Effects

Healthy oral mucosa is essential for normal gustatory function. The mucus membranes of the oral cavity are, however, directly exposed to the external environment, and thus susceptible to toxic effects from a variety of sources, both external and internal. Liquids and gases containing toxic agents that enter the oral cavity may have a direct or indirect negative impact on taste function. Some chemicals may burn the mucosa while others have toxic effects. Microorganisms, fungal growth and viral infections in the oral cavity can lead to localized inflammatory response that can alter taste function [6]. Tobacco and nicotine-containing products [7], or gastric contents from gastroesophageal reflux, may cause inflammation and irritation of the mucosa, and alter the taste of foods. Gastric acid in particular may produce a sour taste sensation typical of many acids. Lastly, a variety of autoimmune disorders such as Behçet's syndrome, pemphigus vulgaris, systemic lupus erythematosus, and scleroderma may alter the oral mucosa and thus taste function [8].

Taste Bud Dysfunction

A number of toxic substances may lead to alterations in taste receptor activation, or may even reduce the number of taste receptors present. Several toxins such as botulinum or tetrodotoxin found in the puffer fish can block normal receptor cell function by blocking receptor sites or conductance through ion channels. Chemotherapeutic agents that alter cell turnover also interfere with the normal turnover of taste receptor cells [9]. The same may be true of other toxins. Tobacco exposure has been shown to increase olfactory sensory neuron death. Continued exposure might thus eventually overwhelm the regenerative capacity of the epithelium, leading to reduction of receptors, and hyposmia [10]. The same effect may occur with continued or repetitive exposure of taste receptors to certain toxins.

Effects on Neural Pathways

Toxins and infections that may have direct effects on the taste nerves include botulism, herpes zoster, Bell's palsy, Guillain-Barré syndrome and poliomyelitis [11, 12]. Agents that produce CNS infection and inflammation such as is the case with herpes encephalitis and multiple sclerosis may also alter taste due to the involvement of central taste pathways [13, 14]. Dysgeusia is a common taste alteration arising from lesions in cortical taste structures [15, 16].

External Toxins

Medications

Medications can affect gustation through any of the mechanisms discussed above. Virtually all medications can elicit a taste response of their own, which can be perceived as unpleasant [17, 18]. Depending on the form of the medication (i.e. solution, tablet, or capsule), this is often through direct stimulation of taste receptors upon dissolution of the medication in the saliva. However, as some blood-borne agents can trigger gustatory sensations, termed intravascular gustation, there is a possibility that some medications may reach taste receptors through a more complex route [19]. Direct application of some medications to the tongue can also acutely alter or reduce taste perceptions elicited by known substances [17, 18]. For example, tricyclic antidepressants applied to the gerbil tongue were shown to acutely reduce chorda tympani responses to a variety of taste-stimulating solutions, including sodium chloride, quinine hydrochloride, sucrose, and citric acid [18]. This effect resolved after complete clearance of the drug. The authors state that this suggests a direct effect of these medications at the level of the peripheral receptors. Another possible mechanism of altered taste induced by medications is reduction of the regenerative capacity of taste receptor cells. This in turn might lead to temporary depletion of receptor cells and taste loss. Chemotherapeutic agents, which are well known to carry a high

Table 1. Common medications associated with taste disturbances

- Antimicrobials [17, 77]
  - Antifungals: amphotericin B, terbinafine
  - Antivirals/protease inhibitors: indinavir, ritonavir, saquinavir
  - β-Lactam antibiotics: penicillin, ampicillin
  - Other: metronidazole, tetracycline
- Anti-inflammatories [17]: diclofenac, nabumetone, sulindac
- Antihyperlipidemics [34]
  - Fibric acid derivatives: gemfibrozil
  - HMG-CoA reductaste inhibitors: atorvastatin, lovastatin, pravastatin, simvastatin
- Antihypertensives [34]
  - Angiotensin-converting enzyme inhibitors: captopril, enalopril, lisinopril, fosinopril
  - Angiotensin II receptor antagonists: losartan
  - Calcium channel blockers: amlodipine, diltiazem, nifedipine
- Antineoplastics [78, 79]: bleomycin, cisplatin, cytosine arabinoside, doxorubicin, 5-fluorouracil, methotrexate
- Antithyroid agents [80]: methimazole, propylthiouracil
- Diuretics [34]
  - Potassium-sparing: amiloride, spironolactone
  - Thiazide: hydrochlorothiazide
- Neurologic medications [81, 82]
  - Anticonvulsants: carbamazepine
  - Antiparkinson agents: levodopa
- Psychiatric medications [18, 33]
  - Antidepressants: amitriptyline, doxepin, imipramine, fluoxetine
  - Antipsychotics: BuSpar, lithium
  - Anxiolytics: buspirone, flurazepam, triazolam

risk of taste disturbances [20], and with their intended effects on cellular proliferation, might lead to such a phenomenon.

Many medications may alter taste due to their impact on saliva production. Agents such as antihistamines [21, 22], antidepressants [23], and diuretics [24] have been shown to affect salivary flow or composition. However, there is conflicting evidence in the literature as to the true impact of reduced salivary flow on taste function [25, 26]. No studies have shown direct effects of medications on the central pathways, although the large number of centrally acting drugs that are associated with taste alterations suggests that this is plausible. Two specific examples are cytosine arabinoside [27] and cisplatin [28, 29], both chemotherapeutic agents with known propensity toward neurotoxicity. As shown in table 1, medications used in the management of seizures, Parkinson's disease, and a variety of psychiatric disorders have all been reported to affect taste. Although in some cases a peripheral mechanism for taste disturbance has been shown [18, 30], it is likely that a portion of the taste effects of these agents may be attributable to alteration of activity in central taste pathways. At the very least, such agents may reduce the hedonic value of taste sensations, thus leading to subjective taste complaints and alterations in appetite [31].

For most medications, quantitative data on the incidence of taste disturbance are lacking. The Physicians' Desk Reference is likely the most widely used reference for medication usage and adverse reactions [32]. Taste disturbances are reported with descriptors including 'altered taste', 'dysgeusia', 'ageusia', 'bad taste', or 'metallic taste'. For some medications, data from controlled trials are available showing specific incidence rates for taste disturbances. However, for many others, the adverse reactions listed are derived from case reports or series in the literature, or from voluntary reports from individual physicians to the publishers. Even in cases of large controlled trials, the incidence of taste disturbance is often derived from patient reports rather than objective testing. As such, the true incidence of taste disturbances with usage of any given agent is difficult to determine. In addition, interactions between medications may also impact the likelihood of taste disturbance in patients on multiple agents.

As a group, the angiotensin-converting enzyme inhibitors, which include captopril, enalapril, fosinopril, and lisinopril, are probably the most common group of agents leading to taste disturbance. This stems both from their efficacy and thus widespread use in the management of hypertension and congestive heart failure, and the relatively high rate of taste disturbances associated with their usage, estimated from 1 to 8% [33]. While the mechanism of this effect is uncertain, one possibility is chelation of zinc at the receptor level [34]. Taste deficits reported from angiotensin-converting enzyme inhibitors include ageusia, dysgeusia, and metallic taste. While in most cases normal taste returned after cessation of the offending agent, rare reports exist of permanent dysgeusia or even ageusia [33]. Another class of medications with both significant rates of taste disturbance and widespread use are the HMG-CoA reductase inhibitors, or 'statin' antihyperlipidemic drugs, including atorvastatin, lovastatin, pravastatin, and simvastatin [34]. Controlled trials have, however, reported rates of taste disturbances less than 1% for these agents [32], and their mechanism of taste disturbance remains unknown.

Foods, Chemicals, Tobacco, Radiation

Taste disturbance secondary to external exposures has only recently begun to be explored. Three sources of toxins have emerged: foods, chemicals and radiation. Most toxic encounters are the result of human activity. Exposures that damage taste function, therefore, are likely to increase as the human impact on the environment grows.

Food containing toxins can possess an unpleasant taste. Occasionally, circulating toxins cause a generalized taste disturbance. Scombroid poisoning is an example [35]. This form of ichthyosarcotoxism is caused by the ingestion of affected dark-meat fish. In addition to parathesias, digestive tract symptoms, and headache, an unusual taste sensation is often reported. The exact mechanism is unknown and, although not a food allergy, patients respond to antihistamine therapy.

Taste alterations caused by cigarette smoking are thought to result from exposure to toxic chemicals. Gromysz-Kalkowska et al. [36] studied the effects of age, gender, and cigarette smoking on 471 randomly selected subjects. Taste testing revealed that smoking had a small, diversified, but in most cases, statistically insignificant effect on taste sensitivity.

Schiffman and Nagle [37] have reviewed the adverse effects of chemical toxins on taste. Frequent contact with pesticides is reported to induce a sustained metallic, bitter taste. Organophosphate-based pesticides, in particular, have been linked to a garlic odor and taste [37]. Evidence exists for both central and peripheral causes of pesticide-related taste disturbances. Organophosphates demonstrate disruption of sensory nerve terminations on taste buds in the experimental model but may also interfere with central neurotransmitters [38].

Metal workers have complained of metallic tastes as well, often coinciding with the particular metal being used. Brass pipe fitters [39] and jewelry workers suffering from lead poisoning [40], for example, frequently report a sweet metallic taste. A possible direct mechanism of action has been proposed since heavy metals are concentrated in saliva and their topical application to the tongue can blunt taste perception [41]. Indeed, an increase in taste threshold has been noted in chromium workers [42]. It is unknown if this effect is reversible. Administration of the chelator D-penicillamine has relieved symptoms in lead-contaminated silver workers [40].

The effects of solvents on taste are not well studied. In 1992, Hotz et al. [43] reported subjective smell and/or taste disturbances in a cross-sectional study of 264 workers exposed to organic solvents. The symptoms were mostly transient and related to concentration peaks rather than duration of exposure supporting an acute depressor effect.

Radiation exposure can occur from an ingested or external source. Radioactive iodine (I^{131}) is used in the treatment of differentiated thyroid cancers. Ingested I^{131} concentrates in salivary glands as well as the thyroid gland. Radiation sialadenitis with associated xerostomia, dental caries, facial nerve involvement, secondary infection, and taste alterations have been reported [44]. The use of sialogogic agents is thought to decrease the transit time of radioactive material in the salivary glands. However, their application and the pursuit of adequate hydration have not yet been proven to be efficacious. External beam head and neck radiation therapy often results in oral sequelae including mucositis, xerostomia, dental caries, and taste loss. Taste loss is not entirely attributable to xerostomia since mucositis and taste loss are usually reversible while hyposalivation persists [45]. Zheng et al. [46] prospectively evaluated 40 patients undergoing head and neck radiation therapy for taste disturbance. Four basic tastes were measured using the whole-mouth method before, during, and after therapy. Bitter taste was the most susceptible, while interference with sweet taste depended on inclusion of the anterior tongue in the radiation field. Salivary function was independent of taste disturbance, supporting the notion that radiation therapy directly damages taste receptors. Fisher et al. [47] have recently provided further support for these findings in a study that randomized radiation therapy patients to receive pilocarpine 5 mg four times daily or placebo during treatment. Although preservation of salivary function was statistically significant in the pilocarpine arm and these patients reported less mouth pain, subjective difficulty with swallowing, hyposalivation, and taste were similar in both groups. Targeted radiation therapy techniques, such as intensity-modulated radiotherapy, that limit taste receptor and salivary gland exposure when feasible, show promise in minimizing taste loss following treatment.

Internal Toxins

Liver Failure

Patients with liver failure from various causes, including cirrhosis, hepatitis, primary biliary cirrhosis, and sclerosing cholangitis, have been shown to have alterations in taste [48]. Taste disturbances tend to be more common than olfactory dysfunction in patients with liver disease. Patients tend to report reduced taste sensitivity, dysgeusia with taste aversions, as well as more taste cravings as compared to controls [48]. Taste disturbances may affect food preferences and thus contribute to the anorexia seen with liver failure, although no consensus exists in the literature [48, 49]. The reversibility of these changes was shown in patients undergoing successful liver transplantation [50]. However, the authors note that complete recovery in all taste modalities did not occur, and that many other factors other than liver function, such as medical regimen, micronutrient deficiencies, patient nutritional status, and psychological state, may have led to the improvements noted.

Several possible mechanisms exist for taste deficits in this population. Effects of altered zinc metabolism have been equivocal. Although plasma zinc levels were found to be low in cirrhotics, this did not correlate with the level of taste acuity, nor did zinc supplementation seem to positively impact taste function [51]. Vitamin A supplementation was found to improve both gustatory and olfactory acuity in patients with alcoholic cirrhosis and vitamin A deficiency, although the mechanism of this effect was unclear [52].

Uremia

Chronic renal failure has long been associated with disturbances of taste. Such disturbances can be significant, contributing to altered food preferences and reduced caloric intake [53], as well as reduced quality of life [54] in patients with renal failure. Impairment of taste has been found to affect patients with uremia, as well as patients on either hemodialysis or peritoneal dialysis [55]. The nature of the disturbances reported include the presence of a foul phantom taste, often termed the 'uremic taste', impaired taste recognition, and elevated taste detection thresholds. The mechanism of these changes is unclear. An immunohistochemical study comparing fungiform papillae taste buds in uremic patients, renal transplant recipients, and healthy controls showed no differences in taste bud architecture and innervation, although the patients with chronic renal failure had fewer taste buds [56]. Patients with chronic renal failure on hemodialysis were shown to have reduced unstimulated salivary flow rates, and increased salivary pH and buffer capacity compared with controls [56]. Many studies have focused on the possible role of abnormal zinc metabolism as a causative factor in dysgeusia in uremic patients [57]. Studies have shown a correlation between plasma zinc levels and taste acuity [58], improvement in taste following oral [59] or dialysate [60] zinc supplementation, and improvement in taste acuity paralleling normalization of zinc metabolism following renal transplantation [61]. Zinc supplementation seemed to have similar effects on nerve conduction velocity and taste acuity, suggesting that taste dysfunction in chronic renal failure might be another manifestation of the 'uremic neuropathy' [57, 60]. One notable but rare mechanism of taste disturbance is amyloid deposition in the tongue. Long-term dialysis is known to lead to amyloid deposition, usually in osteoarticular structures. However, this process may also lead to formation of amyloid nodules in the tongue, as found by Matsuo et al. [62] in about 8% of patients treated with dialysis for over 20 years. Over half of these patients reported lingual dysfunction, including taste disturbances, impaired tongue mobility, or articulation deficits. The effect of medications used by patients with renal failure also cannot be overlooked, as many patients with renal failure may require antihypertensive agents, including angiotensin-converting enzyme inhibitors [34], and diuretics, such as hydrochlorothiazide [63] or amiloride [64], all of which are known to affect gustation.

Diabetes

The prevalence of diabetes mellitus in the general population has been estimated to be between 1 and 5%, a figure that may rise with the increasing prevalence of obesity and increasing life span seen in western populations. Diabetics demonstrate a variety of gustatory and oral manifestations. Diabetics have increased electrical and chemical gustometric thresholds compared to controls [65].

In newly diagnosed diabetics, partial correction of hypogeusia was noted after initiation of antihyperglycemic therapy [66]. Differing opinions exist as to the association between gustatory deficits and neuropathy in diabetics, such that it is unclear that hypogeusia in diabetics is simply another manifestation of diabetic neuropathy [66]. The demonstration of reduced salivary flow rates suggests that this may be another contributory factor to hypogeusia in diabetics [67].

Thyroid Disease

Both hypo- and hyperthyroid states have been associated with disturbances of taste [68]. The spectrum of hypothyroid states ranges from asymptomatic laboratory finding to myxedema coma. In one series of patients with hypothyroidism, almost half had complaints of altered sense of taste, and about 40% had complaints of altered sense of smell [69]. Objective testing showed 83% had decreased acuity for at least one taste (salty, sweet, bitter, sour), although impaired detection or recognition of bitter taste was most common. Most complaints and objective deficits were found to reverse following medical correction of hypothyroidism [69]. Decreased salty and bitter sensations have been reported in hyperthyroid patients [70]. Although the mechanism for this observation is unclear, laboratory work has shown that thyroid hormones, such as thyroxine (T_4) or triiodothyronine (T_3), may have a competitive inhibitory effect on purified taste bud membrane adenosine 3′,5′-monophosphate phosphodiesterase activity [71].

In patients with differentiated thyroid cancer, radioiodine I^{131} administration is commonly used postoperatively to ablate residual thyroid tissue. As iodine is concentrated and secreted in the salivary glands, such treatment can lead to transient or even permanent xerostomia, sialadenitis, and taste disturbances. Mendoza et al. [72] found taste disturbances in approximately 25% of patients receiving radioactive iodine therapy. While 21% experienced acute xerostomia, 35% of patients receiving more than one radioiodine treatment reported xerostomia.

Autoimmune Disease

A number of diseases related to immune dysfunction have been reported to cause disturbances in taste. Due to the systemic nature of these illnesses, autoimmune processes can affect gustation through a number of different mechanisms. One of these disorders is Sjögren's syndrome, which is the second most common autoimmune disease behind rheumatoid arthritis. This disease is HLA-linked, and predominantly afflicts women in their third or fourth decades of life. The disease is characterized by progressive destruction of exocrine glands, including the salivary glands, which results from both lymphocytic infiltration and immune complex deposition. While the classic eccrine gland involvement results in xerophthalmia and xerostomia with taste alterations, virtually any organ system can be involved. Taste complaints seen with Sjögren's syndrome are primarily reduced sensitivity to tastes, which has been confirmed by objective testing [73]. While sensitivity to tastes is generally impaired, suprathreshold taste recognition is not. In addition, there appears to be a rather poor correlation between salivary flow and degree of taste disturbance, suggesting that a mechanism other than just reduced salivary flow may be present [73]. Some possibilities include altered oral bacterial flora, neuropathy [74], and chronic oral mucosal changes such as lingual fissuring [75].

Amyloidosis is defined as the extracellular deposition of proteinaceous material in various sites in the body. This may be localized or systemic, primary (without coexisting disease) or secondary (arising in the setting of chronic inflammatory disease or infection). Amyloid deposition has been noted in the tongue, in addition to the buccal mucosa, palate, and floor of mouth [76]. Lingual involvement may cause taste disturbance, but more urgently can lead to progressive airway compromise necessitating tracheotomy or lingual reductive surgery.

Conclusion

A variety of toxins may alter the sense of taste. Alterations in gustation may occur due to toxic effects on saliva production, the oral mucosa, taste receptor cells, and neural pathways. Common toxins include external factors such as foods, tobacco, chemicals, radiation, or medications, and blood-borne toxins as in liver or renal failure, diabetes, or autoimmune disease. In many cases, further investigation is needed to better understand the precise mechanisms of such toxic injuries to the components of the gustatory system, which might lead to better preventive or restorative therapies.

References

1 Bradley R, Beidler L: Saliva: its role in taste function; in Doty RL (ed): Handbook of Olfaction and Gustation. New York, Dekker, 2003, pp 639–650.
2 Kim M, Chiego DJ Jr, Bradley RM: Morphology of parasympathetic neurons innervating rat lingual salivary glands. Auton Neurosci 2004;111:27–36.
3 Bradley RM: Salivary secretion; in Getchell TV, Doty RL, Bartoshuk LM, Snow JB 2nd (eds): Smell and Taste in Health and Disease. New York, Raven Press, 1991, pp 127–144.
4 De Simone JA, Heck GL: An analysis of the effects of stimulus transport and membrane charge on the salt, acid and water-response of mammals. Chem Senses 1980;5:295–316.
5 Matsuo R, Yamamoto T: Effects of inorganic constituents of saliva on taste responses of the rat chorda tympani nerve. Brain Res 1992;583:71–80.

6 Bromley S, Doty RL: Clinical disorders affecting taste: evaluation and management; in Doty RL (ed): Handbook of Olfaction and Gustation. New York, Dekker, 2003, pp 935–957.
7 Reibel J: Tobacco and oral diseases. Update on the evidence, with recommendations. Med Princ Pract 2003;12(Suppl 1):22–32.
8 Campbell SM, Montanaro A, Bardana EJ: Head and neck manifestations of autoimmune disease. Am J Otolaryngol 1983;4:187–216.
9 Berteretche MV, Dalix AM, d'Ornano AM, Bellisle F, Khayat D, Faurion A: Decreased taste sensitivity in cancer patients under chemotherapy. Support Care Cancer 2004;12:571–576.
10 Vent J, Robinson AM, Gentry-Nielsen MJ, Conley DB, Hallworth R, Leopold DA, et al: Pathology of the olfactory epithelium: smoking and ethanol exposure. Laryngoscope 2004;114:1383–1388.
11 Odaka M, Yuki N, Nishimoto Y, Hirata K: Guillain-Barré syndrome presenting with loss of taste. Neurology 2002;58:1437–1438.
12 Brand N: Taste response and poliomyelitis. Ann Hum Genet 1964;27:233–239.
13 Asnis DS, Micic L, Giaccio D: Ramsay Hunt syndrome presenting as a cranial polyneuropathy. Cutis 1996;57:421–424.
14 Combarros O, Miro J, Berciano J: Ageusia associated with thalamic plaque in multiple sclerosis. Eur Neurol 1994;34:344–346.
15 Finsterer J, Stollberger C, Kopsa W: Weight reduction due to stroke-induced dysgeusia. Eur Neurol 2004;51:47–49.
16 Pritchard TC, Macaluso DA, Eslinger PJ: Taste perception in patients with insular cortex lesions. Behav Neurosci 1999;113:663–671.
17 Schiffman SS, Zervakis J, Westall HL, Graham BG, Metz A, Bennett JL, et al: Effect of antimicrobial and anti-inflammatory medications on the sense of taste. Physiol Behav 2000;69:413–424.
18 Schiffman SS, Zervakis J, Suggs MS, Budd KC, Iuga L: Effect of tricyclic antidepressants on taste responses in humans and gerbils. Pharmacol Biochem Behav 2000;65:599–609.
19 Maruniak JA, Silver WL, Moulton DG: Olfactory receptors respond to blood-borne odorants. Brain Res 1983;265:312–316.
20 Wickham RS, Rehwaldt M, Kefer C, Shott S, Abbas K, Glynn-Tucker E, et al: Taste changes experienced by patients receiving chemotherapy. Oncol Nurs Forum 1999;26:697–706.
21 Van Ganse E, Kaufman L, Derde MP, Yernault JC, Delaunois L, Vincken W: Effects of antihistamines in adult asthma: a meta-analysis of clinical trials. Eur Respir J 1997;10:2216–2224.
22 Sreebny LM, Valdini A, Yu A: Xerostomia. 2. Relationship to nonoral symptoms, drugs, and diseases. Oral Surg Oral Med Oral Pathol 1989;68:419–427.
23 von Knorring L, Mornstad H: Qualitative changes in saliva composition after short-term administration of imipramine and zimelidine in healthy volunteers. Scand J Dent Res 1981;89:313–320.
24 Beal AM: The effect of transport-blocking drugs on secretion of fluid and electrolytes by the mandibular gland of red kangaroos, *Macropus rufus*. Arch Oral Biol 1997;42:705–716.
25 Christensen CM, Navazesh M, Brightman VJ: Effects of pharmacologic reductions in salivary flow on taste thresholds in man. Arch Oral Biol 1984;29:17–23.
26 Mattes RD, Christensen CM, Engelman K: Effects of hydrochlorothiazide and amiloride on salt taste and excretion (intake). Am J Hypertens 1990;3:436–443.
27 Hoffman DL, Howard JR Jr, Sarma R, Riggs JE: Encephalopathy, myelopathy, optic neuropathy, and anosmia associated with intravenous cytosine arabinoside. Clin Neuropharmacol 1993;16: 258–262.
28 Fanning J, Hilgers RD: High-dose cisplatin carboplatin chemotherapy in primary advanced epithelial ovarian cancer. Gynecol Oncol 1993;51:182–186.
29 Wickham RS, Rehwaldt M, Kefer C, Shott S, Abbas K, Glynn-Tucker E, et al: Taste changes experienced by patients receiving chemotherapy. Oncol Nurs Forum 1999;26:697–706.
30 Schiffman SS, Graham BG, Suggs MS, Sattely-Miller EA: Effect of psychotropic drugs on taste responses in young and elderly persons. Ann NY Acad Sci 1998;855:732–737.
31 Treit D, Berridge KC: A comparison of benzodiazepine, serotonin, and dopamine agents in the taste-reactivity paradigm. Pharmacol Biochem Behav 1990;37:451–456.
32 Physicians' Desk Reference, ed 58. Montvale, Thomson PDR, 2004.
33 Ackerman BH, Kasbekar N: Disturbances of taste and smell induced by drugs. Pharmacotherapy 1997;17:482–496.

34 Doty RL, Philip S, Reddy K, Kerr KL: Influences of antihypertensive and antihyperlipidemic drugs on the senses of taste and smell: a review. J Hypertens 2003;21:1805–1813.
35 Muller GJ, Lamprecht JH, Barnes JM, De Villiers RV, Honeth BR, Hoffman BA: Scombroid poisoning. Case series of 10 incidents involving 22 patients. S Afr Med J 1992;81:427–430.
36 Gromysz-Kalkowska K, Wojcik K, Szubartowska E, Unkiewicz-Winiarczyk A: Taste perception of cigarette smokers. Ann Univ Mariae Curie Sklodowska 2002;57:143–154.
37 Schiffman SS, Nagle HT: Effect of environmental pollutants on taste and smell. Otolaryngol Head Neck Surg 1992;106:693–700.
38 Vij S, Kanagasuntheram R: Effect of tri-O-cresyl phosphate (TOCP) poisoning on sensory nerve terminations of slow loris (*Nycticebus coucang coucang*). Acta Neuropathol (Berl) 1972;20: 150–159.
39 Armstrong CW, Moore LW Jr, Hackler RL, Miller GB Jr, Stroube RB: An outbreak of metal fume fever. Diagnostic use of urinary copper and zinc determinations. J Occup Med 1983;25:886–888.
40 Kachru DN, Tandon SK, Misra UK, Nag D: Occupational lead poisoning among silver jewelry workers. Indian J Med Sci 1989;43:89–91.
41 Halpern BP: Environmental factors affecting chemoreceptors: an overview. Environ Health Perspect 1982;44:101–105.
42 Seeber H, Fikentscher R: Taste disorders in chromium-exposed workers. Z Gesamte Hyg 1990;36: 33–34.
43 Hotz P, Tschopp A, Soderstrom D, Holtz J, Boillat MA, Gutzwiller F: Smell or taste disturbances, neurological symptoms, and hydrocarbon exposure. Int Arch Occup Environ Health 1992;63: 525–530.
44 Mandel SJ, Mandel L: Radioactive iodine and the salivary glands. Thyroid 2003;13:265–271.
45 Vissink A, Jansma J, Spijkervet FK, Burlage FR, Coppes RP: Oral sequelae of head and neck radiotherapy. Crit Rev Oral Biol Med 2003;14:199–212.
46 Zheng WK, Inokuchi A, Yamamoto T, Komiyama S: Taste dysfunction in irradiated patients with head and neck cancer. Fukuoka Igaku Zasshi 2002;93:64–76.
47 Fisher J, Scott C, Scarantino CW, Leveque FG, White RL, Rotman M, et al: Phase III quality-of-life study results: impact on patients' quality of life to reducing xerostomia after radiotherapy for head-and-neck cancer – RTOG 97–09. Int J Radiat Oncol Biol Phys 2003;56:832–836.
48 Deems RO, Friedman MI, Friedman LS, Munoz SJ, Maddrey WC: Chemosensory function, food preferences and appetite in human liver disease. Appetite 1993;20:209–216.
49 Madden AM, Bradbury W, Morgan MY: Taste perception in cirrhosis: its relationship to circulating micronutrients and food preferences. Hepatology 1997;26:40–48.
50 Bloomfeld RS, Graham BG, Schiffman SS, Killenberg PG: Alterations of chemosensory function in end-stage liver disease. Physiol Behav 1999;66:203–207.
51 Sturniolo GC, D'Inca R, Parisi G, Giacomazzi F, Montino MC, D'Odorico A, et al: Taste alterations in liver cirrhosis: are they related to zinc deficiency? J Trace Elem Electrolytes Health Dis 1992;6:15–19.
52 Garrett-Laster M, Russell RM, Jacques PF: Impairment of taste and olfaction in patients with cirrhosis: the role of vitamin A. Hum Nutr Clin Nutr 1984;38:203–214.
53 Bellisle F, Dartois AM, Kleinknecht C, Broyer M: Perceptions of and preferences for sweet taste in uremic children. J Am Diet Assoc 1990;90:951–954.
54 Rocco MV, Gassman JJ, Wang SR, Kaplan RM: Cross-sectional study of quality of life and symptoms in chronic renal disease patients: the Modification of Diet in Renal Disease Study. Am J Kidney Dis 1997;29:888–896.
55 Fernstrom A, Hylander B, Rossner S: Taste acuity in patients with chronic renal failure. Clin Nephrol 1996;45:169–174.
56 Astback J, Fernstrom A, Hylander B, Arvidson K, Johansson O: Taste buds and neuronal markers in patients with chronic renal failure. Perit Dial Int 1999;19(Suppl 2):S315–S323.
57 Bolton CF: Peripheral neuropathies associated with chronic renal failure. Can J Neurol Sci 1980;7:89–96.
58 Burge JC, Schemmel RA, Park HS, Greene JA 3rd: Taste acuity and zinc status in chronic renal disease. J Am Diet Assoc 1984;84:1203–1206, 1209.
59 Mahajan SK, Prasad AS, Rabbani P, Briggs WA, McDonald FD: Zinc deficiency: a reversible complication of uremia. Am J Clin Nutr 1982;36:1177–1183.

60 Sprenger KB, Bundschu D, Lewis K, Spohn B, Schmitz J, Franz HE: Improvement of uremic neuropathy and hypogeusia by dialysate zinc supplementation: a double-blind study. Kidney Int Suppl 1983;16:S315–S318.
61 Mahajan SK, Abraham J, Hessburg T, Prasad AS, Migdal SD, Abu-Hamdan DK, et al: Zinc metabolism and taste acuity in renal transplant recipients. Kidney Int Suppl 1983;16:S310–S314.
62 Matsuo K, Nakamoto M, Yasunaga C, Goya T, Sugimachi K: Dialysis-related amyloidosis of the tongue in long-term hemodialysis patients. Kidney Int 1997;52:832–838.
63 Zervakis J, Graham BG, Schiffman SS: Taste effects of lingual application of cardiovascular medications. Physiol Behav 2000;68:405–413.
64 Anand KK, Zuniga JR: Effect of amiloride on suprathreshold NaCl, LiCl, and KCl salt taste in humans. Physiol Behav 1997;62:925–929.
65 Le Floch JP, Le Lievre G, Verroust J, Philippon C, Peynegre R, Perlemuter L: Factors related to the electric taste threshold in type 1 diabetic patients. Diabet Med 1990;7:526–531.
66 Perros P, MacFarlane TW, Counsell C, Frier BM: Altered taste sensation in newly-diagnosed NIDDM. Diabetes Care 1996;19:768–770.
67 Moore PA, Guggenheimer J, Etzel KR, Weyant RJ, Orchard T: Type 1 diabetes mellitus, xerostomia, and salivary flow rates. Oral Surg Oral Med Oral Pathol Oral Radiol Endod 2001;92:281–291.
68 Pittman JA, Beschi RJ: Taste thresholds in hyper- and hypothyroidism. J Clin Endocrinol Metab 1967;27:895–896.
69 McConnell RJ, Menendez CE, Smith FR, Henkin RI, Rivlin RS: Defects of taste and smell in patients with hypothyroidism. Am J Med 1975;59:354–364.
70 Bhatia S, Sircar SS, Ghorai BK: Taste disorder in hypo- and hyperthyroidism. Indian J Physiol Pharmacol 1991;35:152–158.
71 Law JS, Henkin RI: Thyroid hormone inhibits purified taste bud membrane adenosine 3′,5′-monophosphate phosphodiesterase activity. Res Commun Chem Pathol Pharmacol 1984;43:449–462.
72 Mendoza A, Shaffer B, Karakla D, Mason ME, Elkins D, Goffman TE: Quality of life with well-differentiated thyroid cancer: treatment toxicities and their reduction. Thyroid 2004;14:133–140.
73 Weiffenbach JM, Schwartz LK, Atkinson JC, Fox PC: Taste performance in Sjögren's syndrome. Physiol Behav 1995;57:89–96.
74 Urban PP, Keilmann A, Teichmann EM, Hopf HC: Sensory neuropathy of the trigeminal, glossopharyngeal, and vagal nerves in Sjögren's syndrome. J Neurol Sci 2001;186:59–63.
75 Soto-Rojas AE, Villa AR, Sifuentes-Osornio J, Alarcon-Segovia D, Kraus A: Oral manifestations in patients with Sjögren's syndrome. J Rheumatol 1998;25:906–910.
76 Stoopler ET, Sollecito TP, Chen SY: Amyloid deposition in the oral cavity: a retrospective study and review of the literature. Oral Surg Oral Med Oral Pathol Oral Radiol Endod 2003;95:674–680.
77 Schiffman SS, Zervakis J, Heffron S, Heald AE: Effect of protease inhibitors on the sense of taste. Nutrition 1999;15:767–772.
78 Wickham RS, Rehwaldt M, Kefer C, Shott S, Abbas K, Glynn-Tucker E, et al: Taste changes experienced by patients receiving chemotherapy. Oncol Nurs Forum 1999;26:697–706.
79 Bartoshuk LM: Chemosensory alterations and cancer therapies. NCI Monogr 1990;179–184.
80 Genter MB: Evaluation of olfactory and auditory system effects of the antihyperthyroid drug carbimazole in the Long-Evans rat. J Biochem Mol Toxicol 1998;12:305–314.
81 Barajas Garcia-Talavera F: Ageusia in a patient treated with carbamazepine. Neurologia 1988;3: 126–127.
82 Deguchi K, Furuta S, Imakiire T, Ohyama M: Case of ageusia from a variety of causes. J Laryngol Otol 1996;110:598–601.

Evan R. Reiter, MD
Department of Otolaryngology – HNS, 1201 East Marshall Street
Virginia Commonwealth University, Box 980146
Richmond, VA 23298-0146 (USA)
Tel +1 804 828 2766, Fax +1 804 828 3494, E-Mail erreiter@vcu.edu

Hummel T, Welge-Lüssen A (eds): Taste and Smell. An Update.
Adv Otorhinolaryngol. Basel, Karger, 2006, vol 63, pp 278–287

Burning Mouth Syndrome

Miriam Grushka[a], *Victor Ching*[a], *Joel Epstein*[b]

[a]Department of Surgery, William Osler Health Center – Etobicoke Campus, Toronto, Canada; [b]Department of Oral Medicine, Faculty of Dentistry, University of Illinois at Chicago, Chicago, Ill., USA

Abstract

Burning mouth syndrome (BMS) has been considered an enigmatic condition because the intensity of pain rarely corresponds to the clinical signs of the disease. As a result, BMS patients have variously been labelled as depressed, anxious or hypochondriacal and have often been underserviced by the medical and dental communities. Recently, there has been a resurgence of interest in this disorder with the discovery that the pain of BMS may be neuropathic in origin and originate both centrally and peripherally. This chapter discusses some of our recent understandings of the etiology and pathogenesis of BMS as well as the role of pharmacotherapeutic management in this disorder.

Burning mouth syndrome (BMS) is variously referred to as glossopyrosis, glossodynia (when the burning occurs on the tongue only) and syndrome of oral complaints as well as numerous other monikers, although all refer to the same or a similar constellation of symptoms. It is usually described as oral burning pain, sometimes with dysesthetic qualities similar to those present in other neuropathic pain conditions with the absence of clinical and laboratory abnormalities. The dorsal tongue, palate, lips and gingival tissues, individually or in combination, are the most common sites involved. Symptoms are usually bilateral, but can be unilateral as well. In some reports, oral burning pain has been found to be associated with jaw pain [1, 2], taste changes and subjective dry mouth, geographic and fissured tongue [3], painful teeth, loss of a comfortable jaw position, uncontrollable jaw tightness [4–8], headache [5, 9], neck and shoulder pain, increased parafunctional activity, difficulty speaking, nausea, gagging and swallowing difficulties [4].

Although the events preceding the onset of BMS are often not identified, the condition has been reported to follow dental treatment, antibiotic usage and a severe upper respiratory infection [9, 10].

The pain from BMS is constant, progressively increases over the day, and usually decreases during eating. Although it may interfere with onset of sleep, it rarely wakes the patient at night and is at its lowest intensity in the morning [9]. Patients, who are frequently distressed by their unremitting symptoms, may demonstrate psychological abnormalities including anxiety and depression in both questionnaire and psychiatric examinations [11–13]. The presence of emotional issues in BMS appears to be in accord with studies which have demonstrated psychological profiles of distress in the presence of chronic pain [14]. The lack of pathology to account for the pain can be equally frustrating [5, 15].

Prevalence

Epidemiological surveys have reported a prevalence rate of between 0.7–2.6% with an NIH survey estimating close to 1 million burning mouth sufferers in America. Although most prevalent amongst postmenopausal women, men and women of any age can also be affected [16].

Differential Diagnosis

Alternate causes of oral burning pain should be ruled out, including both systemic and peripheral pathology (table 1), before a diagnosis of BMS is entertained. Burning pain can indicate a previously undiagnosed systemic condition. This includes anemia, vitamin B or iron deficiency, untreated diabetes, renal disease, and connective tissue disorders such as Sjögren's syndrome and systemic lupus – both of which can be associated with oral dryness and consequent candidal infections. Some medications such as angiotensin-converting enzyme inhibitors have been reported to be associated with burning pain [10].

Local changes within the oral cavity may cause burning pain and include allergic reactions to dental materials and dentures, and products such as toothpastes, mouth rinses and food constituents such as cinnamon. Candidiasis infection in susceptible patients and painful lesions in the mucosa may also be both causative and treatable. In one small study of the role of viral infection in BMS, 22 subjects complaining of burning mouth were assessed; of these, 9 were

Table 1. Differential diagnosis of BMS

a Systemic causes
Nutritional deficiency: vitamin B, iron, zinc
Allergy: food or dental materials
Esophageal reflux disorder
Uncontrolled diabetes
Acoustic neuroma
Central changes including multiple sclerosis, Parkinson's disease, trigeminal neuralgia
Autoimmune disorders: Sjögren's syndrome
b Local causes
Oral candidiasis infection
Poorly fitting dentures, restorations
Lichen planus and other oral vesiculobullous conditions
Dry mouth: autoimmune disorders, medication
Viral infection: herpes simplex, herpes zoster
Trauma to lingual or mandibular nerve following dental surgery
Oral inflammatory condition: geographic, fissured tongue

diagnosed with BMS and the rest were found to have mucosal changes especially erosive lichen planus [17]. While low salivary flow can coexist with BMS and may exacerbate the pain, there is no indication at this time that xeristomia by itself is a primary causative factor [10].

Diagnosis

History taking is the key to diagnosis of BMS. Both diagnosis and management may be difficult because patients often present with multiple oral complaints, may be focused on their symptoms and may be anxious or depressed, which intensifies the pain experience. The diagnosis is based on clinical characteristics, including either a sudden or intermittent onset of pain, bilateral presentation, a progressive increase in pain during the day and the remission of pain with eating (although some foods may intensify the pain) and sleeping. Salivary flows and taste function should be assessed [18]. Important clinical questions are presented in table 2. Neurological imaging and consultation should be considered when patients present with a more complex symptom array, including both sensory and motor changes, to rule out a neurodegenerative disorder such multiple sclerosis, Parkinson's disease, and stroke.

Table 2. Clinical features that are helpful in the diagnosis of BMS

Unilateral or bilateral burning pain localized to tongue, palate, lips and gingival
Pain that gets worse over the day
Decreased pain on eating
Decreased pain with sleep
Absence of clinical finding
Presence of abnormal or dysgeusic tastes, usually metallic, bitter or sour
Complaint of dry mouth in presence of normal flows
Sensory changes or parasthesias including complaints of areas of roughness or irritation

Table 3. Clinical tests that may be helpful

Hematological tests: CBC, glucose, nutritional factors, autoimmune panel
Oral cultures for fungal, viral or bacterial infections if suspected
MRI to rule out central changes, especially if pain is unilateral, atypical or does not respond to medication
Salivary flows for unstimulated and stimulated whole saliva (<1.5 ml/0.5 min, unstimulated; <4.5 mg/5 min stimulated)
Salivary uptake scan if low salivary flows and Sjögren's syndrome suspected
Allergy testing, if needed, especially to dental panel of allergens
Removal of possibly offending medication including angiotensin-converting enzyme inhibitors

Considerations in differential diagnosis, diagnostic testing, and clinical history are outlined in tables 1–3.

Etiology

Although many etiologies have been suggested in BMS, spanning the range from nutritional factors to dental intolerances, none have been found to account for a substantial portion of patients. Some of the more recent considerations have been the possibility that BMS is a neuropathic disorder as a result of damage to the taste system, possibly by viral infection.

Viral Infection

In view of the relatively quick onset of burning and dysesthetic pain, the relatively high prevalence of BMS (see above), and the previous demonstration that herpesvirus can lead to neuropathies following oropharyngeal infection [19], a possible association between BMS and herpes viral damage was evaluated in a recent study. In this study, 22 subjects complaining of oral burning pain were

Table 4. Percentage of positive Ig findings in serologic tests of the BMS and control groups

	n	HSV, %		CMV, %		HZV, %	
		IgM	IgG	IgM	IgG	IgM	IgG
BMS	9	0	66.7	0	66.7	0	100
Control	13	0	66.7	0	66.7	11.1	88.9

HZV = Herpes zoster virus.

assessed for viral serology, 9 of whom were diagnosed with BMS and the rest were found to have oral mucosal changes and were used as control subjects.

The results of the serologic tests for herpes simplex virus (HSV), cytomegalic virus (CMV) and varicella-zoster virus (VZV) are presented in table 4. No IgM seropositivity for any of the 3 viruses was seen in most patients. All but 1 subject in both the study and control groups were negative for IgM antibody to the herpesviruses tested except for 1 patient with pain compatible with herpes zoster who was found to be positive for varicella-zoster virus antibody. Most subjects in both groups were positive for HSV, CMV and herpes zoster virus IgG and no significant difference was found in the prevalence of the positive findings between the two groups. Although no evidence was found in this preliminary study that would support the presence of an active or past viral infection in BMS subjects, the possibility of a 'hit and run' model for viral damage in BMS cannot be ruled out.

Taste Changes in BMS

Work by Bartoshuk et al. [20] has demonstrated the convergence of taste sensation and pain clinically and experimentally. The chorda tympani nerve leaves the tongue with the lingual nerve (cranial nerve V) and travels through the pterygomandibular space. The inferior alveolar nerve, which conveys sensation from the lower teeth, also passes through the same space. Often, dental anesthesia of the inferior alveolar and lingual nerves required for dental restorations abolishes touch and pain, but also taste on the injected side. The chorda tympani and lingual nerves separate and the chorda tympani passes through the middle ear. Bartoshuk et al. [20], Lehman et al. [21] and Yanagisawa et al. [22] have demonstrated that anesthesia of the chorda tympani behind the tympanic membrane intensifies tastes from the area innervated by the glossopharyngeal nerve at the back of the tongue on the opposite side, supporting a model of central inhibition between the chorda tympani and glossopharyngeal nerves. According to Bartoshuk and fellow workers, reduction of input into the central nervous system from one taste nerve releases inhibition of other taste.

Tie et al. [23] found that anesthesia of the chorda tympani can intensify pain induced by capsaicin on the contralateral anterior tongue suggesting the presence of central inhibitory interactions between taste and oral pain. Furthermore, the intensification of pain was found to be related to an individual's genetic ability to taste PROP (6-n-propylthiouracil), with the greatest intensification found in 'supertasters' who report the most bitter sensation from PROP testing [24].

Based on these taste/pain interactions, it is believed that BMS could also be the clinical manifestation of taste damage, either to the chorda tympani, with release of inhibition in the glossopharyngeal nerve (taste alterations, alterations in touch and pain) or the trigeminal nerve (touch and pain changes). Consistent with this model, severe taste damage has been found in many BMS patients. Notably, the intensity of the peak oral pain was also found to correlate with the density of fungiform papillae and patients with BMS were primarily supertasters [25]. Furthermore, it has also been suggested that interactions between taste and oral pain were not limited to BMS but involved other orofacial pain complaints as well. For instance, patients with atypical odontalgia (pain appearing to originate from healthy teeth) showed taste damage [26].

It should also be noted that although we are not aware of similar studies linking taste and inhibition of the motor component of the trigeminal nerve, based on reports of increased bruxism in BMS patients [27], as well as increased headaches in BMS [9], the possibility that taste also inhibits the motor component of the trigeminal nerve, leading to muscle hyperactivity in the mastication system, is being considered. The anatomical substrate for this inhibition is known to be present in animal studies which demonstrate projections from the gustatory portion of the nucleus of the solitary tract to the oromotor nuclei in the medulla subserving the masticatory muscles [28].

Other abnormalities have also been noted in BMS, including elevated thresholds for temperature and touch [29], altered pain tolerance [30] as well as changes in blink reflex, corneal reflex, jaw jerk, sensory neurography of the inferior alveolar nerve and trigeminal somatosensory evoked potentials [31]. There may also be an increase in sympathetic output which leads to decreased blood flow [32] in the tongue, altered salivary composition [18, 33], high blood pressure, difficulty sleeping and increased esophageal reflux [9].

A hypothesis based on the taste pathways inhibiting other cranial nerves can explain why conditions such as BMS and AO can often encompass sensory, motor and sympathetic abnormalities; why multiple orofacial phenomena are linked together in these pain syndromes; why the onset of these problems can occur suddenly; why there is almost always a lack of associated organic mucosal and dental pathology, and why these conditions may respond to centrally acting drugs, especially those affecting the GABAergic pathways [2, 34, 35] which are involved in taste transmission and in neuroinhibition [28].

GABA is known to be an inhibitory neurotransmitter found in the taste system [36–38] and may be a key target. According to Bartoshuk et al. [20], if taste damage produces a sufficient loss of the inhibition normally exerted on central structures mediating oral pain, then replacement of a GABA agonist such as clonazepam might ameliorate the loss of inhibition and relieve the pain in BMS.

Interestingly, GABA agonists such as clonazepam [39] have also been found to have value in the treatment of nausea, coughing and hiccups [40, 41] and in taste disturbances when associated with BMS [34]. Thus, it is possible that the inhibition produced by the taste system is important in controlling other anatomy associated with eating.

Management

Therapy for BMS involves the use of centrally acting medications for neuropathic pain, such as tricyclic antidepressants, benzodiazepines or gabapentin [42]. Studies support the use of low-dose (0.25–0.75 mg) clonazepam or tricyclic antidepressants (10–40 mg), including amitriptyline, desipramine, nortriptyline, imipramine and clomipramine. Clonazepam is a benzodiazepine used either topically or systemically [1, 34, 35], which appears to have excellent efficacy in the relief of the symptoms related to BMS. In view of only partial or lack of response in some BMS patients taking these medications, other GABA receptor-acting anticonvulsants have been used in combination with clonazepam with apparent success [43]. Topical medications, including clonidine and capsaicin, may be considered for application to local sites. Systemic use of capsaicin has also been suggested [44] as has α-lipoic acid with or without psychotherapy [45].

Polypharmacy in Pain Management

A recent retrospective study of low-dose anticonvulsant medications used in combination for the management of BMS was carried out. Patients were prescribed up to 0.5 mg clonazepam and asked to add as needed up to 1,200 mg of gabapentin (up to 300 mg 4 times a day); 30 mg of baclofen (in 3 divided doses) and then up to 200 mg of lamotrigine (in 2 divided doses) as needed and pain scores were recorded on a modified adjectival/visual analogue scale. Of the

Table 5. Drug combinations used by patients with average doses

	C	G	C/G	C/G/B	L + other	L
Patients	11	4	12	4	4	2
Average dose, mg	0.25	300	0.27/290	0.44/300/25	n/a	50

B = Baclofen; C = clonazepam; G = gabapentin; L = lamotrigine.

45 patients who were diagnosed with BMS and tried the protocol, 1 patient reported an increase in pain after using the protocol and 6 patients did not find any difference; the rest [38] observed some reduction in pain. The average maximum pain rating before treatment was 60.6 and the average maximum pain rating after treatment was 32.1, which was found to be significant ($p < 0.001$) (table 5).

The most common adverse effect reported with the medication protocol was drowsiness followed by dizziness and perceived changes in mood. Eighteen patients reported some side effects at some point of the treatment, and the majority of them were able to resolve the side effects by titrating down the dose of medication. Only 2 patients elected to stop treatment completely because of the side effects.

These results suggest that treatment of BMS may be efficacious with a combination of medications rather than higher doses of a single medication, especially with regard to controlling adverse effects.

References

1 Woda A, Pionchon P: A unified concept of idiopathic orofacial pain: clinical features. J Orofac Pain 1999;13:172–184.

2 Woda A, Pionchon P: A unified concept of idiopathic orofacial pain: pathophysiologic features. J Orofac Pain 2000;14:196–212.

3 Drage LA, Rogers RS: Clinical assessment and outcome in 70 patients with complaints of burning or sore mouth symptoms. Mayo Clin Proc 1999;74:223–228.

4 Marbach JJ: Psychosocial factors for failure to adapt to dental prostheses. Dent Clin North Am 1985;29:215–233.

5 Melis M, Lobo Lobo S, Ceneviz C, Zawawi K, Al-Badawi E, Maloney G, Mehta N: Views and perspectives. Atypical odontalgia: a review of the literature. Headache 2003;43:1060–1070.

6 Marbach JJ: Medically unexplained chronic orofacial pain. Temporomandibular pain and dysfunction syndrome, orofacial phantom pain, burning mouth syndrome, and trigeminal neuralgia. Med Clin North Am 1999;83:691–710.

7 Marbach JJ: Orofacial phantom pain: theory and phenomenology. J Am Dent Assoc 1996;127: 221–229.

8 Vickers ER, Cousins MJ, Walker S, Chisholm K: Analysis of 50 patients with atypical odontalgia. A preliminary report on pharmacological procedures for diagnosis and treatment. Oral Surg Oral Med Oral Pathol Oral Radiol Endod 1998;85:24–32.

9 Grushka M: Clinical features of burning mouth syndrome. Oral Surg Oral Med Oral Pathol 1987;63:30–36.
10 Grushka M, Epstein JB, Gorsky M: Burning mouth syndrome. Am Fam Physician 2002;65:615–620.
11 Soto Araya M, Rojas Alcayaga G, Esguep A: Association between psychological disorders and the presence of oral lichen planus, burning mouth syndrome and recurrent aphthous stomatitis. Med Oral 2004;9:1–7.
12 Nicholson M, Wilkinson G, Field E, Longman L, Fitzgerald B: A pilot study: stability of psychiatric diagnoses over 6 months in burning mouth syndrome. J Psychosom Res 2000;49:1–2.
13 Trikkas G, Nikolatou O, Samara C, Bazopoulou-Kyrkanidou E, Rabavilas AD, Christodoulou GN: Glossodynia: personality characteristics and psychopathology. Psychother Psychosom 1996;65: 163–168.
14 Sternbach RA, Timmermans G: Personality changes associated with reduction of pain. Pain 1975;1:177–178.
15 Zakrewska JM: The burning mouth syndrome remains an enigma. Pain 1995;62:253–257.
16 Ship JA, Grushka M, Lipton JA, Mott AE, Sessle BJ, Dionne RA: Burning mouth syndrome: an update. J Am Dent Assoc 1995;126:842–853.
17 Epstein JB, Grushka M, Gorsky M: Role of herpes simplex virus in BMS, submitted.
18 Nagler RM, Hershkovich O: Sialochemical and gustatory analysis in patients with oral sensory complaints. J Pain 2004;5:56–63.
19 Hashizume K: Herpes zoster and post-herpetic neuralgia. Jap J Clin Med 2001;59:1738–1742.
20 Bartoshuk LM, Chapo A, Duffy VB, Gruhska M, Norgren R, Kveton JF, Pritchard TC, Snyder D: Oral phantoms: evidence for central inhibition produced by taste. Chem Senses 2002;27:A52.
21 Lehman CD, Bartoshuk LM, Catalanotto FC, Kveton JF, Lowlicht RA: The effect of anesthesia of the chorda tympani nerve on taste perception in humans. Physiol Behav 1995;57:943–951.
22 Yanagisawa K, Bartoshuk LM, Catalanotto FA, Karrer TA, Kveton JF: Anesthesia of the chorda tympani nerve and taste phantoms. Physiol Behav 1998;63:329–335.
23 Tie K, Fast K, Kveton J, Cohen Z, Duffy VB, Green B, et al: Anesthesia of chorda tympani nerve and effect on oral pain. Chem Senses 1999;24:609.
24 Bartoshuk LM, Duffy VB, Miller IJ: PTC/PROP tasting: anatomy, psychophysics, and sex effects. Physiol Behav 1994;56:1165–1171.
25 Grushka M, Bartoshuk LM, Chapo AK, Duffy VB, Norgren R, Kveton J, Pritchard TC, Snyder DJ: Oral pain: associated with damage to taste. J Pain 2000;145:P141.
26 Grushka M, Bartoshuk LM, Chapo AK, Duffy VB, Norgren R, Kveton J, Pritchard TC, Snyder DJ: Oral pain: associated with damage to taste. Proc 10th World Congr Pain, San Diego, 2002.
27 Paterson AJ, Lamb AB, Clifford TJ, Lamey PJ: Burning mouth syndrome: the relationship between the HAD scale and parafunctional habits. J Oral Pathol Med 1995;24:289–292.
28 King MS: Distribution of immunoreactive GABA and glutamate receptors in the gustatory portion of the nucleus of the solitary tract in rat. Brain Res Bull 2003;60:241–254.
29 Forssell H, Jaaskelainen S, Tenovuo O: Sensory dysfunction in burning mouth syndrome. Pain 2000;99:41–44.
30 Grushka M, Sessle BJ, Howley TP: Psychophysical assessment of tactile, pain and thermal sensory functions in burning mouth syndrome. Pain 1987;28:169.
31 Jaaskelainen SK: Clinical neurophysiology and quantitative testing in the investigation of orofacial pain and sensory function. J Orofac Pain 2004;18:85–107.
32 Cekic-Arambasin A, Vidas I, Stipetic-Mravak M: Clinical oral test for the assessment of oral symptoms of glossodynia and glossopyrosis. J Oral Rehabil 1990;17:495–502.
33 Chimenos-Kustner E, Marques-Soares MS: Burning mouth and saliva. Med Oral 2002;7:244–253.
34 Grushka M, Epstein J, Mott A: An open-label, dose escalation pilot study of the effect of clonazepam in burning mouth syndrome. Oral Surg Oral Med Oral Pathol Oral Radiol Endod 1998;86:557–561.
35 Gremeau-Richard C, Woda A, Navez ML, Atta N, Bouhassira D, Gagnieu MC, Laluque JF, Picard P, Pioncon P, Tubert S: Topical clonazepam in stomatodynia: a randomized placebo-controlled study. Pain 2004;108:51–54.
36 Davis BJ: GABA-like immunoreactivity in the gustatory zone of the nucleus of the solitary tract in the hamster: light and electron microscopic studies. Brain Res Bull 1993;30:69–77.

37 Smith DV, Li CS: GABA-mediated corticofugal inhibition of taste-responsive neurons in the nucleus of the solitary tract. Brain Res 2000;858:408–415.
38 Wang L, Bradley RM: Influence of GABA on neurons of the gustatory zone of the rat nucleus of the solitary tract. Brain Res 1993;616:144–153.
39 Shindo EF, Emoto M, Mohtai H, Hachisuga T, Kawarabayashi T, Shirakawa K: The prevention of cancer chemotherapy-induced emesis with granisetron and clonazepam. Gan To Kagaku Ryoho 1995;22:233–237.
40 Dicpinigaitis PV, Grimm DR, Lesser M: Baclofen-induced cough suppression in cervical spinal injury. Arch Phys Med Rehabil 2000;81:921–923.
41 Smith HS, Busracamwongs A: Management of hiccups in the palliative care population. Am J Hosp Palliat Care 2003;20:149–154.
42 White TL, Kent PF, Kurtz DB, Emko P: Effectiveness of gabapentin for treatment of burning mouth syndrome. Arch Otolaryngol Head Neck Surg 2004;130:786–788.
43 Ching V, Grushka M, Epstein JB: Clinical efficacy of titrated anticonvulsant analgesics on atypical odontalgia and burning mouth syndrome: retrospective study, in preparation.
44 Petruzzi M, Lauritano D, De Benedittis M, Baldoni M, Serpico R: Systemic capsaicin for burning mouth syndrome: short-term results of a pilot study. J Oral Pathol Med 2004;33:111–114.
45 Femiano F, Gombos F, Scully C: Burning mouth syndrome: open trial of psychotherapy alone, medication with alpha-lipoic acid (thioctic acid), and combination therapy. Med Oral 2004;9:8–13.

Further Reading

Bartoshuk LM, Grushka M, Duffy VB, Fast L, Lucchina L, Putkin J, et al: Burning mouth syndrome: damage to CN VII and pain phantoms in CN V. Chem Senses 1999;24:609–613.

Grushka M, Bartoshuk LM: Burning mouth syndrome and oral dysesthesia: taste injury is a piece of the puzzle. Can J Diagn 2000;17:99–109.

Grushka M, Epstein JB, Gorsky M: Burning mouth syndrome. Am Fam Physician 2002;65:615–620.

Grushka M, Epstein JB, Gorsky M: Burning mouth syndrome and other oral sensory disorders: a unifying hypothesis. Pain Res Manag 2003;8:133–135.

Ship JA, Grushka M, Lipton JA, Mott AE, Sessle BJ, Dionne RA: Burning mouth syndrome: an update. J Am Dent Assoc 1995;126:842–853.

Svensson P, Kaaber S: General health factors and denture function in patients with burning mouth syndrome and matched control subjects. J Oral Rehabil 1995;22:887–895.

Tammiala-Salonen T, Hiidenkari T, Parvinen T: Burning mouth in a Finnish adult population. Community Dent Oral Epidemiol 1993;21:67–71.

Zakrewska JM: The burning mouth syndrome remains an enigma. Pain 1995;62:253–257.

Dr. Miriam Grushka
974 Eglinton W
Toronto, Ont. M6c 2C5 (Canada)
Tel. +1 416 787 2930, Fax +1 416 656 8328, E-Mail mgrushka@yahoo.com

Author Index

Subject Index